AF581622

Wearable and BAN Sensors for Physical Rehabilitation and eHealth Architectures

Wearable and BAN Sensors for Physical Rehabilitation and eHealth Architectures

Editors

Maria de Fátima Domingues
Andrea Sciarrone
Ayman Radwan

MDPI • Basel • Beijing • Wuhan • Barcelona • Belgrade • Manchester • Tokyo • Cluj • Tianjin

Editors
Maria de Fátima Domingues
Campus Universitário de Santiago
Portugal

Andrea Sciarrone
University of Genoa
Italy

Ayman Radwan
Campus Universitário de Santiago
Portugal

Editorial Office
MDPI
St. Alban-Anlage 66
4052 Basel, Switzerland

This is a reprint of articles from the Special Issue published online in the open access journal *Sensors* (ISSN 1424-8220) (available at: https://www.mdpi.com/journal/sensors/special_issues/Wearable_BAN_Sensors?listby=date&view=default).

For citation purposes, cite each article independently as indicated on the article page online and as indicated below:

LastName, A.A.; LastName, B.B.; LastName, C.C. Article Title. *Journal Name* **Year**, *Volume Number*, Page Range.

ISBN 978-3-0365-2812-0 (Hbk)
ISBN 978-3-0365-2813-7 (PDF)

Contents

About the Editors

Maria de Fátima Domingues (Dr.) received her MSc in Applied Physics in 2008 and completed her PhD in Physics Engineering in 2014, both at the University of Aveiro, Portugal. In 2015, Domingues began a Research Fellow position at the Instituto de Telecomunicações – Aveiro and the Consejo Superior de Investigaciones Científicas (CSIC) – Madrid, Spain. At present, Domingues is a Researcher at Instituto de Telecomunicações – Aveiro, and her current research interests embrace new solutions of photonic-based sensors and their application in e-Health scenarios, with a focus in physical rehabilitation architectures. Domingues has authored and co-authored more than 100 journal and conference papers, 5 book chapters, and 3 books.

Andrea Sciarrone (Prof.) is Assistant Professor at the University of Genoa, Italy. He received his bachelor's degree in 2007 and Master of Science in 2009, both in Telecommunication Engineering. In 2014 he was awarded his Ph.D. in Science and Technology for Information and Knowledge at the University of Genoa. He is currently as member of the research staff in the Telecommunication Group and of the Digital Signal Processing (DSP) Laboratory at the DITEN department of the University of Genoa. His main research activities concern e-health applications, signal processing over the Internet of Things, and context and location awareness. Andrea Sciarrone has contributed with many papers to previous e-health tracks of the ComSoc flagship conferences and has several publications in journals and magazines of interest for the relevant research community. Furthermore, he has been involved in several funded e-health projects, both national and international. In 2016, he received the best paper award from the IEEE Communication Society, which was presented during the International Conference on Cloud Networking (CloudNet 2016), 3–5 October 2016, Pisa, Italy. Currently, he is Vice-Chair of IEEE Comsoc e-Health Committee.

Ayman Radwan (Prof.) is Assistant Professor and EU project coordinator with the Instituto de Telecomunicações and Universidade de Aveiro, Aveiro, Portugal. Dr. Radwan has received his Ph.D. from Queen's University (Kingston, ON, Canada), in 2009, and his Master of Applied Science (M. A. Sc.) from Carleton University (Ottawa, ON, Canada). He joined Instituto de Telecomunicações in January 2010. Since 2010, Dr. Radwan has been intensively active in European projects, coordinating and technically managing multiple EU projects. He was also the Coordinator of the CELTIC Project "Green-T" and the CELTIC Plus project "MUSCLES". He is currently acting coordinator of the CELTIC-NEXT "SAFE-HOME". He has extensive experience in coordinating and managing collaborative projects at EU and international levels with multimillion Euro budgets. He is the author of more than 100 scientific works in the field of wireless networks, with emphasis on future generations of mobile networks, RRM, and efficient networking for IoT. He is the author of 2 books and 4 patents. He is a Senior IEEE member and associate editor of IEEE Comm. Letters and IEEE Network.

MDPI

Editorial

Special Issue "Wearable and BAN Sensors for Physical Rehabilitation and eHealth Architectures"

Maria de Fátima Domingues [1,*], Andrea Sciarrone [2] and Ayman Radwan [1]

1 Instituto de Telecomunicações, Campus Universitário de Santiago, 3810-193 Aveiro, Portugal; aradwan@av.it.pt
2 Department of Electrical, Electronic, Telecommunications Engineering, and Naval Architecture (DITEN), University of Genoa, Via dell'Opera Pia 13, 16145 Genoa, Italy; andrea.sciarrone@unige.it
* Correspondence: fatima.domingues@ua.pt

Citation: Domingues, M.F.; Sciarrone, A.; Radwan, A. Special Issue "Wearable and BAN Sensors for Physical Rehabilitation and eHealth Architectures". *Sensors* **2021**, *21*, 8509. https://doi.org/10.3390/s21248509

Received: 14 December 2021
Accepted: 15 December 2021
Published: 20 December 2021

1. Introduction

The demographic shift of the population toward an increased number of elder citizens, together with the sedentary lifestyle we are adopting, is reflected in the increasingly debilitated physical health of the population. The resulting physical impairments require rehabilitation therapies that may be assisted by the use of wearable sensors or body area network sensors (BANs). The use of novel technology for medical therapies can also contribute to reducing the cost of healthcare systems and decrease the patient overflow in medical centers. Sensors are the primary enablers of any wearable medical device, with a central role in eHealth architectures. The accuracy of the acquired data relies on the sensors; hence, when considering wearable and BAN sensing integration, they must prove to be accurate and reliable solutions.

This Special Issue (SI) focuses on the current state-of-the-art BANs and wearable sensing devices for the physical rehabilitation of impaired or debilitated citizens. Both original research papers and review articles describing the current state-of-the-art were considered for publication. We believe that this SI will provide the reader with an overview of the present status and a future outlook of the aforementioned topics.

The contributions to this SI resulted in a collection of 10 published manuscripts reporting on the advances in research related to different sensing technologies (optical or electronic) and body area network sensors (BANs); their design and implementation; advanced signal processing techniques and the application of these technologies in areas such as physical rehabilitation, robotics, medical diagnostics and therapy.

A short overview of the collection of papers accepted for publication in this SI is presented in Section 2.

The guest editors would like to show their token of appreciation to all the authors that contributed to the success of this SI, by providing a set of original papers with a comprehensive and up-to-date overview of a variety of topics, under the umbrella of "Wearable and BAN Sensors for Physical Rehabilitation and eHealth Architectures".

Furthermore, the work and support of the academic editors and reviewers is highly appreciated. They were a key factor to guarantee the high quality and the scientific rigor of the published manuscripts and, consequently, of this Special Issue.

2. Contributed Papers

The manuscripts accepted for publication in this SI mirror the relevance of the topic for the research community, and the vast field of research that still exists to be explored to enhance the wearable and BAN solutions for physical rehabilitation applications and eHealth architectures.

The authors of [1] presented the design and study of different mobility aids (smart walker) configurations, targeting the population who suffers from visual and mobility impairments. In this study, the authors explored different technologies and software configurations to evaluate the performance of the different solutions and reach the conclusion that there is not one configuration that will be suitable for all. Instead, they found that multiple and different choices of sensors can provide a similar user experience. Nevertheless, emphasis should be given to the fact that active sensors (ultrasonic distance sensors or infrared depth cameras) provide a better accuracy for the localization of objects/obstacles [1]. The authors also reach the conclusion that it is necessary to perform a holistic evaluation of the walker in terms of its end-to-end performance, and that the user interface is of big importance to the overall performance of a smart walker [1].

Bezuidenhout et al. presented a study on the reliability of Actigraph GT3X+ (AG) accelerometers to detect gait parameters. The devices were worn on the hip and on the ankle by thirty healthy individuals walking in a straight line and turning at different speeds. As a reference, a Stepwatch (SW) activity monitor was used, which was attached to the right ankle [2]. The authors found that the AG placed on the ankle provided the best accuracy for the detection of steps at speeds less than 0.6 m/s, and for speeds above this value, the detection of steps was only possible by applying a low frequency extension filter (LFEF). The hip worn AG presented accuracy above 87% for gait speeds <0.1 m/s, which was considerably degraded with an increase in the gait speed. The authors' findings suggest that the location where the sensor is placed, together with the type of data filters used, are key factors that influence the accuracy of the step counts [2].

In the third contribution to this SI, the authors Di Tocco et al. presented their study on the development of wearable solutions for unobtrusive cardio-respiratory monitoring [3]. The proposed solution is based on four conductive textiles sensors, which are placed on the user's chest. The deformation induced on the sensors, by the expansion and contraction of the rib cage due to the respiratory cycle, provides reliable information, from which the users breathing activity can be inferred [3]. As for the heart rate, the authors used an IMU placed on the left-hand side of the chest. In the trials performed with the wearable system based on a multi-sensor configuration, the authors found that it provided consistent measures for the respiratory and heart rate for all the subjects and scenarios tested [3].

The authors from [4] presented a study on the long- and short-term effects of a scapular exercise on the function and pain in individuals with rotator-cuff-related pain syndrome (RCS) and anterior shoulder in-stability (ASI) [4]. The results presented were the outcome of a study performed in one hundred and eighty-three patients, from which 171 suffered from RCS and 66 from ASI. The assessment of the shoulder pain and function was performed during the implantation of the structure exercise protocol at its beginning, 4th week and at the 2-year follow up [4]. The authors found a substantial improvement in the 4-week assessment, and not a major difference between the 4th week and the 2nd year follow up, which is a valuable indicator of the positive impact of the exercise protocol implemented in the short and long term [4].

The authors of [5] presented a study on the physiological parameters (with particular focus on the heart rate variability (HRV)) that can be extracted from wearable devices to detect stress levels in car drivers. The authors developed a predictive model based on different machine learning (ML) methodologies such as K-Nearest Neighbor (KNN), Random Forest (RF), among others that is able to classify the stress level extracted from ECG-derived HRV features [5]. The techniques proposed by the authors show that the HRV features can act as markers for stress level detection, achieving a recall of 80% with the ML models proposed [5].

The contribution by Rutkowski et al., a study focusing on the use of physical activity sensors (such as the SenseWear armband) in patients with chronic obstructive pulmonary disease (COPD), was presented to monitor their activity level in day-to-day life and for the duration and intensity of physical activity. The approach implemented by the authors allowed them to understand the daily activity of the patients and if they undertake the prescribed unsupervised physical activity, and additionally, to understand the strengths and weaknesses of the selected type of sensors [6]. Based on the sensors' feedback, in terms of resting time, number of steps, physical activity level and energy expenditure (kcal), the authors did not find a significant difference (statistically) between the non-supervised and supervised physical activity days. Furthermore, the authors found that the use of this type of sensor may improve the patient's self-esteem and motivate them to continue physical activity and, in that way, improve their health condition [6].

Another work devoted to the human activity recognition (HAR), using smartwatch inertial sensors, was presented by the authors of [7]. The authors study the performance of three algorithms for the out-of-distribution (OOD) detection of activity classes data that are not present in the training data of the ML [7]. The authors collected a new data set (SPARS9x) from inertial smartwatch sensors worn by 20 volunteers, first performing supervised physical exercises and, after, performing other unrelated physical movements (OOD). From this analysis, the authors showed that traditional algorithms outperform deep learning algorithms for this particular case of OOD detection for HAR [7].

The valuable contribution by Liu et al., for the success of this SI, was also focused on the use of wearable inertial sensors, but for the ambulatory detection of the human gait phase [8]. The analysis of gait parameters, such as its phase, is of extreme importance in the diagnose of diseases (e.g., Alzheimer's, Parkinson's) or post-surgery rehabilitation evolution. The authors proposed a methodology to infer the gait phase, based on the angular velocity provided by inertial sensors, associated to a Hidden Markov Model (HMM) used to segment the gait phases. The outcome of the experiments implemented by the authors demonstrate that their model is able to accurately recognize the gait phase segmentation [8].

The authors of [9] presented their study on the use of multiple sensing technology (mostly miniature wearable inertial sensor nodes) allied to the extended Kalman filter (EKF) method, to evaluate the training performance (stroke posture, rhythm) of kayakers. The authors, based on the kinematic information retrieved by the sensors, resort to ML algorithms to distinguish the stroke cycle phases, providing a comprehensive evaluation of the kayaker's motion on a real scenario, with a stroke phase match of up to 98% (validated by videography) [9]. The techniques, proposed by these authors, can supply the needed quantitative data for coaches and athletes to improve their physical performance [9].

The review paper presented by Vilela et al. discusses the innovative and relevant topic of fog-computing in the area of eHealth [10]. The authors present a review of eHealth applications using fog-computing. The paper focuses on the existing solutions in the literature that use fog-cloud computing with very tight requirements in terms of latency, security and energy efficiency. An architectural overview of communication technologies is elaborated. The paper concentrates on highlighting the gains in the performance of fog networking, in terms of delay and the amount of data. Finally, the authors shed light on challenges in the area for future research efforts [10].

3. Outlook and Prospects

The set of papers published in this SI is just a small representation of the current research interest regarding the use of wearable and BAN sensors for physical rehabilitation and activity monitoring. As the field for wearable sensors evolves, improving its range of detection, resolution and accuracy, new applications and higher accuracy detection levels can be achieved by widening the application of these technologies even more. When allied to ML algorithms, other emerging fields of applications can be sought, such as the digital twin features, where there is a vast area of research still to be pursued.

Author Contributions: Writing—original draft preparation, M.F.D.; writing—review and editing, M.F.D., A.S. and A.R. All authors have read and agreed to the published version of the manuscript.

Funding: This work is funded by FCT/MCTES through national funds and when applicable co-founded by EU funds under the UIDB/50008/2020-UIDP/50008/2020. This work is also funded by FCT/MEC through national funds and when applicable co-funded by the FEDER-PT2020 partnership agreement under the project UID/EEA/50008/2019.

Acknowledgments: M. Fátima Domingues acknowledges the REAct (FCT-IT-LA) scientific action.

Conflicts of Interest: The authors declare no conflict of interest.

References

1. Mostofa, N.; Feltner, C.; Fullin, K.; Guilbe, J.; Zehtabian, S.; Bacanlı, S.S.; Bölöni, L.; Turgut, D. A Smart Walker for People with Both Visual and Mobility Impairment. *Sensors* **2021**, *21*, 3488. [CrossRef] [PubMed]
2. Bezuidenhout, L.; Thurston, C.; Hagströmer, M.; Moulaee Conradsson, D. Validity of Hip and Ankle Worn Actigraph Accelerometers for Measuring Steps as a Function of Gait Speed during Steady State Walking and Continuous Turning. *Sensors* **2021**, *21*, 3154. [CrossRef] [PubMed]
3. Di Tocco, J.; Raiano, L.; Sabbadini, R.; Massaroni, C.; Formica, D.; Schena, E. A Wearable System with Embedded Conductive Textiles and an IMU for Unobtrusive Cardio-Respiratory Monitoring. *Sensors* **2021**, *21*, 3018. [CrossRef] [PubMed]
4. Dos Santos, C.; Jones, M.A.; Matias, R. Short- and Long-Term Effects of a Scapular-Focused Exercise Protocol for Patients with Shoulder Dysfunctions—A Prospective Cohort. *Sensors* **2021**, *21*, 2888. [CrossRef] [PubMed]
5. Dalmeida, K.M.; Masala, G.L. HRV Features as Viable Physiological Markers for Stress Detection Using Wearable Devices. *Sensors* **2021**, *21*, 2873. [CrossRef] [PubMed]
6. Rutkowski, S.; Buekers, J.; Rutkowska, A.; Cieślik, B.; Szczegielniak, J. Monitoring Physical Activity with a Wearable Sensor in Patients with COPD during In-Hospital Pulmonary Rehabilitation Program: A Pilot Study. *Sensors* **2021**, *21*, 2742. [CrossRef] [PubMed]
7. Boyer, P.; Burns, D.; Whyne, C. Out-of-Distribution Detection of Human Activity Recognition with Smartwatch Inertial Sensors. *Sensors* **2021**, *21*, 1669. [CrossRef]
8. Liu, L.; Wang, H.; Li, H.; Liu, J.; Qiu, S.; Zhao, H.; Guo, X. Ambulatory Human Gait Phase Detection Using Wearable Inertial Sensors and Hidden Markov Model. *Sensors* **2021**, *21*, 1347. [CrossRef] [PubMed]
9. Liu, L.; Wang, H.-H.; Qiu, S.; Zhang, Y.-C.; Hao, Z.-D. Paddle Stroke Analysis for Kayakers Using Wearable Technologies. *Sensors* **2021**, *21*, 914. [CrossRef] [PubMed]
10. Vilela, P.H.; Rodrigues, J.J.P.C.; Righi, R.d.R.; Kozlov, S.; Rodrigues, V.F. Looking at Fog Computing for E-Health through the Lens of Deployment Challenges and Applications. *Sensors* **2020**, *20*, 2553. [CrossRef] [PubMed]

Review

Looking at Fog Computing for E-Health through the Lens of Deployment Challenges and Applications

Pedro H. Vilela [1,†], Joel J. P. C. Rodrigues [2,3,4,*,†], Rodrigo da R. Righi [5,†], Sergei Kozlov [4,†] and Vinicius F. Rodrigues [5,†]

1 National Institute of Telecommunications, Santa Rita do Sapucaí, MG 37540-000, Brazil; pedrov@mtel.inatel.br
2 Federal University of Piauí (UFPI), Teresina, PI 64049-550, Brazil
3 Instituto de Telecomunicações, 6201-001 Covilhã, Portugal
4 International Institute of Photonics and Optoinformatics, ITMO University, 197101 St. Petersburg, Russia; kozlov@mail.ifmo.ru
5 Applied Computing Graduate Program, Universidade do Vale do Rio dos Sinos, Unisinos, São Leopoldo, RS 93022-750, Brazil; rrrighi@unisinos.br (R.d.R.R.); vfrodrigues@unisinos.br (V.F.R.)
* Correspondence: joeljr@ieee.org
† These authors contributed equally to this work.

Received: 13 March 2020; Accepted: 26 April 2020; Published: 30 April 2020

Abstract: Fog computing is a distributed infrastructure where specific resources are managed at the network border using cloud computing principles and technologies. In contrast to traditional cloud computing, fog computing supports latency-sensitive applications with less energy consumption and a reduced amount of data traffic. A fog device is placed at the network border, allowing data collection and processing to be physically close to their end-users. This characteristic is essential for applications that can benefit from improved latency and response time. In particular, in the e-Health field, many solutions rely on real-time data to monitor environments, patients, and/or medical staff, aiming at improving processes and safety. Therefore, fog computing can play an important role in such environments, providing a low latency infrastructure. The main goal of the current research is to present fog computing strategies focused on electronic-Health (e-Health) applications. To the best of our knowledge, this article is the first to propose a review in the scope of applications and challenges of e-Health fog computing. We introduce some of the available e-Health solutions in the literature that focus on latency, security, privacy, energy efficiency, and resource management techniques. Additionally, we discuss communication protocols and technologies, detailing both in an architectural overview from the edge devices up to the cloud. Differently from traditional cloud computing, the fog concept demonstrates better performance in terms of time-sensitive requirements and network data traffic. Finally, based on the evaluation of the current technologies for e-Health, open research issues and challenges are identified, and further research directions are proposed.

Keywords: fog computing; cloud computing; e-health; healthcare; Internet of Things

1. Introduction

During the past few years, the healthcare industry noticed the potential of how Internet services could help to enhance the patient's life quality by offering analysis and processing of data in real-time. An efficient model able to provide storage and application processing over the Internet is the concept of cloud computing [1]. This model can be described as a service provided by large data centers that offer part of their infrastructure—both hardware and software—to third parties (corporations and/or individuals) and public organizations. Once these clients adhere to this type of service, they will have computational resources with increasing capacity without the need for significant investments of financial capital for the acquisition, maintenance, and management of such resources [2,3].

Complementing this concept, fog computing expands the services offered by the traditional cloud model to the network boundary [4]. Fog computing has as its main characteristics low latency, the greater geographic distribution of data, mobility, large numbers of nodes in the network, predominantly wireless access, execution of real-time applications, and device heterogeneity [4–6]. Its purpose is to enhance efficiency, performance, and reduce the volume of data sent to the cloud for processing, analysis, and storage. Nonetheless, the data collected by the sensors is gathered, processed, and stored in a temporary database instead of handing it to the cloud, thus avoiding round-trip delays and network traffic. This feature is particularly crucial for electronic-Health (e-Health) applications that transmit data over the Internet for remote real-time processing, such as remote ECG monitoring [7]. Such applications aim to monitor patients and/or processes actively [8], thus producing valuable information for decision making.

One of the reasons for the emergence and implementation of fog computing was the need to create a platform that would support the recent paradigm inherited from ubiquitous computing, the Internet of Things (IoT) [9], where any object may act as a sensor node and offer a particular service, such as data processing. In this context, CISCO [10] predicts that in 2020 the volume of data generated by IoT devices may reach six hundred zeta bytes per year. This scenario implies significant challenges to how data is exchanged among devices and the cloud, due to the high demand for bandwidth and network latency. In the context of e-Health, such issues pose some challenges regarding how to handle an increasing amount of data to maintain low latency for real-time applications. Although cloud and fog computing offer similar services, there are differences when considering the context of fog computing. With the massive amount of data arising from the end-devices, using remote cloud networks to transport data may become impractical or resource-prohibitive [11,12].

E-Health applications are a group of software and services focused on the acquisition and transmission of medical information used to deliver healthcare services [13,14]. Typically, these applications require higher levels of security and quality of service (QoS) from the system infrastructure. Currently, a few studies focus on presenting literature surveys and reviews in the scope of fog computing and e-Health applications [15–20]. However, they focus mostly either on presenting characteristics of strategies and challenges that might be a target for future research, or specific health scenarios, such as smart homes [20]. These studies do not describe a comprehensive taxonomy regarding the main characteristics of e-Health applications employing fog computing, and also do not perform an analysis considering a comparison between cloud and fog environments. With that in mind, the main objective of this paper is to update the current state-of-the-art on fog computing, focusing our discussions and contributions on the application requirements, their challenges, and open gaps still existing in the literature.

In this context, the current research presents details of fog computing and e-Health applications, analyzing the main strategies present in the literature. Based on these papers, we propose a taxonomy for the joint combination of e-Health and fog computing, thus defining the characteristics of the main applications encountered in fog computing deployments. The most relevant contributions of this study are listed as follows:

(i) An analysis of how e-Health applications benefit from the fog computing architecture in terms of deployment, communication protocols, data security, and infrastructure details;
(ii) A mapping regarding the focus of e-Health systems employing the concept of fog computing from the point of view of application requirements and main tasks;

(iii) A definition of the concerns and challenges regarding the joint combination of e-Health and fog computing, also giving directions for further research and developments.

The remainder of this paper is organized as follows. Section 2 introduces relevant concepts addressed in the paper. Section 3 presents a short overview of research initiatives related to the current research. Section 4 describes the method followed to perform the study. Section 5 presents the results of the survey; in particular, Section 5.1 describes the main architecture used in e-Health applications and identifies the most relevant fog computing features, Section 5.2 outlines the most relevant e-Health scenarios and applications under use, and Section 5.3 discusses lessons that emerged during this study. Finally, Section 6 presents the final considerations, again highlighting our contributions.

2. Background

This section presents the current state of e-Health applications regarding the fog computing concept, describes the practical approaches already under use and explains the need for a new architectural model to deal with the massive amount of generated data, especially by IoT smart devices. The focus of this section is to depict the healthcare scenario involving the IoT environment and outline its major application scenarios. It also presents the definition of some computing paradigms, such as cloud and fog computing, and compares them in terms of their computational capacity and critical network metrics.

2.1. E-Health Scenarios

Many countries are facing a considerable challenge to manage a rapidly growing aging population and the increase in chronic diseases [21,22]. The demand for medical care has risen in recent years due to the popularization of IoT smart devices, which opens a field of study for new models of delivering medical services that improve the way health information is handled. Currently, the conventional method consists of patients visiting their doctors only when they fall sick. To check their health conditions, they frequently visit hospitals or clinics to meet their physicians. As a consequence, most of the time, health parameters are manually monitored and transferred to healthcare systems, which leads to inefficient use of resources, and sometimes to higher costs. With the employment of e-Health smart sensors and medical devices, many manual tasks could be released from caregivers, since patients' conditions can be automatically monitored and analyzed remotely [23]. This new technique may revolutionize the way diagnostics and treatments are performed. Another important fact is the possibility to use healthcare systems for patients' health monitoring to keep them out of hospitals and, thus, improve hospital resource management. Furthermore, employing IoT devices for monitoring patients remotely enables them to receive medical care ubiquitously [24]. By analyzing the e-Health scenarios, we can affirm that each one can be enclosed in at least one of the goals: (i) mobility support [25]; (ii) ambient assisted living (AAL) [26]; and (iii) in-hospital treatment [8].

For the first group, with the popularization of the Internet in mobile devices, a new paradigm of e-Health has emerged: mobile-Health (m-Health) [25]. Although there is no standard definition for this concept yet, according to the World Health Organization (WHO) [27], m-Health may be understood as the offering of medical services through mobile devices such as mobile phones, sensors, and other wearable devices. This scenario also encompasses the connectivity in an ambulance or any medical air transport. In turn, the concept of AAL, as stated in [26], intends to link the usage of biomedical system monitoring with other environmental smart device sensors to provide more efficient assistance to people who live alone or have a particular disability/chronic disease. For that means, AAL performs a fundamental role in healthcare by detecting possible accidents or indicating evidence of abnormality, which helps people with their daily routine and, at the same time, helps to reduce government spending on elderly healthcare. Finally, in-hospital treatment encompasses all medical equipment used to monitor and analyze vital sign parameters from patients within hospital care [8]. Compared to the scenarios mentioned above, the devices in this context are more complex and are usually owned and maintained by the hospital itself.

2.2. Cloud Computing

Cloud computing, as stated in [28], is a prominent concept of delivering computational and storage as a service to a shared pool of selectable resources. Users can request and/or decline any computational processing or storage they may require automatically. Differently from traditional services, this approach enables the development of elastic applications that are charged accordingly to their use. As the authors in [29] mention,one of the attributes of cloud computing applied to healthcare is the possibility to consume resources from the cloud whenever needed and pay only for the used resources. Another great advantage is the capability of sharing information among health professionals, caregivers, and patients in a more structured way, reducing the risks of losing documents such as exams and medical records. However, the majority of the cloud data centers are geographically centralized and located far from its end-users. In that event, real-time applications sensitive to delays suffer from some issues, such as high round-trip delay and network congestion.

2.3. Fog Computing

With the arrival of the IoT, much has been planned on how to execute all information processing brought about by smart devices correctly, and that is precisely what the concept of fog computing tries to solve. Its purpose is to make the processing of generated data occur directly at the device, or next to it, at the network boundary, in more powerful equipment, with no need to send it to the cloud [30]. The term "fog computing" was first introduced by CISCO, and it is defined as an architecture that extends the computational and storage capacity of the cloud to the edge of the network [29]. In other words, it is a cloud infrastructure closer to its end-users. Consequently, it allows data to be collected and processed locally, reducing network latency as well as bandwidth usage. The main benefits brought by the fog computing paradigm found in the literature are: (i) reduced latency [31,32]; (ii) enhanced privacy [33]; (iii) lower need of bandwidth [10]; (iv) dependability [34]; (v) energy efficiency [35]; and (vi) data security.

First, handling data at the network's border reduces the latency when compared to other cloud-based architectures, since the physical distance is shorter. Therefore, potential data center delays may be avoided. Another advantage brought by the fog computing concept is the possibility to move computation-intensive tasks from devices with limited resources to a more powerful node [31,32]. Second, differently from the cloud model, the privacy of user data may be enhanced once the fog approach enables the analysis and processing of data on a local gateway, instead of sending the information to the cloud [33]. In other words, the number of hops that the user's data are transmitted over the network is smaller, reducing the data's exposure to external routers and networks. Third, since the fog model enables the data to be collected and processed closer to end-users, the volume of data transferred to the cloud is reduced, avoiding network traffic expenditures. This is possible because only a small part of the data is sent to a remote cloud data center for storage, whereas the rest of the data is analyzed and processed at the user location [10]. Fourth, the fog paradigm can improve the system's dependability by sharing the same functionality among the many different fog nodes. Thus, it enhances data redundancy. Further, since the computational resources are placed closer to end-users, the system may be less dependent on network connection availability [34]. Finally, energy is a primordial item that must be carefully analyzed in the IoT environment. When talking about sensor devices, reducing energy consumption is very important, since most of the sensors are battery-driven. In such cases, the overall energy efficiency can be improved by employing fog nodes acting as gateways. Such gateways can handle requests or update processes while the sensor is in a sleep state. Whenever it wakes up, it takes control of the whole application. Additionally, tasks requiring more intensive power processing can be offloaded from energy-constrained devices to nearby gateways [35].

3. Related Work and Motivation

This section presents current studies that present surveys or review articles in the scope of fog computing and healthcare in the last few years. More precisely, this section aims to demonstrate related research with this article in a way that it is possible to demonstrate its contributions compared to current research. In the fog and health scope, a few articles propose systematic literature reviews employing different methodologies [15–18,20,36]. In [18], the authors present a literature review on pervasive health applications, focusing on identifying characteristics of such applications that might benefit from fog computing environments. The authors discuss which computational tasks from e-Health sensors can be moved to fog computing infrastructures and where these tasks can be executed. Combined with this, the authors examine the trade-off of placing the identified computational tasks in the network. The study consists of a systematic literature review based on three research questions. Although the article focuses specifically on fog computing, their inclusion criteria consider a broad set of terms from the wireless sensor networks field. However, this broad set of articles is used to extract possible environments in which fog computing might play an important role. The authors identified in the literature five deployment scenarios that can benefit from fog computing: mobile, home, hospital, non-hospital, and transport. With a different scope, in [36], the authors perform a systematic literature review in the context of resource management in fog environments answering six research questions. Their inclusion criteria consider articles a set of words related to resource management plus fog or edge computing. They reviewed a group of 100 articles with publishing years ranging from 2014 to 2018. One of the main outcomes of their study regards a taxonomy of resource management approaches with the respective article for each topic.

The authors in [17] performed a literature review to summarize the main domains and issues related to fog computing and healthcare. Their primary focus was to provide an overview of fog issues tackled by the literature versus the application domain. Their method consisted of employing a systematic literature review to answer three research questions in the specific field of fog-health: the statistical publication trends according to publisher and date; the application domains; and the most discussed issues. For the first issue, they organized the articles to demonstrate the research interest in the subject in the previous few years. Their main contribution relies on an evaluation of the relationship between domains and issues in this specific research field. From their literature corpus, they found that data analysis and response time issues for remote health monitoring are the main studied characteristics. By analyzing the literature, the authors demonstrated that many applications present strategies regarding time-delay in the health scenario due to its critical importance for the treatment of patients. Silva and Júnior [16] presented a literature review on fog computing for healthcare, focusing on the state-of-the-art and challenges in this field. The paper seeks to answer what are the types of applications used, the aimed-at diseases, the characteristics of the fog solutions, the reasons for each research, and what are the main challenges. Among the authors' main findings, the most important regards the lack of a well-defined fog architecture in the context of healthcare. The search strategy employed by the authors considered studies that use fog computing in healthcare scenarios to answer six research questions. Although the authors did not define specific years for the studied articles, their strategy resulted in a short year range (2015–2018) that considers less than four years since the research was done in early 2018. From the resulting corpus, for each research question, the authors classified the articles according to their focus.

In [15], the authors presented a systematic literature review of healthcare IoT applications employing fog computing with shared resources. The article performed a systematic literature review with main inclusion criteria that encapsulate cloud computing, fog computing, and edge computing. Their focus was to present the current literature research status in fog computing in the healthcare applications field, show the performance evaluation objectives of the state-of-the-art, define the main methods they employ, and find some future directions in this research topic. From the resulting corpus, the authors focused specifically on technical details from strategies of fog computing in healthcare IoT systems. This literature strategy resulted in a broad set of articles that the authors classified within their three-category taxonomy: methods, system development, and review and survey. According to the authors, these categories were derived from previous studies and aimed to cover articles that focus on models, implemented architectures, and literature reviews. In [20], the authors presented a literature review focused on fog computing for smart home scenarios. The authors' strategy consisted of a systematic literature review that seeks to answer six research questions regarding fog-based smart home applications. To do so, their inclusion criteria considered the terms "fog" and "smart" combined to a set of words referring to building and homes. Based on the reviewed articles, the authors proposed a taxonomy categorizing the solutions that are focused on resource management or service management. Differently from the other studies, In [19], the authors did not employ a systematic literature review in their research process. Instead, they discussed characteristics of edge and fog computing environments and how they can be applied for healthcare applications. The authors aimed at describing the principal concepts of edge and fog computing that are important in the context of healthcare. Their main contributions focus on presenting strategies for how to combine such technologies to distribute computing tasks that are currently performed in the cloud computing layer. The authors presented an analysis of health IoT applications integrating the fog computing layer between the hospital infrastructure and the cloud environment. This layer allows applications to respond quickly in case of a medical crisis, since the fog processing capabilities are closer than the cloud.

Table 1 summarizes the six related works presented in this section. Half of the articles focus only on fog computing as technology, while the other half also include at least one more technology in their evaluation. Considering if they present a taxonomy in their research or not, the last three articles propose a taxonomy with their findings [15,20,36]. On the one hand, the authors in [15] presented a taxonomy dividing the strategies into three categories in which they classified if the articles present methods or systems, or perform literature reviews. On the other hand, in [20], the authors built a taxonomy dividing the articles into two categories that classify the articles according to whether they are service-based or resource-based. Besides them, the authors in [36] presented a taxonomy focusing specifically on resource management approaches and the respective articles tackling each category. The current related work demonstrates that the studies focus mainly on presenting fog computing as a new layer between the end-users and cloud infrastructures. Additionally, the studies present some contributions by defining the main characteristics of solutions employing fog computing to healthcare scenarios. Considering all the studies presented in this section, they fall into at least one of the following affirmations: (i) papers do not describe a comprehensive taxonomy regarding the main characteristics of healthcare applications employing fog computing; and (ii) articles do not perform an analysis considering a comparison between cloud and fog environments. The next sections of the current article focus on exploring these issues.

Table 1. Summary of surveys and reviews related to the current research. IoT: Internet of things; WSN: wireless sensor networks.

Paper	Year	Technologies	Range	Taxonomy	Description
[18]	2017	Fog, IoT, WSN	2005–2016	-	Systematic review of wireless sensor networks applications that might benefit from fog computing in healthcare.
[16]	2018	Fog	2015–2018	-	Review classifying articles according to applications, diseases, research, characteristics, motivations, and challenges.
[17]	2018	Fog	2010–2018	-	Systematic review of fog computing issues for the applications domain in the healthcare scenario.
[19]	2019	Fog, Edge	-	-	Description of the edge and fog concepts, and proposal of an architecture that combines them for healthcare applications.
[15]	2019	Fog, IoT	2007–2017	✓	Review of methods, systems, and surveys on IoT-based healthcare applications.
[36]	2020	Fog, IoT	2014–2018	✓	Systematic literature review of resource management strategies in fog environments.
[20]	2020	Fog	2016–2019	✓	Systematic literature review of fog computing applications for smart homes.

4. Literature Review Methodology

This study presents a literature review based on the principles of systematic literature reviews [37] in order to make it reproducible and achieve high-quality results. This section outlines the research methodology, presenting the strategy used for the collection and selection of the most appropriate contributions and corresponding papers. The goal of this research is to summarize and update the current state-of-the-art of e-Health applications employing the concept of fog computing and present its major characteristics in terms of its computing tasks. Therefore, Table 2 defines a set of research questions (RQ) that guide the review process of this study. The importance of the research questions is to provide a better understanding of the impact of fog computing on e-Health applications. Furthermore, it helps to identify the major characteristics of systems that have already deployed the fog concept concerning architectural models, network metrics for performance evaluation, security issues, and so on.

Table 2. Research questions (RQs).

ID	Question
RQ1	How do e-Health applications benefit from the fog computing architecture?
RQ2	What is the focus of e-Health systems employing the concept of fog computing?
RQ3	What are the current issues related to fog computing on e-Health?

The literature selection was based on journal and conference papers with the aid of electronic database resources . The following databases were queried: the Institute of Electrical and Electronics Engineers (IEEE) IEEExplore digital library (https://ieeexplore.ieee.org/); the Elsevier journal directory (https://www.elsevier.com/); and ResearchGate social networking (https://www.researchgate.net/). The inclusion criteria consisted of querying these databases with different keywords to collect the raw literature corpus. In each database, the following set of keywords was applied in the search string: "fog computing", "edge computing", "fog-based system", "fog-health", and "fog-cloud computing". Figure 1 presents the whole process of collecting and selecting the candidate articles for the review. The resulting raw corpus of articles contained 18,490 articles, to which a title analysis was applied as a first filter to decrease the number of candidates. This process resulted in 1314 articles that were joined for the last filter phase, which consisted of exclusion criteria.

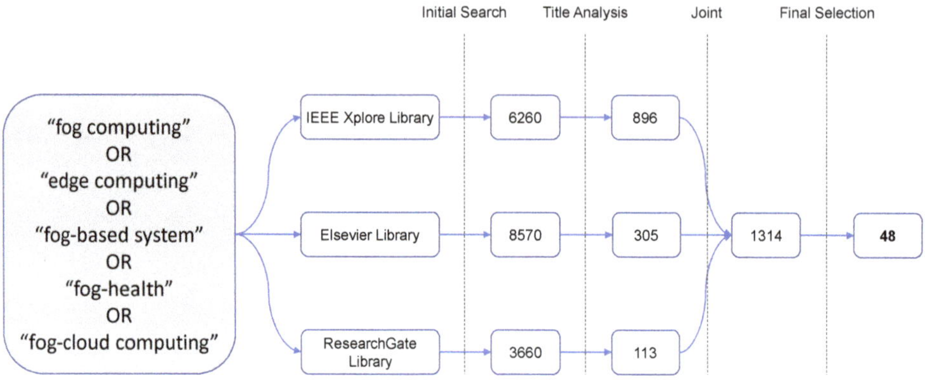

Figure 1. Summary of the systematic literature review process.

The exclusion criteria process was employed through a first reading phase in order to filter the most relevant research to be included in this study. The exclusion consisted of the following removal steps: (i) removal of papers written in languages other than English; (ii) removal of redundant and/or unwanted papers that are not related to fog computing or e-Health applications; (iii) removal of books, manuals, theses, and papers not related to fog computing; and (iv) removal of papers mentioning fog computing but not applied to healthcare. Finally, at the end of the process, 48 articles were selected to be reviewed and included in this study. This resulting corpus of studies was carefully analyzed to identify their main aspects. The next three sections are organized in a way to reflect the three research questions, and the articles are reviewed in them aiming at answering the RQs presented in Table 2.

5. Results

This section presents the answers for the three questions detailed in Table 2.

5.1. RQ1: How Do E-Health Applications Benefit from the Fog Computing Architecture?

To answer that question, it is first necessary to understand the main characteristics of fog computing deployments present in the literature. Therefore, Sections 5.1.1–5.1.4 present the main characteristics of such deployments present in the literature regarding architectural model, communication, infrastructure, and security, respectively. Then, Section 5.1.5 discusses the most important details and presents a resulting taxonomy that depicts the main findings regarding fog computing in the scope of health applications.

5.1.1. Architecture Model

As a new approach for computation, fog computing supplements the classic cloud computing and its services closer to its end-users. For this purpose, it is capable of providing computation and storage resources in a decentralized model. The main concept of fog computing architecture is a promising subject in the telecommunication research field. Recently, many architectures for fog computing have been presented, where the three-tier architecture is considered to be the most predominant structure [4]. The basic architecture deployed in the fog computing paradigm is illustrated in Figure 2 and is composed of the following layers: edge, fog, and cloud. The edge (device) layer is the closest layer to end-users/devices. It consists of several devices, such as sensors and mobile phones. The devices in this layer are responsible for collecting data from physical objects and sending them to the upper layer through short-range radio frequency technologies. Located at the edge of the network, the fog layer is comprised of network devices, such as routers, gateways, and base stations, among several others. Such devices/nodes are responsible for tasks such as scheduling, storing, and managing distributed computation. The third layer is the tier with sufficient storage and computational resources, responsible for extensive data analysis and permanent storage. Different from traditional cloud architectures, this layer accesses the cloud core network in a periodical and controlled way, improving the utilization of available resources [38].

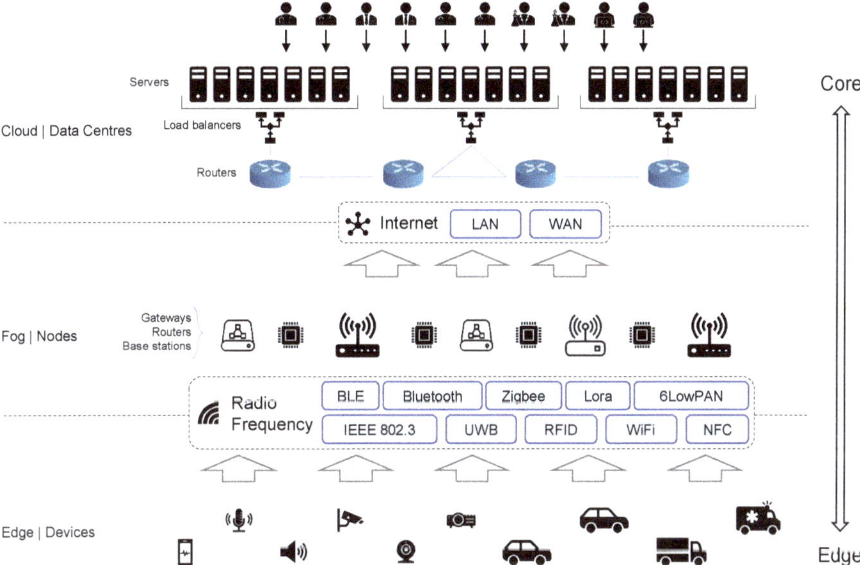

Figure 2. Illustration of a basic three-tier fog architecture presenting some sensors, devices, and communication protocols present in the model.

In [39], the authors present an architectural design for IoT systems. In this design, the cloud works as an extension of the fog layer in an assistive way. The fog gateway is placed between the cloud and the user devices, such as sensors and actuators, to meet network requirements as well as to manage and provide resources to several distributed fog nodes. According to [40], offering computation and storage at the edge of the network also helps to reduce bandwidth usage and mitigates security and privacy concerns. Several approaches place their computation task on a single node of a personal area network (PAN) or local area network (LAN) [41,42]. The data collected at this level are processed and sent to the upper level and, sometimes, to the cloud. Other approaches employ two or more fog nodes linked between the device sensing and the cloud access points [43,44].

5.1.2. Communication Protocols

IoT protocols are very relevant when performance and energy efficiency are required. In the study presented in [34], a detailed research was carried out for evaluating some of the most important protocols applied to the e-Health system, in terms of performance and energy consumption. The main protocols used in e-Health applications can be classified into three groups: (i) constrained application protocol (CoAP); (ii) Internet protocol version 6 over low power wireless personal area network (6LoWPAN); and (iii) message queuing telemetry transport (MQTT). CoAP is an Internet application protocol defined by the RFC 7252 [45] that is designed for resource-constrained sensor devices [46]. It is most explored in wireless sensor networks (WSN) in which devices have energy constraints and need a lightweight communication protocol to perform their operations. It implements methods similar to hypertext transfer protocol (HTTP), where data is encoded in a simple binary format, often based on JavaScript object notation (JSON). CoAP uses the representational state transfer (REST) style to make the resources available over the user datagram protocol (UDP) [42,46–48]. One of the the main ideas behind CoAP is to allow machine-to-machine (M2M) communication while keeping the message overhead small to avoid packet fragmentation, thus increasing the packet probability delivery [45]. WSNs are typically deployed in e-Health systems for data collection. By this means, according to [47], the utilization of lightweight protocols helps to speed up the response time once the amount of data exchanged from the application and the back-end system is smaller.

6LoWPAN is a protocol that allows data transmission between low power devices over IEEE 802.15.4 networks [49]. The IEEE 802.15.4 is a communication standard that aims devices with low-data-rate and low-power capabilities, allowing them short-range radio frequency transmissions in low-rate wireless personal area networks (WPANs) [50]. 6LoWPAN is designed on top of this standard to provide such devices with the ability to use the Internet protocol capabilities. Its specifications can be found in different RFC definitions, since each one deals with specific subjects: RFC 4919 [51], RFC 4944 [52], RFC 6282 [53], RFC 6775 [54], and RFC 7668 [55].

Finally, MQTT is a very common protocol, defined by the ISO/IEC 20922 [56], present in IoT environments [46]. Its main goal is to provide M2M communication through the exchange of bi-directional messages so that remote nodes can communicate using the MQTT infrastructure. It is an extremely lightweight and simple protocol designed to operate in hardware-constrained devices and offer low bandwidth consumption. MQTT architecture is based on transmission control protocol/internet protocol (TCP/IP), and its messages are exchanged through the publish/subscribe paradigm. Such a paradigm is composed of a component called a broker, which is responsible for receiving, queuing, and dispatching messages from publishers to subscribers [41,46,57].

5.1.3. Infrastructure Technologies

Infrastructure details involve both communication and hardware technologies used to set up fog environments in the health scope. E-Health systems typically employ a combination of networks to interconnect medical devices to the cloud. Among several networks, some commonly appear in the literature: wireless body area networks (WBANs), WPANs, LANs, and wide area networks (WANs). Using a particular technology for radio frequency, WBANs and WPANs can be seen as networks that enable the connection of devices acquiring personal information close to the patients. While WBAN refers to wearable sensors attached to patients [58], WPANs offer a higher layer on top of WBAN, allowing sensors and devices to communicate among themselves and to access local networks [59]. On the other hand, LAN and WAN are the already-spread networks for local and long-range communication. In a mobile scenario, the majority of sensors are connected to an Internet access point via Wi-Fi communication [60,61]. Another way to connect sensors is through wireless sensor and actuator networks (WSANs), which consist of a group of sensors and actuators wirelessly connected that perform distributed sensing and actuating tasks. A fog-based gateway architecture for WSAN is presented in [62], where the vitalized model is a joining of a gateway and a micro server connected via Ethernet. The communication of the interface between the gateway nodes may be wired (e.g., Ethernet) or wireless, such as third generation (3G), and long term evolution (LTE).

Regarding hardware technologies, there are several development platforms found in the literature for both research and commercial purposes. Such platforms are based on mobile computer boards that allow programmers to develop and quickly prototype their systems. The Arduino is one of the most popular platforms used in many applications due to its ease of programming and its low cost. It is a printed circuit board consisting of ATMEL microcontroller input/output circuits, which can be easily connected to a computer and programmed in the Assembly, C, and C++ languages [63]. Intel Galileo is a 32-bit open-source development board specially designed for IoT applications. It is compatible with Arduino and its respective shields. The board supports blacktooth, Wi-Fi, radio frequency, and Ethernet communication. Additionally, Intel Galileo supports a variety of sensors, such as temperature, electrocardiography, and oxygen concentration, among many others [64]. Another development board is the Raspberry Pi, which is a complete plug-and-play computer with an integrated processor and random access memory (RAM) chips used for IoT development. Despite its higher cost compared to the others, the small device is trendy because of its massive number of open-source packages and libraries, which can be easily implemented using Python, for instance. Besides, the board can be equipped with Ethernet, Wi-Fi, or blacktooth interfaces [65]. Finally, Pandaboard is based on a small advanced RISC machine (ARM) computer from Texas Instruments. It is commonly used for research development purposes because of its minimal energy consumption as well as its low cost. It works with several network standards, such as Wi-Fi, blacktooth, and Ethernet [66].

The literature does not only include physical development platforms but also simulation tools that permit researches to evaluate the effect of their software solutions. An evaluation environment for real-time applications employing fog computing is necessary to enhance the innovation and development of new technologies. As testbeds in the real world are, most of the time, very expensive, the development of software for simulations proves to be an efficient tool to address these problems. Focusing on cloud computing environments, the authors in [67] proposed a framework called *cloudSim* for the simulation of public/private cloud environments. The platform allows the user to model its cloud environment to perform several tests and evaluate its performance before deploying a production environment. In the fog scenario, in [68] the authors proposed a platform called *ifogSim*, which is a toolkit for the modeling and simulation of resource management written in JAVA that intends to minimize latency, energy consumption, bandwidth usage and operational costs. Currently available in the cloud, *ifogSim* provides a simulation environment in which it is possible to model and test massive scenarios of IoT and fog environments.

5.1.4. Security Issues

IoT applications in healthcare must be able to keep medical data private and safe from unauthorized access. Any part of the system exposed to a hacker or malicious software may cause critical consequences. Many researchers are concerned about how to solve security issues in IoT systems since security cannot be 100% guaranteed. Thus, health professionals and caregivers must define an acceptable risk limit of IoT applications. An evaluation of the kind of data that is being processed is a critical task. Depending on its purpose, failures may have a huge impact on the patient's life. Hence, it is an essential requirement that needs to be considered to keep the system resilient against security threats. The literature presents some studies that address different security aspects: (i) authentication [69]; (ii) privacy [43,70–73]; and (iii) data encryption [9]. Because of the huge amount of devices within a fog environment, user authentication plays an important role in keeping the system safe from unwanted access to healthcare services. On the other hand, privacy is one attribute of extreme importance in e-Health applications. Patients expect their private information to be kept confidential. However, IoT health systems must allow information sharing, which is necessary for delivering high-quality care and at the same time ensure its privacy. Policies and techniques must be improved to share health data with authorized users only. In addition to the first two security issues, data encryption is also necessary, since it does not matter for the infrastructure. Patients' data are very sensitive to leaks. Although fog computing normally processes data locally, on some occasions data must be forwarded to the cloud. Therefore, fog nodes must guarantee that those data are adequately encrypted before sending them to remote environments.

In [69], the authors propose a security policy management approach to address the major challenges that are important to keep data sharing and its collaboration secure within a fog computing environment. Furthermore, the authors in [43] reinforce that risk management with all stakeholders is key for achieving optimal safety and performance of medical devices. Currently, there are a few legal commitments to preserve private data in medical applications employing IoT smart devices. In [70,71], some of these legal obligations are presented. In [72], the authors present an architecture for autonomic security management that can assess risks in healthcare information systems applying a cost-efficient self-protecting approach with little or no human intervention at all. This framework also offers prevention mechanisms for monitoring and management solutions helping decision making actions based on security issues. The authors in [73] came up with a framework concerning how to keep private health information safe from eavesdropping or malicious manipulation. Based on this framework, the authors developed a medical expert system to tackle low effectiveness due to manual operations and privacy breaches caused by the participation of doctors in the medical information process. With encryption in mind, in [9] an encryption layer is implemented in the proposed fog architecture.

5.1.5. Discussion

In the field of healthcare, one of the most critical metrics to handle critical situations is time. Considering the health status of patients, quickly diagnosing anomalies in health parameters may improve the physicians' time response to it and, consequently, save lives. One of the main characteristics of fog computing is the new layer between health sensors and the cloud data centers that process data [4]. This new layer is designed to be physically close to the sensors, and it also provides processing of data for quick responses. From the network point of view, this infrastructure closer to sensors allows data communication over the network to have lower latency levels. Consequently, by inserting this new feature to the e-Health application infrastructure, it is possible to decrease the time response of the system and gain time, which, as mentioned before, is important with regards to patients' health. Therefore, the main contribution fog computing provides for e-Health applications is this improved network infrastructure for rapid data transfer and processing.

Figure 3 depicts the proposed taxonomy showing the primary architecture model employed in the fog computing concept. It also presents the most used light-weighted protocols and platforms used in the development of medical applications in the current literature. Security and privacy issues are presented in terms of how the fog concept can enhance the privacy breaches of health information in the current model due to the vast exchange of data among IoT sensors and the cloud.

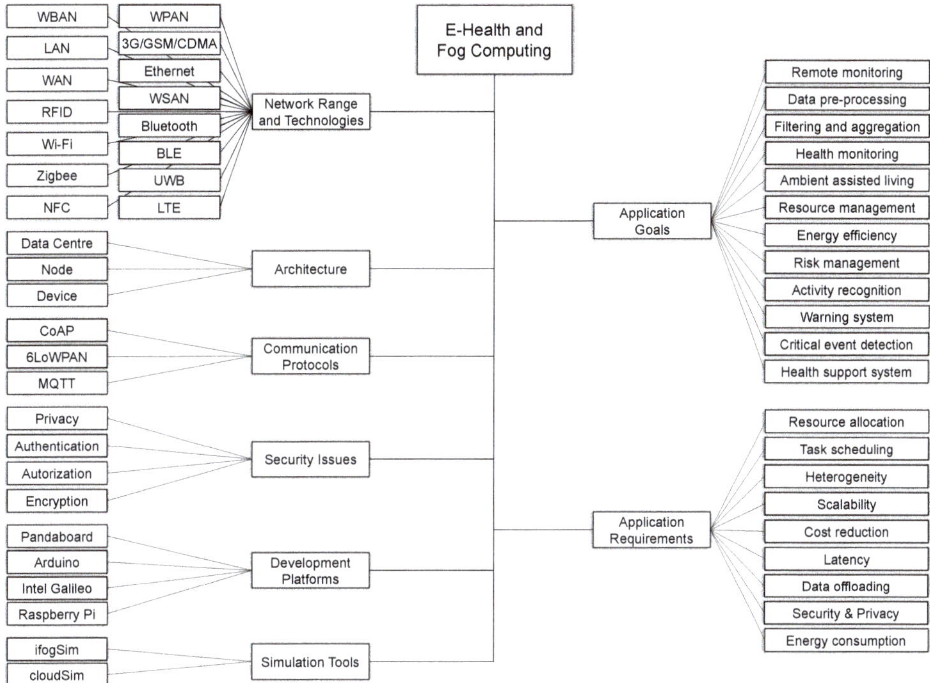

Figure 3. Proposal of a fog computing taxonomy in the scope of e-Health applications.

5.2. RQ2: *What Is the Focus of E-Health Systems Employing the Concept of Fog Computing?*

To answer the above question, this section describes the e-Health applications published in the current literature related to the fog computing paradigm. Such applications are classified in groups accordingly to their main computing task: data collection, data processing, critical data analysis, and real-time feedback. The next four sections individually describe articles that fall into these classifications. Then, Section 5.2.5 discusses the main findings, also presenting a comparison between all article reviews for each class.

5.2.1. Data Collection

As more and more IoT devices are connected to the Internet, the huge volume of data generated will require real time responses. Therefore, this amount of data implies high bandwidth costs. If false data is introduced in these IoT devices, besides compromising the accuracy of the data, it may increase the use of communication resources. Some studies focus on data collection strategies in order to decrease bandwidth needs and improve response time. Among different strategies, these studies fall in at least one of the following: (i) remote monitoring [74,75]; (ii) pre-processing [76]; (iii) compression [77]; and (iv) filtering and aggregation [76–78]. Remote monitoring systems focus on tracking the health conditions of patients outside health environments. In [74], the authors propose the eWall project, which is a system to monitor and supervise patients with mild dementia and chronic obstructive pulmonary disease (COPD) at home. The fog concept applied to this project aims at speeding up data processing in real-time for emergencies. Additionally, in the remote monitoring scope, the authors in [75] proposed an evaluation of a fog-based smart monitoring system using long-range (LoRa) radio communication in remote locations where there is no Internet connectivity. The suggested system helps minimize the power consumption with the implementation of a network using long-range wireless radio communication.

Despite remote monitoring strategies, the other strategy employs transformations to the data to improve its transmission performance. Pre-processing solutions offer some intelligence locally to process the data and extract higher levels of information. For instance, in [76], the authors proposed a three-tier structure for a real-time epileptic seizure detection system that contains a mobile device placed in the middle layer responsible for filtering, pre-processing, and extracting the electroencephalogram (EEG) characteristics. Filtering and aggregation strategies combine multiple samples of data to avoid sending redundant information. In this scope, the authors in [78] presented an aggregation scheme of lightweight privacy-preserving data. The researchers performed several experiments showing that by applying new filters at the edge of the network, communication resources are saved, avoiding false data injection. In turn, compression, as the name suggests, consists of shrinking the amount of data to be transmitted. Employing compression strategies, the authors in [77] demonstrated a medical processing system, which is responsible for inter-device communication and interpreting many wireless protocols used in e-Health applications. The system has a gateway node able to process data locally by applying data compression or fusion, and it also offers customized filtering and local storage.

5.2.2. Data Processing

Differently from data collection strategies, some studies present strategies that focus on methods to process collected data. Such studies aim at the processing level layer in the fog architecture in which data is analyzed before transmission to remote locations. In the literature, articles focusing on this issue range between different e-Health applications. However, most of them concentrate efforts to develop health monitoring systems [79–82]. In this context, the authors in [79] come up with a new architecture model to minimize dependency on cloud storage and analytics for e-Health remote monitoring. The model resulted in a better system responsiveness and a lower bandwidth requirement due to the shorter distance between the data acquisition and data processing modules. In turn, in [80], the authors presented a fog-based monitoring system focusing on the detection and prevention of mosquito-borne diseases. The goal of the system is to analyze the physical sensed data and diagnose/differentiate the several types of mosquito-borne diseases at an early stage. In [81], the authors presented a robust infrastructure for electrocardiogram (ECG) monitoring applications, where primary and backup servers responsible for processing data are placed at the edge of the network. They proposed a processing architecture for optimizing the placement of server nodes to reduce energy consumption and networking equipment. The authors in [82] suggested an innovative framework based on IPv6 as a means of mitigating the difficulty of medical applications requiring low response time through the implementation of resource scheduling techniques. Such techniques were developed in a three-layer architecture, where the physiological data is collected at the body sensing layer and transferred to the fog layer for real-time processing.

Although health monitoring catches the attention of many studies, there are also some studies that focus on different aspects. Fog nodes are composed of hundreds of fog devices that can handle storage and small computational processes. Nonetheless, reaching the resource capacity offered by the conventional cloud is still a very tough challenge. For that reason, efficient management of resources is essential for fog environment operation. In this scope, in [83], the authors proposed a dynamic resource management solution for fog environments. The main goal of their strategy is to verify parameters from the active IoT devices and then estimate the best amount of resources required by a particular node, consequently avoiding the waste of resources. In the context of AAL, the authors in [84] proposed an AAL fog architecture containing a gateway node in charge of processing speech data from patients with Parkinson's disease. The node works as an interface that processes the raw data collected by a smartwatch and sends it to the cloud infrastructure. Finally, the authors in [71] presented a study of the IoT paradigm for e-Health applications, in which they focus on the Network of Things concept. They introduced a framework for describing such environments for designing and implementing IoT solutions in healthcare.

5.2.3. Critical Data Analysis

A group of articles focuses on the critical data analysis class, which considers strategies in which the main focus is to analyze data from critical conditions. In this scope, the authors in [85] demonstrated concerns regarding the security of patients data. They designed a fog-enabled architecture for healthcare services with a focus on risk assessment and information sharing. Additionally, the cooperation with the fog device enhances the privacy of medical data due to the implementation of robust operations of cryptography. Differently, other articles vary their strategies between ECG monitoring [86] to activity recognition for e-Health services [87]. Related to the first topic, the authors in [86] proposed an IoT-based application architecture that benefits from fog to enhance the quality of service of medical systems for local and remote patients. Instead of replacing the cloud infrastructure, a fog gateway is placed at the edge of the network to collaborate with the cloud, sharing the weight of handling all information from biosensors. In turn, related to the second topic, in [87], the authors presented a blockchain fog computing scheme for human activity recognition regarding e-Health services. The use of local servers placed at the network border adds preliminary data filtering, which improves the performance of complex operations and provides faster responses to relevant events.

For fog architectures that collect sensitive data, maintaining performance and decreasing costs are a key point for such systems. Although energy is consumed during the sensing procedure, the majority of energy consumption is related to computational tasks and data transmission. In order to quantify the effect of a great volume of data in mobile applications, the study in [44] demonstrated a systematic method based on whether offloading to the fog is better than to the cloud. In their approach, two concepts of applications were studied regarding latency and energy utilization. In addition, they focused on data transmission over WiFi and 4G LTE networks, which are common for mobile devices such as smartphones. Also focusing on efficiency, the authors in [42] considered task scheduling as a way of offloading traffic from the network core by strategically allocating services among fog nodes while minimizing the cost of resources. A three-layer hierarchy system is able to manage available resources located at both cloud and fog nodes and provide the most appropriate scheduling for the workflow. Finally, in the same direction, the authors in [31] investigated a resource management fog computing strategy applied to medical cyber-physical systems. The authors focused on the challenging environment for these systems that suffer from transmission instabilities between medical devices and the cloud data center. The model presented helps to share the burden of offloading traffic from the core network by distributing tasks through base station association, where virtual machines are deployed.

5.2.4. Real-Time Feedback

Fog computing may benefit applications that need to detect unwanted events in real-time and respond to that event quickly. The real-time feedback class encompasses solutions that aim at providing quick responses to critical situations in healthcare environments. Several articles target the most variate set of applications in that scope: (i) critical event detection [61,88]; (ii) warning systems [89]; (iii) security issues [49,90]; and (iv) breath support systems [91]; In the first topic, in [88], the authors implemented a real-time signal processing algorithm for fall detection, which is able to deliver information to caregivers. These algorithms are executed at the network's border by fog servers, which collect and process all health information. Similarly, ref. [61] suggested a solution to cerebrovascular accident mitigation , where the authors created a real-time analytic system for monitoring falls caused by strokes. The fall detection allocates tasks among smart devices and the cloud in a collaborative way. In the scope of the second topic, the authors in [89] presented a remote medical monitoring system as a reliable and efficient IoT-based approach that combines both machine learning algorithms and automated management components to provide monitoring and notification services. The system is responsible for deciding whether the information goes to the cloud when processing power is needed or to the local gateway, helping in decision making.

Although IoT devices play an important role in delivering services more efficiently in the health environment, many open issues related to privacy and security are still not addressed in the literature. In this context, the authors of [90] conceived a protection framework for medical applications based on fog computing able to detect privacy leakages. The advantage of this method is the capability to analyze data closer to the source of information instead of forwarding it to the cloud for processing. In such a case, the detection and blockage of privacy leakages become a lot faster than the currently used models. Likewise, in [49], the authors defined a new scheme applying fog computing that addresses the distribution of digital certificates in the IoT environment. The new scheme approach ensures that the revoked certificates can be immediately sent to the fog nodes, mitigating the risk of accepting a revoked certificate. Finally, regarding the last topic, the authors of [91] analyzed the effects caused by the implementation of fog-to-cloud computing models in health services. To exemplify the scenario, they proposed a breath support system for patients with pulmonary problems.

5.2.5. Discussion

Current studies employ concepts of fog computing for e-Health applications with different goals. Table 3 summarizes and classifies these research initiatives according to their main tasks, which are: (i) data collection [74–78]; (ii) data processing [71,79–84]; (iii) critical data analysis [31,42,44,85–87]; and (iv) real-time feedback [49,61,88–91]. Such functionalities in medical services focus on the enhancements of the security and privacy of sensitive information. The studies are concentrated in fog-based architectures to enable faster response times for real-time applications or resource scheduling techniques to minimize the total network bandwidth usage and consequently energy consumption. Data collection covers strategies that aim at acquiring data from health sensors quickly. In turn, data processing can be seen as the next level of data collection in which strategies focus on processing data acquired from health sensors. This process results in significant information that can be monitored in real-time. On the other hand, critical data analysis comprises studies focused on improving the healthcare processes by defining new architectures for such systems. More specifically, these solutions are concentrated on critical patients' data from both the health and security points of view. Finally, real-time feedback is composed of solutions that monitor real-time data from patients focusing on diagnosing critical situations. In particular, this last group contains initiatives that provide real-time feedback for critical situations, helping physicians to improve response times for such situations. Figure 4 depicts the number of papers that focus on each one of these issues. The figure demonstrates that there is a high concentration of papers dealing with cost, latency, and data offloading. This indicates that most of the strategies are focused on improving healthcare services by focusing on data transfer but also by paying attention to costs reduction. Besides that, the figure also demonstrates that some studies focus on managing computational resources, which benefits not only the system performance but also energy efficiency. By improving energy efficiency it is possible to reduce costs at the same time.

Table 3. Reviewed e-Health applications.

	Reference Number	Application	Resource Allocation	Task Scheduling	Heterogeneity	Scalability	Cost Reduction	Latency	Data Offloading	Security & Privacy	Energy Consumption
			Application Requirement								
Data Collection	[74]	COPD monitoring system						✓	✓		
	[75]	Health monitoring system based on LoRa							✓		✓
	[76]	ECG monitoring system					✓	✓			
	[77]	Smart e-Health gateway			✓	✓					✓
	[78]	Enhanced aggregation privacy scheme					✓			✓	
Data Processing	[82]	IPv6-based framework for e-Health applications			✓		✓	✓			
	[84]	Parkinson speech device	✓						✓		
	[71]	Proposed framework for healthcare						✓	✓		
	[83]	Resource management implementation	✓				✓				
	[79]	Novel e-Health architecture for edge-IoT ecosystem						✓	✓		
	[80]	Fog-based system for mosquito diseases	✓				✓				
	[81]	Infrastructure for health monitoring applications									✓
Critical Data Analysis	[86]	Proposed architecture for Enhanced QoS		✓			✓		✓		
	[44]	Data offloading method						✓	✓		✓
	[42]	Service allocation strategy		✓			✓		✓		
	[31]	Concept of fog in medical system		✓					✓		
	[87]	Block-chain fog computing monitoring framework	✓				✓	✓			
	[85]	Fog-assisted data sharing scheme					✓			✓	✓
Real-Time Feedback	[88]	Real-time fall detection system						✓			
	[61]	Fall monitoring system	✓			✓		✓			
	[89]	Remote medical monitoring application	✓	✓			✓				
	[91]	Breath support system					✓			✓	
	[49]	New security approach for fog nodes		✓						✓	
	[90]	Privacy leakage method for medical system						✓		✓	✓

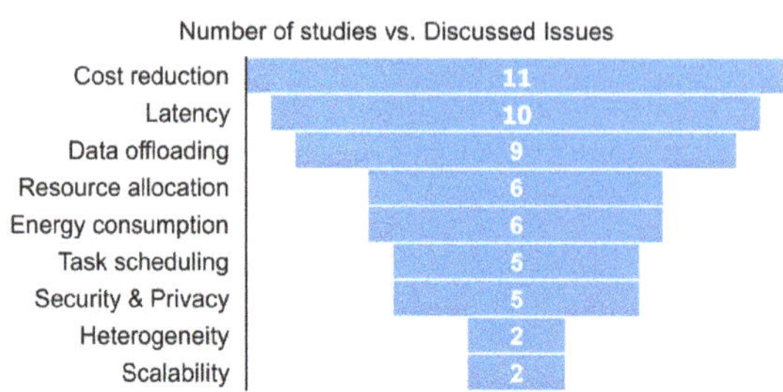

Figure 4. Amount of studies that focus on each issue.

5.3. RQ3: What Are the Current Open Issues Related to the Fog Computing on e-Health?

The fog computing approach plays an important role in healthcare by enabling applications to have fast data processing with low latency. Additionally, it addresses the security and confidentiality issues required by medical systems. Additionally, as fog nodes are closer to their end-users, fog computing helps to reduce bandwidth usage by offloading data traffic from the network core. In terms of e-Health systems, the concept of fog computing enhances data security and privacy, since all information is handled by fog devices and kept within the local network. It also helps to minimize medical expenses by reducing the amount of data exchanged between a particular health application and the cloud infrastructure. In addition, as the resources are placed closer to end-users, fog computing is able to provide instant responses for applications requiring real-time control. Although fog computing demonstrates better performance, it cannot totally replace cloud computing. Nonetheless, fog and cloud architectures will coexist while keeping their own advantages.

5.3.1. Fog versus Cloud

Apart from the similarity between fog and cloud computing, there are many differences between the two architecture models. Applications hosted in the cloud are scalable and cheaper than in the fog model, due to the cloud's huge storage capacity and hardware abstraction. Such abstraction is hard to be achieved in a fog network, since the edge devices must be acquired by the network owner. As a result of the heterogeneous nature of fog networks, a particular resource cannot be guaranteed in all fog devices [92]. Thus, keeping the system reliable requires a lot more effort, regarding complexity and cost, in fog when compared to the traditional cloud model. Ensuring a minimum of privacy and security in the fog model is a lot harder than in the cloud, since fog devices are normally maintained by many different service providers [93]. A single breach in a particular fog node may cause the system to operate incorrectly. In terms of connectivity in the fog environment, despite having hardware in full operation, if a single node loses access to the network, the entire system may be put at risk. Because fog is a decentralized model, ordinary hardware checks are much more complex and costly than in cloud infrastructures.

Nowadays, with the adoption of the IoT paradigm, current architectural approaches are unsustainable to provide services to the massive number of applications being developed. Many IoT devices are facing challenges related to latency, network bandwidth, dependability, privacy, and security, which cannot be handled in a cloud computing model. As a result, a fog computing approach has been presented as a propitious architecture to tackle these issues. The concept of fog computing is an extension of the cloud computing model to the network border. Both of them provide computational resources, storage, and network services to their end-users. Nonetheless, the fog paradigm differs from the cloud model in terms of network type, computing capacity, storage, physical location, and so on. As a new architecture, fog computing suggests several features that make it more complex than cloud [94]. According to [95–98], a comparison between these two models is presented in Table 4.

Table 4. Fog computing vs. cloud computing.

Requirements	Fog Computing	Cloud Computing
Geographical distribution	Distributed	Centralised
Location of server	At the edge of network	Within the Internet
Distance between client and server	Close (one hop)	Far (several hops)
Coverage	Extensive	Global
Latency	Low	High
Bandwidth	Low	High
Response time	short	long
Hardware	Limited resources	Scalable resources
Data storage	Temporary	Permanent
Flexibility	High	Limited

5.3.2. Trends, Future Directions, and Open Issues

The emergence of the fog computing paradigm is relatively new when comparing to other well-established technologies such as cloud and IoT. Therefore, there is enough space for future developments and the integration of this paradigm into the e-Health scenario and other technologies. Some of the challenges faced by e-Health systems that can be addressed by fog computing can be classified in three groups of interest: (i) latency; (ii) power consumption; and (iii) heterogeneity and interoperability.

As some applications are commonly sensitive to delay, fog networks must guarantee that the response time of a certain request is within a limit. In other words, the fog system must employ mechanisms to verify if a specific task can be concluded or not regarding its defined metrics. In case a fog device is not able to deliver the service based on that metric, the service request must be rejected or forwarded to the cloud. With that in mind, defining a proper task allocation is of extreme importance. This scope encompasses a variety of application fields that can benefit from delay improvements, such as medical cyber-physical systems, ultra low latency applications, and tactile internet. In medical cyber-physical systems, devices monitoring physical parameters from both the environment and patients are integrated through the network [99]. Traditionally, medical devices produce data that must be transmitted to remote stations that process these data (for instance in the cloud). In such environments, the transmission of sensitive data among nodes with proper delay is crucial to improve the quality of medical services. From the ultra-low latency applications point of view, there are almost no initiatives that focus on applications that require constant low delay communication [100]. In such a context, ultra-low-latency networks are optimized to process a very high volume of data with a low tolerance for delay. These networks are designed to support real-time applications and react quickly to changes in the data streams. In addition, recent studies are focusing on the new tactile internet paradigm [101,102]. The goal is to bring the internet to a lower level of granularity: the human senses, such as touch. This will produce a new level of data generation that might scale rapidly. Therefore, collecting and transmitting this amount of data may be challenging, requiring communication infrastructures capable of handling it with low delay.

Unlike cloud systems, fog networks are composed of several decentralized fog devices commonly connected by battery or through inefficient communication interfaces. Defining a more efficient protocol to deal with resource allocation within the fog network may help to minimize the energy consumption in the fog environment. In that context, WSNs environments are promising, since in such architectures, the devices are characterized by low energy consumption [103]. WSNs are composed of a set of small nodes that employ low-energy protocols for communication and data exchange. In particular, combining fog and

WSN has been explored by some researches in the literature; however there is still space for developing new strategies in this context. In the context of heterogeneity and interoperability, systems must inter-operate among several types of devices and vendors. Standards that ensure interoperability among such devices have not yet been addressed in the literature. For instance, it is possible to explore web services interfaces to allow data exchange between different modules and systems [104]. REST APIs (application programming interfaces) [105] can be used as a way of providing data for different systems in fog environments. Additionally, the development of new techniques for the announcements of available resources among fog devices as well as their communication protocols is fundamental to enable heterogeneous network interconnection.

6. Conclusions

In the present study, a comprehensive review of the fog computing framework was conducted, highlighting its major characteristics, such as its main architecture model, used network technologies, platforms for development, simulation tools, and its main security issues. To achieve this, a systematic literature review was performed that resulted in 48 articles analyzed carefully in this paper. Several application deployment cases regarding data collection and analysis were introduced before explaining the fog computing approach in the context of healthcare. Fog computing can be analyzed in the context of healthcare from two different points of view. The first one regards the fog infrastructure itself. Currently, studies agree that fog is a new layer between sensor devices and traditional cloud infrastructure. Such a model comprises some specific communication protocols and also several communication technologies, ranging from WPANs to LTE channels. This demonstrates that this new paradigm is used as an extension of current platforms using current technologies. The other point of view is the e-Health landscape, which comprises the use of computational applications to provide better healthcare services. Current studies focus on several issues that can affect the performance of systems and applications. In fact, one of the main issues these strategies target is cost and latency. On the one hand, cost reduction is one issue that is targeted in several areas and this is no different in healthcare scenarios. On the other hand, latency has a direct relation with time, which is a very important parameter in healthcare. Studies propose monitoring systems for data collection and processing that generate valuable feedback regarding patients' health parameters. Fog computing can improve the time it takes to reach these results, since its hardware infrastructure is closer to the users.

The main contribution of the current research has two points of view. From the technical point of view, the article presented the main characteristics of fog computing architectures focused on the health scenario. In addition, the study defined the main tasks applications perform when employing such solutions. From the society point of view, the study demonstrated some possible paths for study combining fog and healthcare. More specifically, it demonstrated how future research might explore this field to provide solutions for healthcare applications in order to increase patients' safety and security. Finally, considering future research, the topics include applying the research methodology from this survey to different paper databases. In the future, it is also possible to consider different areas and compare how fog computing is employed in each one of them. Additionally, the current research focused specifically on e-Health applications.

Author Contributions: Conceptualization, P.H.V. and J.J.P.C.R.; methodology, V.F.R., P.H.V., J.J.P.C.R. and S.K.; software, P.H.V. and J.J.P.C.R.; validation, P.H.V., R.d.R.R. and J.J.P.C.R.; formal analysis, V.F.R., P.H.V., S.K. and J.J.P.C.R.; investigation, V.F.R., P.H.V. and J.J.P.C.R.; resources, S.K. and J.J.P.C.R.; data curation, V.F.R., P.H.V. and J.J.P.C.R.; writing—original draft preparation, P.H.V. and J.J.P.C.R.; writing—review and editing, V.F.R., J.J.P.C.R. and R.d.R.R.; visualization, S.K., J.J.P.C.R. and R.d.R.R.; supervision, J.J.P.C.R.; project administration, J.J.P.C.R.; funding acquisition, J.J.P.C.R. and S.K. All authors have read and agreed to the published version of the manuscript.

Funding: This work is supported by FCT/MCTES through national funds and when applicable co-funded by EU funds under the project UIDB/EEA/50008/2020; by the Government of the Russian Federation, Grant 08-08; and by the Brazilian National Council for Scientific and Technological Development (CNPq) via Grants No. 431726/2018-3 and 309335/2017-5.

Conflicts of Interest: The authors declare no conflict of interest.

References

1. Bonomi, F.; Milito, R.; Natarajan, P.; Zhu, J. Fog computing: A platform for internet of things and analytics. In *Big Data and Internet of Things: A Roadmap for Smart Environments*; Springer: Cham, Switzerland, 2014; pp. 169–186.
2. Arunkumar, G.; Venkataraman, N. A novel approach to address interoperability concern in cloud computing. *Procedia Comput. Sci.* **2015**, *50*, 554–559. [CrossRef]
3. Atlam, H.F.; Alenezi, A.; Alharthi, A.; Walters, R.J.; Wills, G.B. Integration of cloud computing with internet of things: Challenges and open issues. In Proceedings of the 2017 IEEE International Conference on Internet of Things (iThings) and IEEE Green Computing and Communications (GreenCom) and IEEE Cyber, Physical and Social Computing (CPSCom) and IEEE Smart Data (SmartData), Exeter, UK, 21–23 June 2017; pp. 670–675.
4. Stojmenovic, I.; Wen, S. The fog computing paradigm: Scenarios and security issues. In Proceedings of the 2014 Federated Conference on Computer Science and Information Systems (FedCSIS), Warsaw, Poland, 7–10 September 2014; pp. 1–8.
5. Bonomi, F.; Milito, R.; Zhu, J.; Addepalli, S. Fog computing and its role in the internet of things. In Proceedings of the First Edition of the MCC Workshop on Mobile Cloud Computing, Helsinki, Finland, 17 August 2012; ACM: New York, NY, USA, 2012; pp. 13–16.
6. Mukherjee, M.; Shu, L.; Wang, D. Survey of Fog Computing: Fundamental, Network Applications, and Research Challenges. *IEEE Commun. Surv. Tutor.* **2018**, *20*, 1826–1857. [CrossRef]
7. Ozkan, H.; Ozhan, O.; Karadana, Y.; Gulcu, M.; Macit, S.; Husain, F. A Portable Wearable Tele-ECG Monitoring System. *IEEE Trans. Instrum. Meas.* **2020**, *69*, 173–182. [CrossRef]
8. Habib, C.; Makhoul, A.; Darazi, R.; Couturier, R. Health risk assessment and decision-making for patient monitoring and decision-support using Wireless Body Sensor Networks. *Inf. Fusion* **2019**, *47*, 10–22. [CrossRef]
9. Aazam, M.; Huh, E.N. Fog computing and smart gateway based communication for cloud of things. In Proceedings of the 2014 International Conference onFuture Internet of Things and Cloud (FiCloud), Barcelona, Spain, 27–29 August 2014; pp. 464–470.
10. Networking, C.V. *Cisco Global Cloud Index: Forecast and Methodology, 2015–2020;* White Paper; Cisco Public: San Jose, CA, USA, 2016.
11. Chen, S.; Zhang, T.; Shi, W. Fog Computing. *IEEE Internet Comput.* **2017**, *21*, 4–6. [CrossRef]
12. Ai, Y.; Peng, M.; Zhang, K. Edge computing technologies for Internet of Things: A primer. *Digit. Commun. Netw.* **2018**, *4*, 77–86. [CrossRef]
13. Svensson, P.G. eHealth applications in health care management. *Ehealth Int.* **2002**, *1*, 5. [CrossRef]
14. Bisio, I.; Estatico, C.; Fedeli, A.; Lavagetto, F.; Pastorino, M.; Randazzo, A.; Sciarrone, A. Brain Stroke Microwave Imaging by Means of a Newton Conjugate Gradient Method in L^p Banach Spaces. *IEEE Trans. Microw. Theory Tech.* **2018**, *66*, 3668–3682. [CrossRef]
15. Mutlag, A.A.; Ghani, M.K.A.; Arunkumar, N.; Mohammed, M.A.; Mohd, O. Enabling technologies for fog computing in healthcare IoT systems. *Future Gener. Comput. Syst.* **2019**, *90*, 62–78. [CrossRef]
16. da Silva, C.A.; de Aquino Júnior, G.S. Fog Computing in Healthcare: A Review. In Proceedings of the 2018 IEEE Symposium on Computers and Communications (ISCC), Natal, Brazil, 25–28 June 2018; pp. 1126–1131. [CrossRef]
17. Ibrahim, W.N.H.; Selamat, A.; Krejcar, O.; Chaudhry, J.A. Recent Advances on Fog Health—A Systematic Literature Review. In *New Trends in Intelligent Software Methodologies, Tools and Techniques—Proceedings of the 17th International Conference SoMeT_18, Granada, Spain, 26–28 September 2018*; Fujita, H., Herrera-Viedma, E., Eds.; Frontiers in Artificial Intelligence and Applications; IOS Press: Amsterdam, The Netherlands, 2018; Volume 303, pp. 157–170. [CrossRef]
18. Kraemer, F.A.; Braten, A.E.; Tamkittikhun, N.; Palma, D. Fog Computing in Healthcare—A Review and Discussion. *IEEE Access* **2017**, *5*, 9206–9222. [CrossRef]

19. Dash, S.; Biswas, S.; Banerjee, D.; Rahman, A.U. Edge and Fog Computing in Healthcare—A Review. *Scalable Comput. Pract. Exp.* **2019**, *20*, 191–206. [CrossRef]
20. Rahimi, M.; Songhorabadi, M.; Kashani, M.H. Fog-based smart homes: A systematic review. *J. Netw. Comput. Appl.* **2020**, *153*, 102531. [CrossRef]
21. Forsman, B.; Svensson, A. Frail Older Persons' Experiences of Information and Participation in Hospital Care. *Int. J. Environ. Res. Public Health* **2019**, *16*, 2829. [PubMed]
22. Hansson, A.; Sevensson, A.; Ahlström, B.H.; Larsson. L.G.; Forsman, B.; Alsén, P. Flawed communications: Health professionals' experience of collaboration in the care of frail elderly patients. *Scand. J. Public Health* **2017**, *46*, 680–689. [CrossRef] [PubMed]
23. Shuwandy, M.L.; Zaidan, B.B.; Zaidan, A.A.; Albahri, A.S. Sensor-Based mHealth Authentication for Real-Time Remote Healthcare Monitoring System: A Multilayer Systematic Review. *J. Med. Syst.* **2019**, *43*, 33. [CrossRef] [PubMed]
24. Mohapatra, S.; Mohanty, S.; Mohanty, S. Smart Healthcare: An Approach for Ubiquitous Healthcare Management Using IoT. In *Big Data Analytics for Intelligent Healthcare Management*; Dey, N., Das, H., Naik, B., Behera, H.S., Eds.; Advances in Ubiquitous Sensing Applications for Healthcare; Academic Press: Cambridge, MA, USA, 2019; Chapter 7, pp. 175–196. [CrossRef]
25. Gagnon, M.P.; Ngangue, P.; Payne-Gagnon, J.; Desmartis, M. m-Health adoption by healthcare professionals: A systematic review. *J. Am. Med. Inform. Assoc.* **2016**, *23*, 212–220. [CrossRef]
26. Silva, B.M.; Rodrigues, J.J.; Simoes, T.M.; Sendra, S.; Lloret, J. An ambient assisted living framework for mobile environments. In Proceedings of the 2014 IEEE-EMBS International Conference on Biomedical and Health Informatics (BHI), Valencia, Spain, 1–4 June 2014; pp. 448–451.
27. Kay, M.; Santos, J.; Takane, M. mHealth: New horizons for health through mobile technologies. *World Health Organ.* **2011**, *3*, 66–71.
28. Mell, P.; Grance, T. *The NIST Definition of Cloud Computing*; Computer Security Division, Information Technology Laboratory, National Institute of Standards and Technology Gaithersburg: Gaithersburg, MD, USA, 2011.
29. Griebel, L.; Prokosch, H.; Köpcke, F.; Toddenroth, D.; Christoph, J.; Leb, I.; Engel, I.; Sedlmayr, M. A scoping review of cloud computing in healthcare. *BMC Med. Inform. Decis. Mak.* **2015**, *15*, 17. [CrossRef]
30. Mouradian, C.; Naboulsi, D.; Yangui, S.; Glitho, R.H.; Morrow, M.J.; Polakos, P.A. A Comprehensive Survey on Fog Computing: State-of-the-Art and Research Challenges. *IEEE Commun. Surv. Tutor.* **2018**, *20*, 416–464. [CrossRef]
31. Gu, L.; Zeng, D.; Guo, S.; Barnawi, A.; Xiang, Y. Cost Efficient Resource Management in Fog Computing Supported Medical Cyber-Physical System. *IEEE Trans. Emerg. Top. Comput.* **2017**, *5*, 108–119. [CrossRef]
32. Choi, N.; Kim, D.; Lee, S.J.; Yi, Y. A fog operating system for user-oriented iot services: Challenges and research directions. *IEEE Commun. Mag.* **2017**, *55*, 44–51. [CrossRef]
33. Kapsalis, A.; Kasnesis, P.; Venieris, I.S.; Kaklamani, D.I.; Patrikakis, C.Z. A Cooperative Fog Approach for Effective Workload Balancing. *IEEE Cloud Comput.* **2017**, *4*, 36–45. [CrossRef]
34. Mun, D.H.; Dinh, M.L.; Kwon, Y.W. An Assessment of Internet of Things Protocols for Resource-Constrained Applications. In Proceedings of the 2016 IEEE 40th Annual Computer Software and Applications Conference (COMPSAC), Atlanta, GA, USA, 10–14 June 2016; Volume 1, pp. 555–560. [CrossRef]
35. Ni, J.; Zhang, K.; Lin, X.; Shen, X.S. Securing fog computing for internet of things applications: Challenges and solutions. *IEEE Commun. Surv. Tutor.* **2017**, *20*, 601–628. [CrossRef]
36. Ghobaei-Arani, M.; Souri, A.; Rahmanian, A.A. Resource management approaches in fog computing: A comprehensive review. *J. Grid Comput.* **2019**, 18, 1–42. [CrossRef]
37. Kitchenham, B.; Charters, S. *Guidelines for Performing Systematic Literature Reviews in Software Engineering*; Technical Report; Keele University: Keele, UK, 2007.
38. Sarkar, S.; Misra, S. Theoretical modelling of fog computing: A green computing paradigm to support IoT applications. *IET Netw.* **2016**, *5*, 23–29. [CrossRef]

39. Nikoloudakis, Y.; Panagiotakis, S.; Markakis, E.; Pallis, E.; Mastorakis, G.; Mavromoustakis, C.X.; Dobre, C. A Fog-Based Emergency System for Smart Enhanced Living Environments. *IEEE Cloud Comput.* **2016**, *3*, 54–62. [CrossRef]
40. Liu, H.; Eldarrat, F.; Alqahtani, H.; Reznik, A.; de Foy, X.; Zhang, Y. Mobile Edge Cloud System: Architectures, Challenges, and Approaches. *IEEE Syst. J.* **2018**, *12*, 2495–2508.
41. Caro, N.D.; Colitti, W.; Steenhaut, K.; Mangino, G.; Reali, G. Comparison of two lightweight protocols for smartphone-based sensing. In Proceedings of the 2013 IEEE 20th Symposium on Communications and Vehicular Technology in the Benelux (SCVT), Namur, Belgium, 21 November 2013; pp. 1–6. [CrossRef]
42. Pham, X.Q.; Huh, E.N. Towards task scheduling in a cloud-fog computing system. In Proceedings of the 2016 18th Asia-Pacific Network Operations and Management Symposium (APNOMS), Kanazawa, Japan, 5–7 October 2016; pp. 1–4. [CrossRef]
43. Sametinger, J.; Rozenblit, J.; Lysecky, R.; Ott, P. Security challenges for medical devices. *Commun. ACM* **2015**, *58*, 74–82. [CrossRef]
44. Hu, W.; Gao, Y.; Ha, K.; Wang, J.; Amos, B.; Chen, Z.; Pillai, P.; Satyanarayanan, M. Quantifying the Impact of Edge Computing on Mobile Applications. In Proceedings of the 7th ACM SIGOPS Asia-Pacific Workshop on Systems (APSys '16), Hong Kong, China, 4–5 August 2016; ACM: New York, NY, USA, 2016. [CrossRef]
45. Shelby, Z.; Hartke, K.; Bormann, C.; Frank, B. RFC 7252: The constrained application protocol (CoAP). *Internet Eng. Task Force* **2014**, 7252, 1–112. Available online: https://tools.ietf.org/html/rfc7252 (accessed on 29 April 2020).
46. Kayal, P.; Perros, H. A comparison of IoT application layer protocols through a smart parking implementation. In Proceedings of the 2017 20th Conference on Innovations in Clouds, Internet and Networks (ICIN), Paris, France, 7–9 March 2017; pp. 331–336. [CrossRef]
47. Khattak, H.A.; Ruta, M.; Sciascio, E.D.; Sciascio, D. CoAP-based healthcare sensor networks: A survey. In Proceedings of the 2014 11th International Bhurban Conference on Applied Sciences Technology (IBCAST), Islamabad, Pakistan, 14–18 January 2014; pp. 499–503. [CrossRef]
48. Bormann, C.; Castellani, A.P.; Shelby, Z. CoAP: An Application Protocol for Billions of Tiny Internet Nodes. *IEEE Internet Comput.* **2012**, *16*, 62–67. [CrossRef]
49. Chen, Q.; Lambright, J.; Abdelwahed, S. Towards Autonomic Security Management of Healthcare Information Systems. In Proceedings of the 2016 IEEE First International Conference on Connected Health: Applications, Systems and Engineering Technologies (CHASE), Washington, DC, USA, 27–29 June 2016; pp. 113–118. [CrossRef]
50. *IEEE Standard for Low-Rate Wireless Networks*; IEEE Std 802.15.4-2015 (Revision of IEEE Std 802.15.4-2011); IEEE: Piscataway, NJ, USA, 2016; pp. 1–709.
51. Kushalnagar, N.; Montenegro, G.; Schumacher, C. Rfc 4919: Ipv6 over low-power wireless personal area networks (6lowpans): Overview. *Assumpt. Probl. Statement Goals* **2007**, *31*, 45–75.
52. Montenegro, G.; Kushalnagar, N.; Hui, J.; Culler, D. RFC 4944: Transmission of IPv6 packets over IEEE 802.15. 4 networks. *Req. Comments* **2007**, *4944*, 1–30. Available online: https://tools.ietf.org/html/rfc4944 (accessed on 29 April 2020).
53. Hui, J.; Thubert, P. RFC 6282: Compression format for IPv6 datagrams over IEEE 802.15. 4-based networks. *Internet Eng. Task Force* **2011**, 6282, 1–24. Available online: https://tools.ietf.org/html/rfc6282 (accessed on 29 April 2020).
54. Shelby, Z.; Chakrabarti, S.; Nordmark, E.; Bormann, C. RFC 6775-Neighbor Discovery Optimization for IPv6 over Low-Power Wireless Personal Area Networks (6LoWPANs). *IETF RFC 6775* **2012**, 6775, 1–55. Available online: https://tools.ietf.org/html/rfc6775 (accessed on 29 April 2020).
55. Nieminen, J.; Savolainen, T.; Isomaki, M.; Patil, B.; Shelby, Z.; Gomez, C. RFC 7668-IPv6 over Bluetooth Low Energy. *IETF RFC 7668* **2015**, *7668*, 1–21. Available online: https://tools.ietf.org/html/rfc7668 (accessed on 29 April 2020).
56. Information Technology—Message Queuing Telemetry Transport (MQTT) v3. 1.1., ISO/IEC20922, 2016. Available online: https://www.iso.org/standard/69466.html (accessed on 29 April 2020).

57. Bandyopadhyay, S.; Bhattacharyya, A. Lightweight Internet protocols for web enablement of sensors using constrained gateway devices. In Proceedings of the 2013 International Conference on Computing, Networking and Communications (ICNC), San Diego, CA, USA, 28–31 January 2013; pp. 334–340. [CrossRef]
58. Khan, J.Y.; Yuce, M.R. Wireless Body Area Network (WBAN) for Medical Applications. In *New Developments in Biomedical Engineering*; Campolo, D., Ed.; IntechOpen: Rijeka, Croatia, 2010; Chapter 31. [CrossRef]
59. Misic, J.; Misic, V. *Wireless Personal Area Networks: Performance, Interconnection, and Security with IEEE 802.15.4*; John Wiley & Sons: Hoboken, NJ, USA, 2008.
60. Chen, H.; Liu, H. A remote electrocardiogram monitoring system with good swiftness and high reliablility. *Comput. Electr. Eng.* **2016**, *53*, 191–202. [CrossRef]
61. Cao, Y.; Chen, S.; Hou, P.; Brown, D. FAST: A fog computing assisted distributed analytics system to monitor fall for stroke mitigation. In Proceedings of the 2015 IEEE International Conference on Networking, Architecture and Storage (NAS), Boston, MA, USA, 6–7 August 2015; pp. 2–11.
62. Lee, W.; Nam, K.; Roh, H.G.; Kim, S.H. A gateway based fog computing architecture for wireless sensors and actuator networks. In Proceedings of the 2016 18th International Conference on Advanced Communication Technology (ICACT), Pyeongchang, Korea, 31 January–3 February 2016; pp. 210–213. [CrossRef]
63. Doukas, C. *Building Internet of Things with the Arduino*; CreateSpace Independent Publishing Platform: North Charleston, SC, USA, 2012.
64. Ramon, M.C. Intel Galileo and Intel Galileo Gen 2. In *Intel® Galileo and Intel® Galileo Gen 2: API Features and Arduino Projects for Linux Programmers*; Apress: Berkeley, CA, USA, 2014; pp. 1–33. [CrossRef]
65. Upton, E.; Halfacree, G. *Raspberry Pi User Guide*; John Wiley & Sons: Hoboken, NJ, USA, 2014.
66. Gia, T.N.; Thanigaivelan, N.K.; Rahmani, A.M.; Westerlund, T.; Liljeberg, P.; Tenhunen, H. Customizing 6LoWPAN networks towards Internet-of-Things based ubiquitous healthcare systems. In Proceedings of the NORCHIP, Tampere, Finland, 27–28 October 2014; pp. 1–6.
67. Calheiros, R.N.; Ranjan, R.; Beloglazov, A.; De Rose, C.A.; Buyya, R. CloudSim: A toolkit for modeling and simulation of cloud computing environments and evaluation of resource provisioning algorithms. *Softw. Pract. Exp.* **2011**, *41*, 23–50. [CrossRef]
68. Gupta, H.; Dastjerdi, A.V.; Ghosh, S.K.; Buyya, R. iFogSim: A Toolkit for Modeling and Simulation of Resource Management Techniques in Internet of Things, Edge and Fog Computing Environments. *arXiv* **2016**, arXiv:1606.02007.
69. Dsouza, C.; Ahn, G.J.; Taguinod, M. Policy-driven security management for fog computing: Preliminary framework and a case study. In Proceedings of the 2014 IEEE 15th International Conference on Information Reuse and Integration (IRI), Redwood City, CA, USA, 13–15 August 2014; pp. 16–23.
70. Laplante, P.A.; Laplante, N. The internet of things in healthcare: Potential applications and challenges. *IT Prof.* **2016**, *18*, 2–4. [CrossRef]
71. Laplante, P.A.; Kassab, M.; Laplante, N.L.; Voas, J.M. Building Caring Healthcare Systems in the Internet of Things. *IEEE Syst. J.* **2018**, *12*, 3030–3037. [CrossRef]
72. Linthicum, D.S. Connecting Fog and Cloud Computing. *IEEE Cloud Comput.* **2017**, *4*, 18–20. [CrossRef]
73. Huang, H.; Gong, T.; Ye, N.; Wang, R.; Dou, Y. Private and Secured Medical Data Transmission and Analysis for Wireless Sensing Healthcare System. *IEEE Trans. Ind. Inform.* **2017**, *13* , 1227–1237. [CrossRef]
74. Fratu, O.; Pena, C.; Craciunescu, R.; Halunga, S. Fog computing system for monitoring Mild Dementia and COPD patients—Romanian case study. In Proceedings of the 2015 12th International Conference on Telecommunication in Modern Satellite, Cable and Broadcasting Services (TELSIKS), Nis, Serbia, 14–17 October 2015; pp. 123–128. [CrossRef]
75. Kharel, J.; Reda, H.T.; Shin, S.Y. Fog computing-based smart health monitoring system deploying lora wireless communication. *IETE Tech. Rev.* **2019**, *36*, 69–82. [CrossRef]
76. Hosseini, M.P.; Hajisami, A.; Pompili, D. Real-time epileptic seizure detection from eeg signals via random subspace ensemble learning. In Proceedings of the 2016 IEEE International Conference on Autonomic Computing (ICAC), Wurzburg, Germany, 17–22 July 2016; pp. 209–218.

77. Rahmani, A.M.; Thanigaivelan, N.K.; Gia, T.N.; Granados, J.; Negash, B.; Liljeberg, P.; Tenhunen, H. Smart e-Health Gateway: Bringing intelligence to Internet-of-Things based ubiquitous healthcare systems. In Proceedings of the 2015 12th Annual IEEE Consumer Communications and Networking Conference (CCNC), Las Vegas, NV, USA, 9–12 January 2015; pp. 826–834. [CrossRef]
78. Lu, R.; Heung, K.; Lashkari, A.H.; Ghorbani, A.A. A Lightweight Privacy-Preserving Data Aggregation Scheme for Fog Computing-Enhanced IoT. *IEEE Access* **2017**, *5*, 3302–3312. [CrossRef]
79. Ray, P.P.; Dash, D.; De, D. Edge computing for Internet of Things: A survey, e-healthcare case study and future direction. *J. Netw. Comput. Appl.* **2019**, *140*, 1–22. [CrossRef]
80. Vijayakumar, V.; Malathi, D.; Subramaniyaswamy, V.; Saravanan, P.; Logesh, R. Fog computing-based intelligent healthcare system for the detection and prevention of mosquito-borne diseases. *Comput. Hum. Behav.* **2019**, *100*, 275–285. [CrossRef]
81. Isa, I.S.M.; Musa, M.O.I.; El-Gorashi, T.E.H.; Elmirghani, J.M.H. Energy Efficient and Resilient Infrastructure for Fog Computing Health Monitoring Applications. In Proceedings of the 2019 21st International Conference on Transparent Optical Networks (ICTON), Angers, France, 9–13 July 2019; pp. 1–5.
82. Hu, J.; Wu, K.; Liang, W. An IPv6-based framework for fog-assisted healthcare monitoring. *Adv. Mech. Eng.* **2019**, *11*. [CrossRef]
83. Aazam, M.; Huh, E.N. Dynamic resource provisioning through Fog micro datacenter. In Proceedings of the 2015 IEEE International Conference on Pervasive Computing and Communication Workshops (PerCom Workshops), St. Louis, MO, USA, 23–27 March 2015; pp. 105–110. [CrossRef]
84. Monteiro, A.; Dubey, H.; Mahler, L.; Yang, Q.; Mankodiya, K. Fit: A Fog Computing Device for Speech Tele-Treatments. In Proceedings of the 2016 IEEE International Conference on Smart Computing (SMARTCOMP), St. Louis, MO, USA, 18–20 May 2016; pp. 1–3. [CrossRef]
85. Tang, W.; Zhang, K.; Zhang, D.; Ren, J.; Zhang, Y.; Shen, X.S. Fog-Enabled Smart Health: Toward Cooperative and Secure Healthcare Service Provision. *IEEE Commun. Mag.* **2019**, *57*, 42–48. [CrossRef]
86. Gia, T.N.; Jiang, M.; Rahmani, A.M.; Westerlund, T.; Liljeberg, P.; Tenhunen, H. Fog Computing in Healthcare Internet of Things: A Case Study on ECG Feature Extraction. In Proceedings of the 2015 IEEE International Conference on Computer and Information Technology; Ubiquitous Computing and Communications, Dependable, Autonomic and Secure Computing, Pervasive Intelligence and Computing, Liverpool, UK, 26–28 October 2015; pp. 356–363. [CrossRef]
87. Islam, N.; Faheem, Y.; Din, I.U.; Talha, M.; Guizani, M.; Khalil, M. A blockchain-based fog computing framework for activity recognition as an application to e-Healthcare services. *Future Gener. Comput. Syst.* **2019**, *100*, 569–578. [CrossRef]
88. Craciunescu, R.; Mihovska, A.; Mihaylov, M.; Kyriazakos, S.; Prasad, R.; Halunga, S. Implementation of Fog computing for reliable E-health applications. In Proceedings of the 2015 49th Asilomar Conference on Signals, Systems and Computers, Pacific Grove, CA, USA, 8–11 November 2015; pp. 459–463. [CrossRef]
89. Azimi, I.; Anzanpour, A.; Rahmani, A.M.; Liljeberg, P.; Salakoski, T. Medical warning system based on Internet of Things using fog computing. In Proceedings of the 2016 International Workshop on Big Data and Information Security (IWBIS), Jakarta, Indonesia, 18–19 October 2016; pp. 19–24. [CrossRef]
90. Gu, J.; Huang, R.; Jiang, L.; Qiao, G.; Du, X.; Guizani, M. A Fog Computing Solution for Context-Based Privacy Leakage Detection for Android Healthcare Devices. *Sensors* **2019**, *19*, 1184. [CrossRef] [PubMed]
91. Masip-Bruin, X.; Marín-Tordera, E.; Alonso, A.; Garcia, J. Fog-to-cloud Computing (F2C): The key technology enabler for dependable e-health services deployment. In Proceedings of the 2016 Mediterranean Ad Hoc Networking Workshop (Med-Hoc-Net), Vilanova i la Geltru, Spain, 20–22 June 2016; pp. 1–5. [CrossRef]
92. Hao, Z.; Novak, E.; Yi, S.; Li, Q. Challenges and software architecture for Fog Computing. *IEEE Internet Comput.* **2017**, *21*, 44–53. [CrossRef]
93. Mukherjee, M.; Matam, R.; Shu, L.; Maglaras, L.; Ferrag, M.A.; Choudhury, N.; Kumar, V. Security and privacy in fog computing: Challenges. *IEEE Access* **2017**, *5*, 19293–19304. [CrossRef]
94. Shi, Y.; Ding, G.; Wang, H.; Roman, H.E.; Lu, S. The fog computing service for healthcare. In Proceedings of the 2015 2nd International Symposium on Future Information and Communication Technologies for Ubiquitous HealthCare (Ubi-HealthTech), Beijing, China, 28–30 May 2015; pp. 1–5. [CrossRef]

95. Firdhous, M.; Ghazali, O.; Hassan, S. Fog computing: Will it be the future of cloud computing? In Proceedings of the Third International Conference on Informatics & Applications, Kuala Terengganu, Malaysia, 8–10 October 2014; pp. 8–15.
96. Dastjerdi, A.V.; Buyya, R. Fog computing: Helping the Internet of Things realize its potential. *Computer* **2016**, *49*, 112–116. [CrossRef]
97. Luan, T.H.; Gao, L.; Li, Z.; Xiang, Y.; Wei, G.; Sun, L. Fog computing: Focusing on mobile users at the edge. *arXiv* **2015**, arXiv:1502.01815.
98. Kai, K.; Cong, W.; Tao, L. Fog computing for vehicular Ad-hoc networks: Paradigms, scenarios, and issues. *J. China Univ. Posts Telecommun.* **2016**, *23*, 56–96. [CrossRef]
99. Dey, N.; Ashour, A.S.; Shi, F.; Fong, S.J.; Tavares, J.A.M. Medical Cyber-Physical Systems: A Survey. *J. Med. Syst.* **2018**, *42*, 1–13. [CrossRef] [PubMed]
100. Shih, Y.; Chung, W.; Pang, A.; Chiu, T.; Wei, H. Enabling Low-Latency Applications in Fog-Radio Access Networks. *IEEE Netw.* **2017**, *31*, 52–58. [CrossRef]
101. Saddik, A.E. Multimedia and the Tactile Internet. *IEEE MultiMedia* **2020**, *27*, 5–7. [CrossRef]
102. Aijaz, A.; Sooriyabandara, M. The Tactile Internet for Industries: A Review. *Proc. IEEE* **2019**, *107*, 414–435. [CrossRef]
103. Yick, J.; Mukherjee, B.; Ghosal, D. Wireless sensor network survey. *Comput. Netw.* **2008**, *52*, 2292–2330. [CrossRef]
104. Richardson, L.; Ruby, S. *RESTful Web Services*; O'Reilly Media, Inc.: Sebastopol, CA, USA, 2008.
105. Masse, M. *REST API Design Rulebook: Designing Consistent RESTful Web Service Interfaces*; O'Reilly Media, Inc.: Sebastopol, CA, USA, 2011.

MDPI

Article

Paddle Stroke Analysis for Kayakers Using Wearable Technologies

Long Liu [1,2], Hui-Hui Wang [1], Sen Qiu [2,*], Yun-Cui Zhang [3] and Zheng-Dong Hao [2]

1 Dalian Neusoft University of Information, Dalian 116023, China; liulong@neusoft.edu.cn (L.L.); wanghuihui@neusoft.edu.cn (H.-H.W.)
2 The Laboratory of Intelligent System, Dalian University of Technology, Dalian 116024, China; bayin@mail.dlut.edu.cn
3 The Research Institute of Photonics, Dalian Polytechnic University, Dalian 116023, China; zhang_yc@dlpu.edu.cn
* Correspondence: qiu@dlut.edu.cn; Tel.: +86-0411-8470-3623

Abstract: Proper stroke posture and rhythm are crucial for kayakers to achieve perfect performance and avoid the occurrence of sport injuries. The traditional video-based analysis method has numerous limitations (e.g., site and occlusion). In this study, we propose a systematic approach for evaluating the training performance of kayakers based on the multiple sensors fusion technology. Kayakers' motion information is collected by miniature inertial sensor nodes attached on the body. The extend Kalman filter (EKF) method is used for data fusion and updating human posture. After sensor calibration, the kayakers' actions are reconstructed by rigid-body model. The quantitative kinematic analysis is carried out based on joint angles. Machine learning algorithms are used for differentiating the stroke cycle into different phases, including entry, pull, exit and recovery. The experiment shows that our method can provide comprehensive motion evaluation information under real on-water scenario, and the phase identification of kayaker's motions is up to 98% validated by videography method. The proposed approach can provide quantitative information for coaches and athletes, which can be used to improve the training effects.

Keywords: paddle stroke analysis; motion reconstruction; inertial sensor; data fusion

Citation: Liu, L.; Wang, H.-H.; Qiu, S.; Zhang, Y.-C.; Hao, Z.-D. Paddle Stroke Analysis for Kayakers Using Wearable Technologies. *Sensors* **2021**, *21*, 914. https://doi.org/10.3390/s21030914

Academic Editor: Maria de Fátima Domingues

Received: 30 December 2020
Accepted: 19 January 2021
Published: 29 January 2021

1. Introduction

As a multi-cycle high-intensity water sports project, kayak includes single boat, double boat, four-person boat and obstacle slalom formats. The athlete sits in the boat, facing forward, holding the paddle with oar handle on the middle position, relying on the feet to steer the rudder to control the course. kayaking is closely related to athletes' professional skills, physical fitness, psychological state and other aspects, and the key to win in the competition is to complete the paddling movement efficiently and without any mistakes under tense conditions. Professional teams and amateur clubs are constantly looking for advanced methods to help athletes improve their athletic performance [1–3].

Several different methods have been used for rowing technique testing. Video-based analysis by researchers to quantify the stroke performance of rowers is one of them [4], but this approach is restricted by the experimental site, which suffers from visual blind field, and it does not observe the behavior accurately. Other studies have been devoted to the creation of instrumented boats to assess rowers' performance by measuring oar's power and motion [5]. Franz Gravenhorst et al. assessed rowing technology by continuously monitoring rowers' seat positions [6]. Henry et al. used strain gauges and potentiometers to measure the forces on the oars and their angular positions to assess rowing performance by power output [7]. Although boat speed, stroke frequency, stroke force and power output of athletes are evaluated, the standardization and normalization of rowing motion are not studied to fundamentally improve the kayaker's technique.

All the above studies are to test kayaking equipment by instruments, so as to study the rowing performance. However, kayaking is a cooperative movement of athletes' arms, torso, upper limbs and body along a certain movement track, which is a combination of factors of athletes' muscle activity, joint flexion/extension angle and limbs activity [8,9]. Therefore, wearable sensors can evaluate the skills of rowing sports based on athletes movements capture. At present, it is a new trend to use inertial sensors on evaluating rowing performance [10,11]. M.Tesconi et al. developed a tight wearable sensor system, but it is only tested in the laboratory without extensive practical testing. In fact, the effort of balance control in on-water scenario results in clumsiness and change in the motor part of the action, and further leads to discredit on the simulated indoor experiments [12]. Rachel C. King et al. introduced a kinematic monitoring system which combines inertial sensors and other body sensor network nodes. However, the rotation of the back and femur in the sagittal plane is mainly monitored, and the flexion and extension of the upper limbs are not studied. The rowing action consists of shoulder abduction/adduction and elbow flexion/extension, and a lack of analysis in coronal plane is worth consideration [13]. Ruffaldi et al. put forward a sensor fusion model which integrates wearable inertial measurement with physiological sensors and marks the buttocks, sternum, acromion, outside humerus, medial epicondyle, ulna and radial styloid process. This method can support the human motion tracking of rowing in indoor and outdoor environments. However, the experimental results and discussions are defined within the indoor training, and the absence of real rowing data is a major barrier for true evaluation [14]. Taken together, the quantitative analysis of kayaking athletes' movement in the above research is relatively incomplete, and there is limited research on monitoring and analyzing athletes' whole-body movement [15].

To improve athletes' rowing skills and provide the comprehensive technical guidance of kayaking sprint, we put forward a method of motion reconstruction and analysis based on inertial measurement units (IMUs). In our study, the athlete body is regarded as a set of rigid models, including several segments with self-defined length, and each body segment is modeled as a line which is connected by the friction-free joints. The attitude information is described by means of quaternions. Based on the quaternion-driven rotation, the joint angle of flexion and extension movement of each human body part is fully described.

The main contributions of this work are as follows:

- We use extend Kalman filter to fuse inertial sensor data and reconstruct real-time postures of kayakers in different paddling positions and capture the movements of single kayakers under realistic conditions.
- The validity and accuracy of the proposed posture estimation algorithm is verified using an optical motion capture system.
- Stroke quality of single kayaker is analyzed based on joint angles obtained from motion reconstruction.
- We use machine learning algorithms for phase partitioning of a stroke cycle.

The article is structured as follows. Section 2 introduces the hardware and software platform. The experimental methodology is described in Section 3. The results of this study is described are Section 4. Finally, discussions and conclusions are given in Section 5 and Section 6, respectively.

2. Systematic Data Collection and Participants

2.1. System Platform

The IMUs-based motion tracking system used to obtain attitude of rowers was self-designed in our lab. The total system consists of several tiny inertial measurement units, a transceiver and a set of self-designed software. The physical device is shown in Figure 1. Each inertial measurement unit contains a triaxial accelerometer, a triaxial gyroscope and a triaxial magnetometer to measure the three-shaft acceleration, three-shaft angular velocity, and three-shaft magnetic field intensity. The detailed specifications are shown in Table 1. We used an STM32 as the micro control unit to record information from the inertial sensor units and store the raw data in a TransFlash card. The motion captured process by IMUs is controlled wirelessly by the Lora signal sent by the computer. In every sensor node, there is a miniature Lora wireless module soldered to the printed circuit board of the slave node. The main controller communicates with the RF module via SPI interface, and it is always waiting for an interrupt to start acquisition of inertial measurement. Once the master transceiver connected to the computer by USB interface receives the "START" signal from the self-made software, it delivers in the broadcasting signal. Upon receiving the public signal, the slave receivers start to collect data simultaneously. In addition, after the initial stage is completed, the participants are required to perform specific movements (stomp and shake hands) for synchronization of different sensor nodes. The system acquires the raw data at a sampling rate of 400 Hz. When the operation of the capture process is completed, the data stored in the memory card are exported for subsequent kinematic data analysis. A 3.7 V (400 mAh) battery is selected to power the whole system. At the end of each experiment, the collected data are immediately copied to the personal computer. The apparatus is preceded by practice trials, and the test battery lasted approximately 2 h. Each experiment period is about 10 min per volunteer. The composition of motion capture system is depicted in Figure 2.

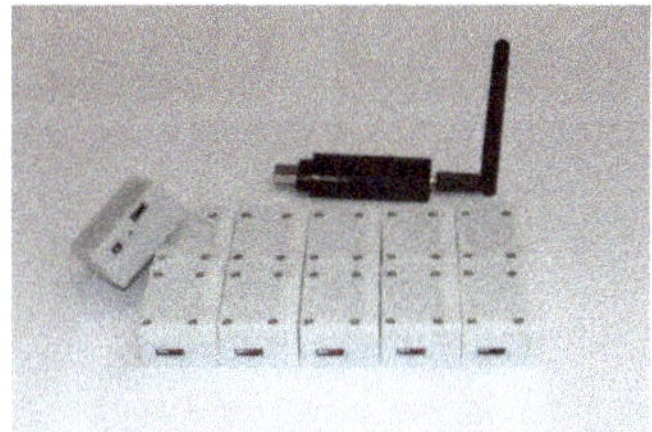

Figure 1. Diagram of apparatus components.

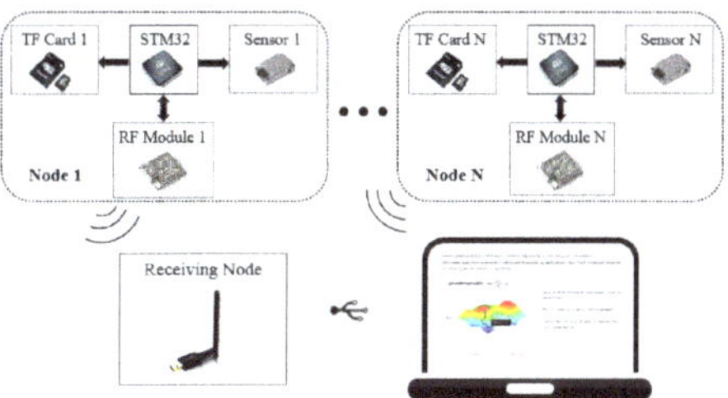

Figure 2. Schematic illustration of the self-designed motion capture system structure.

Table 1. Sensor detailed Specifications.

Unit	Accelerometer	Gyroscope	Magnetometer
Dimensions	3 axis	3 axis	3 axis
Sensitivity(/LSB)	0.833 mg	0.04 deg/s	142.9 uguass
Dynamic Range	±18 g	±1200 deg/s	±1.9 gauss
−3 dB Bandwidth(Hz)	330	330	25
Nonlinearity(%FS)	0.2	±0.1	0.1
Misalignment(deg)	0.2	0.05	0.25

2.2. Participants and Experimental Sites

Six kayakers recruited from the provincial sprint team participates in the preliminary study. The training duration for each participant is approximately more that three years. The participants train six times per week, with daily training sessions of 5–6 h. They have an average weight of 72.4 ± 6.4 kg and height of 1.76 ± 0.33 m. They are all female athletes. All participants had their height and weight recorded, and they were fully informed and consent was obtained. The experiment was conducted at the Sports Training Center in Dalian, Liaoning Province, China (latitude 121°15.194′ N and longitude 38°55.467′ E).

3. Methods

3.1. Rigid Body Model

In the kinematic analysis of this paper, the body is defined as a rigid structure, and the skeletal structural model of the rower can be defined by 17 rigid segments (feet, calves, thighs, pelvis, waist, chest, arms, thighs and head), as shown in Figure 3a. The kayaker's pelvis is set as the zero point. The length of each segment is approximately proportionally determined by the height of the athlete. As the line segments are connected by joint, if the orientation described by Euler angle or quaternion of each segment is obtained, the positions of the other lines can be determined by the length of the skeletal vector in the attitude in Figure 3a by iterative calculation [16]. To capture the limbs kinematical information, the nine-axis inertial measurement is placed on each of the back, waist, thigh, arm, calf and the limb segments, which is used to obtain raw acceleration, angular velocity and magnetometer information during the acquisition process. The placements of inertial sensors are shown in Figure 3b. The collection of kinematic data from head and feet was not the focus of this research, and the sensor nodes for head and feet motion capture were removed. The solutions of motion reconstruction data are replaced by the neighboring nodes. Thus, capturing full-body human motion needs 10 sensor nodes, as depicted in Figure 3b, and only six sensor nodes are required for capturing upper limbs.

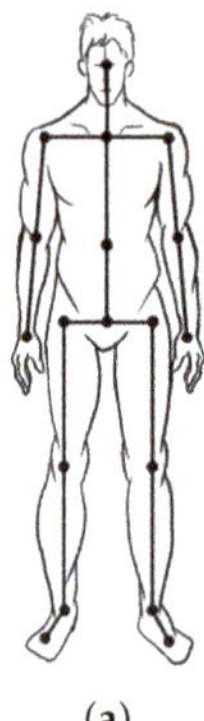

(**a**)

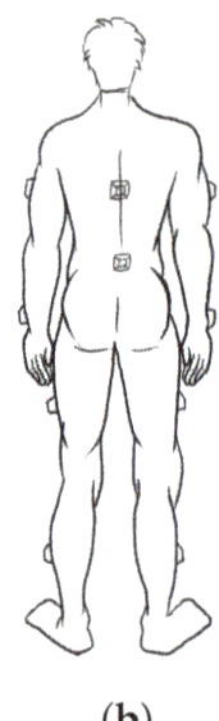

(**b**)

Figure 3. Schematic of the whole body structure definition with rigid body model and the distributed representation of the sensor's location: (**a**) a rigid-body model with 17 human body segments connected via revolute joints, and the segment number can be adjusted according to the specific situation; and (**b**) the location and method of fixation for the sensors.

The joint angles are defined as the angle between the vectors connecting adjacent body segments. The changes in posture of kayakers corresponds to the flexion and extension angle of each joint. The joint angles are depicted in Figure 4. The kayaking movement referred to the athlete who sits in the kayak and the paddling movement is primarily achieved by the upper limbs, so the flexion/extension of the shoulder joint (SF) and flexion/extension of the elbow joint (EF) are the main aspects of our approach [17].

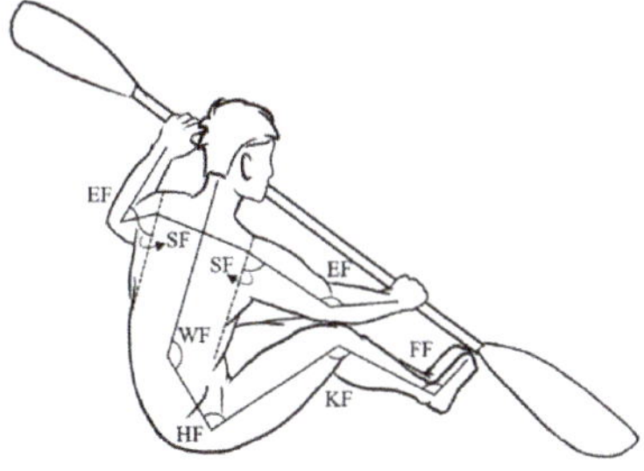

Definition	Abbreviation
Waist Flexion Angle	WF
Hip Flexion Angle	HF
Shoulder Flexion Angle	SF
Elbow Flexion Angle	EF
Knee Flexion Angle	KF
Foot Flexion Angle	FF

Figure 4. The definition of body joint angle.

3.2. Precondition of Rehabilitation Motion Design

As shown in Figure 5, the whole motion capture system contains three coordinate systems, and each three-dimensional coordinate system is based on the standard right-handed 3D cartesian coordinate system. During the experiment, the inertial sensors are secured to each body segment by Velcro straps to acquire information related to the bodily motions of the participants.

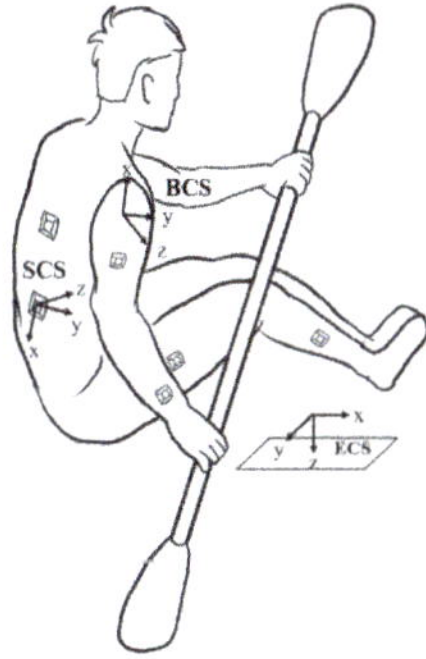

Figure 5. Definition of the coordinate systems used in this work, and including Earth Coordinate System (ECS), Body Coordinate System (BCS) and Sensor Coordinate System (SCS).

The sensor returns signal in local coordinates, which is called Sensor Coordinate System (SCS). It is defined as the coordinates of a sensor node placed on the human body. However, the motion is observed in the Earth Coordinate System (ECS). The trajectories of each body segment and joint are measured with respect to the ECS, as opposed to the Body Coordinate System (BCS). Therefore, the raw inertial data time series are each transformed from SCS to ECS using the rotation matrix.

The x and y axes are not aligned to Earth coordinate system since magnetometer data are always disturbed by metal constructions in experimental environment facilities, which further affects the accuracy of motion tracking. To correct the magnetometer offset, sensitivity and axis-misalignment, some researches put forward calibration methods [14]. The ellipsoid fitting based on least square method is adopted in this paper for magnetometer calibration with data recorded on-site the day of the event [18]. The remaining angular velocity and acceleration signals are also filtered using second-order digital filter with cut-off frequency at 100 Hz [19]. At the end of signal preprocessing of magnetometer, accelerometer and gyroscope, the data fusion algorithm is used to estimate accurate pose of all human body segment.

3.3. Motion Reconstruction Based on Quaternion

Taking into account the gimbal locking problem of Euler angle and its computational complexity of the rotation matrix, the final posture parameters are obtained by fusing acceleration, angular velocity and magnetometer measurement data with the quaternion-based interpretation of body segments rotations [20]. The quaternion is defined by Equation (1).

$$q = \mathrm{q}_0 + \mathrm{q}_1\mathrm{i} + \mathrm{q}_2\mathrm{j} + \mathrm{q}_3\mathrm{k} \tag{1}$$

with the three imaginary units i,j,k, which satisfy the equation $\mathrm{i}^2 = j^2 = k^2 = ijk = -1$ [21] In the initial state, the athlete is asked to face north and stand with his arms down for a few seconds, and the sensor nodes are fixed with a belt on the surface of the body. The initial Euler angles of each sensors nodes are obtained by using Equations (2)–(6). Herein, ϕ, θ and φ, respectively, represent pitch, roll and yaw angle. a_x, a_y and a_z are the linear acceleration of the device in three directions, while m_x^b and m_y^b are the local magnetic intensity around the test sites after calibration. The rotation quaternion ${}_s^eq(0)$ between SCS and ECS at the initial state is also obtained by using Equation (7) [22].

$$\phi = \mathrm{asin}\frac{a_y}{\sqrt[2]{a_x^2 + a_y^2 + a_z^2}} \tag{2}$$

$$\theta = \mathrm{atan}\,2\frac{-a_x}{a_z} \tag{3}$$

$$m_x^a = m_x^b cos(\theta) + m_z^b sin(\theta) \tag{4}$$

$$m_y^a = m_x^b sin(\theta) sin(\phi) + m_y^b cos(\phi) - m_z^b * cos(\theta) sin(\phi) \tag{5}$$

$$\varphi = \mathrm{atan}\,2\left(m_y^a, m_x^a\right) \tag{6}$$

$${}_s^eq(0) = \begin{bmatrix} \cos\left(\frac{\phi}{2}\right)\cos\left(\frac{\theta}{2}\right)\cos\left(\frac{\varphi}{2}\right) - \sin\left(\frac{\phi}{2}\right)\sin\left(\frac{\theta}{2}\right)\sin\left(\frac{\varphi}{2}\right) \\ \cos\left(\frac{\phi}{2}\right)\cos\left(\frac{\theta}{2}\right)\sin\left(\frac{\varphi}{2}\right) - \cos\left(\frac{\phi}{2}\right)\sin\left(\frac{\theta}{2}\right)\sin\left(\frac{\varphi}{2}\right) \\ \cos\left(\frac{\phi}{2}\right)\cos\left(\frac{\theta}{2}\right)\sin\left(\frac{\varphi}{2}\right) + \cos\left(\frac{\phi}{2}\right)\sin\left(\frac{\theta}{2}\right)\sin\left(\frac{\varphi}{2}\right) \\ \cos\left(\frac{\phi}{2}\right)\cos\left(\frac{\theta}{2}\right)\sin\left(\frac{\varphi}{2}\right) + \cos\left(\frac{\phi}{2}\right)\sin\left(\frac{\theta}{2}\right)\sin\left(\frac{\varphi}{2}\right) \end{bmatrix} \tag{7}$$

In the initial state, the rotation quaternion ${}_s^bq(0)$ between SCS and BCS is equal to ${}_s^eq(0)$ because the BCS and ECS overlap, and, since the sensors are placed at fixed position, ${}_s^bq(t)$ is always equal to ${}_s^eq(0)$ throughout the measurement. The main purpose of the project is to analyze the movement of kayak in terms of the Earth coordinate system. The rotation quaternion of each limb from BCS to ECS can be calculated by Equation (8):

$${}_b^eq(t) = {}_s^e q(t) \otimes {}_b^s q(t) \tag{8}$$

where ${}_b^sq(t)$ is the conjugate of ${}_s^bq(t)$ and ${}_s^eq(t)$ is constantly updated with data fusion algorithm over time. In this paper, the bar-shaped human body model conformed to the rigid body model is defined for representing the human pose. The participant's pelvis in the ECS is set as the initial position; the posture of each segment is obtained by the iteration of the relationship.

Taking the adjacent segment of the upper limb as an example to explain how we iteratively compute the attitude of human body, the upper arm and forearm body segments are modeled as two segments in the elbow joint, as illustrated in Figure 6. $S_{u1}(t)$ and $S_{f1}(t)$ are the end position of two segments, while $S_{u0}(t)$ and $S_{f0}(t)$ are the start position of two segments. The length vectors of the upper arm and forearm are $d_u(t)$ and $d_f(t)$ Thus, the position of each segment is obtained from Equations (9) and (10). When all segments' postures of the rigid body model are obtained from the relative skeletal segment iteration calculation, the flexion and extension joint angle can also be solved by inverse cosine between two adjacent skeletal segment vectors.

$$S_{u1}(t) = S_{u0}(t) + ({}^{e}_{b,u}q(t)) \otimes [0 d_u(t)] \otimes ({}^{e}_{b,u}q(t))^* \tag{9}$$

$$S_{f1}(t) = S_{f0}(t) + ({}^{e}_{b,f}q(t)) \otimes [0 d_f(t)] \otimes ({}^{e}_{b,f}q(t))^* \tag{10}$$

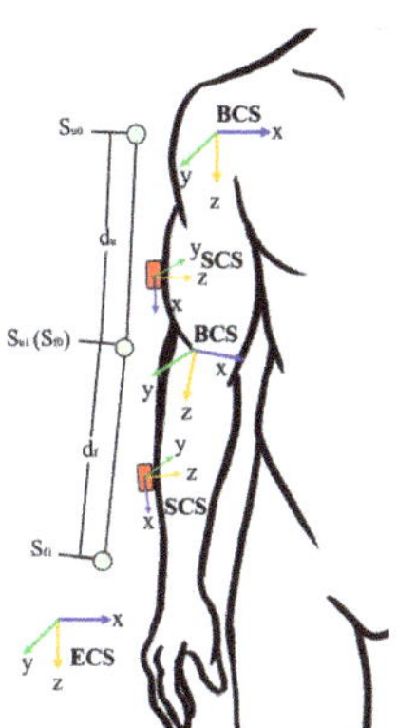

Figure 6. Skeleton structure of upper limb.

3.4. Data Fusion Algorithm

There are many data fusion algorithms for reconstructing human motion, such as complementary filter (CF) [23], gradient descent method (GD) [24] and extended Kalman filter (EKF) [25]. In this article, the EKF method is used for multi-sensor data fusion. The EKF model adopted in this paper is depicted in Equations (11) and (12).

$$x(t) = f(x(t-1), u, t-1) + w(t-1) \tag{11}$$

$$z(t) = h(x(t), t) + v(t) \tag{12}$$

where $x(t)$ stands for state vector at time t; $z(t)$ represents the observation vector at time t; u indicates the measured values of gyroscope; $w(t)$ and $v(t)$ are the process noise of the state variable and system measured noise; and $Q(t)$ and $R(t)$ denote their covariance matrices, respectively. The state variables are defined as follows, where $\boldsymbol{q} = \begin{bmatrix} q_0 & q_1 & q_2 & q_3 \end{bmatrix}^{\mathrm{T}}$ represents the pose quaternions and $\boldsymbol{b}_\omega = \begin{bmatrix} b_{\omega x} & b_{\omega v} & b_{\omega z} \end{bmatrix}^{\mathrm{T}}$ are the measurements biases of the gyroscope. The state variable vector is written in Equation (13).

$$X = \begin{bmatrix} q_0 & q_1 & q_2 & q_3 & b_{\omega x} & b_{\omega y} & b_{\omega z} \end{bmatrix}^T \tag{13}$$

The updated pose quaternions can be solved by differential equations, which are expressed as Equations (14) and (15).

$$\dot{q} = \frac{1}{2}\hat{q} \otimes \omega \tag{14}$$

$$w = \begin{bmatrix} 0 & -w_x & -w_y & -w_z \\ w_x & 0 & w_z & -w_y \\ w_y & -w_z & 0 & w_x \\ w_z & w_y & -w_x & 0 \end{bmatrix} \tag{15}$$

Furthermore, the state transition matrix is obtained by Equations (16)–(18), where T denotes the sample periods.

$$F_q^q(t|t-1) = \begin{bmatrix} 0 & -\frac{1}{2}(\omega_x - b_{\omega x})T & -\frac{1}{2}(\omega_y - b_{\omega y})T & -\frac{1}{2}(\omega_z - b_{\omega z})T \\ \frac{1}{2}(\omega_x - b_{\omega x})T & 0 & \frac{1}{2}(\omega_z - b_{\omega z})T & -\frac{1}{2}(\omega_y - b_{\omega y})T \\ \frac{1}{2}(\omega_y - b_{\omega y})T & -\frac{1}{2}(\omega_z - b_{\omega z})T & 0 & \frac{1}{2}(\omega_x - b_{\omega x})T \\ \frac{1}{2}(\omega_z - b_{\omega z})T & \frac{1}{2}(\omega_y - b_{\omega y})T & \frac{1}{2}(\omega_x - b_{\omega x})T & 0 \end{bmatrix} \tag{16}$$

$$F_\omega^q(t|t-1) = \begin{bmatrix} \frac{1}{2}q_1T & \frac{1}{2}q_2T & \frac{1}{2}q_3T \\ -\frac{1}{2}q_0T & \frac{1}{2}q_3T & -\frac{1}{2}q_2T \\ -\frac{1}{2}q_3T & -\frac{1}{2}q_0T & \frac{1}{2}q_1T \\ \frac{1}{2}q_2T & -\frac{1}{2}q_1T & \frac{1}{2}q_0T \end{bmatrix} \tag{17}$$

$$F(t|t-1) = \begin{bmatrix} F_q^q & F_\omega^q \\ O_{3\times 4} & O_{3\times 3} \end{bmatrix} \tag{18}$$

Then, the state transition function is obtained after model linearizing, as shown in Equation (19),

$$\Phi(t \mid t-1) = I_{7\times 7} + F(t \mid t-1) \times T. \tag{19}$$

Here, let $W(t) = [W_\omega(t), W_{\text{bias}}(t)]^T$ be the vector composed of the noise of gyroscope and the gyroscope bias migration noise. We then proceed to obtain the expressions of the derivatives of two variables. We get $Q(t) = W(\text{t}-1) \times T \times Q_{\text{w}} \times T \times W^T(\text{t}-1)$. The covariance matrix Q_{w} is defined in Equation (20).

$$Q_w = \begin{bmatrix} I_{3\times 3}\sigma_\omega & O_{3\times 3} \\ O_{3\times 3} & I_{3\times 3}\sigma_{\omega b} \end{bmatrix} \tag{20}$$

where σ_ω and $\sigma_{\omega b}$ are the initial value of bias and migration noise of gyroscope. Then, the prior state estimate can be calculated from state transition function (Equation (21)):

$$P(t \mid t-1) = \Phi(t \mid t-1)P(t-1 \mid t-1)\Phi^T(t \mid t-1) + Q(t) \tag{21}$$

In this paper, the acceleration vector and magnetic field vector are selected as the observation variables, which are defined as Equation (22), where a_x, a_y and a_z are the measurements of three-axis accelerations and m_x, m_y and m_z represent the measurements of three-axis magnetometer on the horizontal plane.

$$Z = \begin{bmatrix} a_x & a_y & a_z & m_x & m_y & m_z \end{bmatrix}^T \tag{22}$$

The observation matrix of the accelerometer H_a is calculated on the projection of the gravitation on the carrier coordinate system by Equation (23).

$$H_a = {}_e^bR \begin{bmatrix} 0 \\ 0 \\ -g \end{bmatrix} = \begin{bmatrix} -2g(q_1q_3 - q_0q_2) \\ -2g(q_2q_3 + q_0q_1) \\ -g(q_0^2 - q_1^2 - q_2^2 + q_3^2) \end{bmatrix} \tag{23}$$

where ${}_e^bR$ indicates the rotation matrix between SCS and ECS. The observation matrix H_{mag} is computed in the same fashion as H_a, and the magnetic field values $\begin{bmatrix} h_x & 0 & h_z \end{bmatrix}^T$ also need to be projected into the carrier coordinate system. The observation matrix H is calculated by Equation (24).

$$H = \begin{bmatrix} H_{\text{mag}} & 0_{3\times 3} \\ H_a & 0_{3\times 3} \end{bmatrix} \tag{24}$$

Therefore, the state gain matrix can be expressed as Equation (25).

$$K(t) = P(t \mid t-1)H^T(t)\left[H(t)P(t \mid t-1)H^T(t) + R(t)\right]^{-1} \tag{25}$$

Thus, the estimate value of the state vector at time t can be calculated from Equation (26), and the state error covariance matrix is updated by Equation (26). $K(t)$ represents the gain factor, $X(t \mid t)$ is the posteriori state estimate and $P(t)$ is the posterior covariance matrix, as shown in Equation (27).

$$X(t \mid t) = X(t-1 \mid t) + K(t)[Z(t) - h(X(t \mid t-1))] \tag{26}$$

$$P(t) = [I - K(t)H(t)]P(t \mid t-1) \tag{27}$$

3.5. Evaluation Method for Sprint Kayak Technique

The kayakers under different competitive levels vary substantially in trunk rotation, leg motion, stroke width, stroke rate, overall motion of the kayak, blade and water-contact time and other factors [26]. From these important factors, the most influential is the stroke frequency, stroke phases and stroke variation. After motion reconstruction, the details about the action implementation are also captured despite the visual occlusion conditions. The joint angles were estimated by computing the reverse cosine of the angle between the adjacent segment vectors. Then, the motion information is retained for further analysis.

There are many parametric ways to estimate the technical level of athlete based on the joint angles series. To achieve the best performance, the efficient and consistent stroke cycle is deemed necessary. The information about stroke rhythm can be calculated by searching the joint angle signal peak values. The consistency and variation are also identified by the movement range of joint at the extremities. In addition to these, the stroke phase is a more practical indicator used by coaches. A stroke cycle is defined as the periodic movement. When analyzing rowing technique, the stroke cycle is usually broken down into 2–4 phases, and there are many observational models to distinguish movement phases. In this paper, the model adopted for the categorization and analysis of kayaking movement is shown in Figure 7. The first-level phase is defined as the period of the whole stroke cycle, which corresponds to before and after the same-side paddle blade enteris the water. For detailed division of the stroke cycle, the propulsion phase is separated based on water contact of the paddle, which is considered for the greater visibility of the position. The four sub-phases are defined by the instant of the blade catching, immersing, extracting and releasing from the water. The sub-phases are divided into paddle entry, pull phase, paddle exit and recovery phase within a larger phase. The duration of pull phase and the ratio of propulsion duration to recovery duration have significant effects on the rowing performance [27].

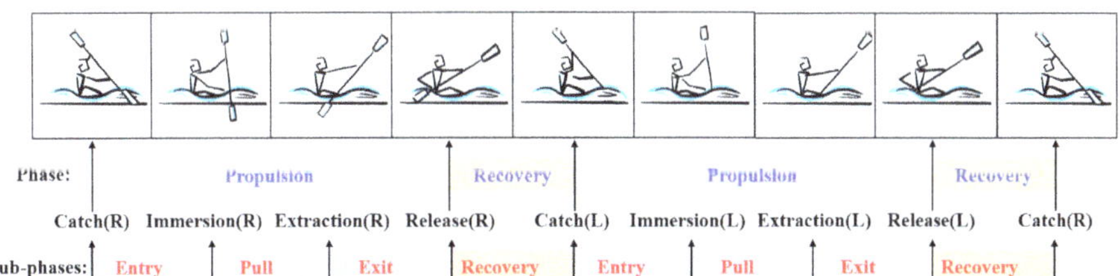

Figure 7. The model for kayak motion analysis including two levels: phases and sub-phases. The phases defining positions are entry, pull, exit and return. R, right side; L, left side.

3.6. Feature Extraction Method

To make predictions on the joint angle sequence data based on the phase partitioning method described above, the present study proposes several machine learning models to predict the movement phases (entry, pull, exit and recovery) based on the feature matrix of upper limbs joint angles. Statistical descriptive feature extraction is a widely used method to calculate the statistical features on the sample record [28]. However, the duration, amplitude and orientation vary among athletes. To obtain more comprehensive information for further phase recognition, the wavelet scattering transform is proposed in this paper for feature extraction [29]. Its algorithm principle is as follows.

The input signal $x = [x_1, x_2, \ldots, x_n]$ is a n-dimensional vector whose length is the number of joint angle series under analysis. $\psi_(x)$ is the chosen wavelet mother function which is used for multiscale and oriented filter bank. The definition is described as follows:

$$\psi_{\lambda_i}(x) = 2^{-2j_i}\psi(\lambda_i x) \tag{28}$$

where $\theta \in \mathcal{R}$ represents the rotation matrix in the finite, discrete rotation group R and $\lambda_i = 2^{-j_i}\theta_i, i = 1, \ldots, m$ represents the joint scaling and rotation operators. $\phi(x)$ is denoted as the low-pass filter, and its definitions is described as follows.

$$\phi_J(x) = 2^{-2J}\phi\left(2^{-J}x\right) \tag{29}$$

In the rest of this paper, j and J are integers within $j \leq J$, and j is the level of scattering. The wavelet scattering transform has important advantages of invariant, stable and informative signal feature representation. The following methods is used to recover the information by the operation of the invariant modulus part. We denote $S_i x, i = 1, 2, \ldots, m$ as the wavelet scattering coefficients of each layer. The output of each layer is written by the function of the modulus, and the low-pass averaging function is described as follows:

$$S_0 x = x * \phi_J(x) \tag{30}$$

$$S_1 x(x, \lambda) = |x * \psi_\lambda| * \phi_J(x) \tag{31}$$

$$S_m x(x, \lambda_1, \ldots, \lambda_m) = |\ldots|x * \psi_{\lambda_1}|\ldots * \psi_{\lambda_m}| * \phi_J(x) \tag{32}$$

The final wavelet scattering coefficients are the whole output of the transform from the 0th to mth order, as expressed in Equation (33).

$$Sx = \{S_0 x, S_1 x, \ldots S_m x\} \tag{33}$$

When the scattering transform of all slide-windowed joint angle record are obtained, the feature matrixes based on wavelet scattering transform are fed into machine learning model for training and predicting movement phases.

4. Results

4.1. Performance Comparison between Self-Made System and Standard Optical System

To verify the performance of the self-made inertial motion capture system (IMC), we set a contrast experiment with the commercial optical motion capture system (OMC). The OMC is mainly considered as the reference standard tool for dynamic measurement of upper and lower limb joint angles. The experimental arrangement is schematically shown in Figure 8. The participant was instructed to wear inertial sensor nodes and the markers of OMC simultaneously and make stretching exercises of the trunk and extremities. Due to the limitations of OMC, the participants wore tight-fighting shorts for the data collection to avoid occlusions. The whole-body movements were recorded using the 3D optical motion capture system (OptiTrack, American) and inertial sensor nodes. The experiment were approved by The University of Dalian Technology at LiaoNing province China and all participants provided written informed consent.

The coordinate system between the OMC and IMC is inconsistent, thus the raw data of OMC need to be converted. The results of comparison between OMC and IMC are presented as follows. Figure 9 shows a comparison graph of the elbow flexion extension for the two systems. First, consider the situation in the left half of Figure 9. The IMC can trace the motion curve accurately compared to OMC. The corresponding correlation coefficients are 0.995 and 0.996 of the two curves, respectively. The measured values of the optical system are used as a standard reference. The descriptive statistical histograms of the relative errors are also calculated for the right half of Figure 9, and the measurement errors are well controlled. All results show that the device we developed is reliable [30].

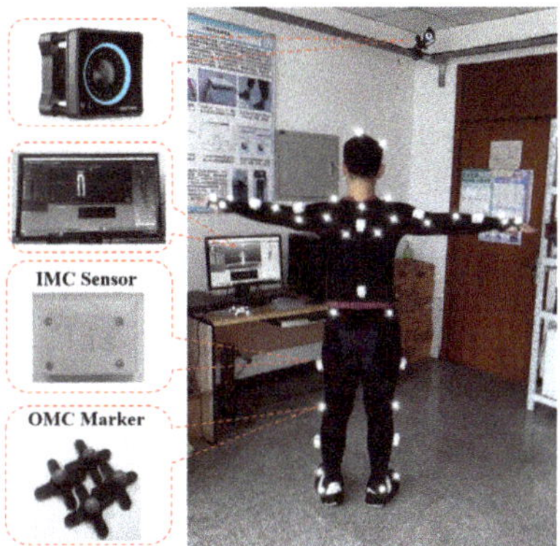

Figure 8. Inertial motion capture (IMC) sensor and optical motion capture (OMC) marker setup shown in the room.

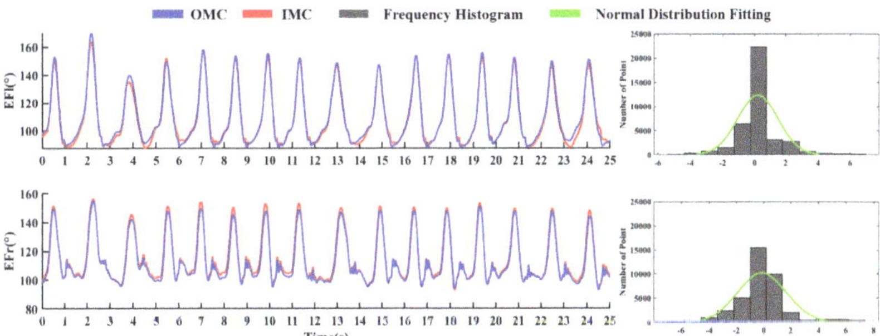

Figure 9. Contrast curve of elbow flexion and extension angle and error statistics.

4.2. Kinematic Statistical Analysis after Motion Reconstruction

The kayaker sprint involves the motion of the whole body. The paddling movement involves arm and trunk muscles, but it needs to activate the bilateral extensors and flexors of the hips and knees to simultaneously twist the body. Only in this way, the power output will be increased, and a higher velocity is obtained. Thus, the motion of whole body was captured in our experiments. All experiments were repeated at least three times for lengths of 200 m each time. At the same time, the entire process as recorded with a motion camera from coronal plane at 240 frame rate. Figure 10 shows the results of motion reconstruction using IMC equipment. To guarantee safe conditions, the observation of camera-based motion capture maintained constant viewing distance. It is inevitable that the vision of lower extremities is occluded by the hull and sometimes the upper limbs are occluded by the paddle. However, the details of motion expression are captured by the IMC, which is proven to be an efficient method to overcome the difficulty caused by the visual method [24].

The elbow and shoulder joint angle curves of a stroke cycle are shown in Figures 11 and 12, where the blue solid lines represent the time-normalized group mean and the black dotted lines represent the time-normalized maximum (MAX) and minimum mean (MIN). The gray shaded area represents the range of motion (ROM) between MAX and MIN. The mean value and standard deviation of MAX, MIN and ROM of elbow and shoulders are also calculated, and the results are shown in Table 1. A stroke cycle is completed in two stages of right-side and left-side stroke. During the two sub-phases, the paddlers try to maintain similar extremity rotation and flexion angles. It can be seen in the graph that joint angles of elbow and shoulder in right-side stroke is the opposite of the extremities movements in left-side stroke. This also accounts for the similarity between EFl in Figure 11 and EFr in Figure 12, and the same applies for EFr, SFr and SFl.

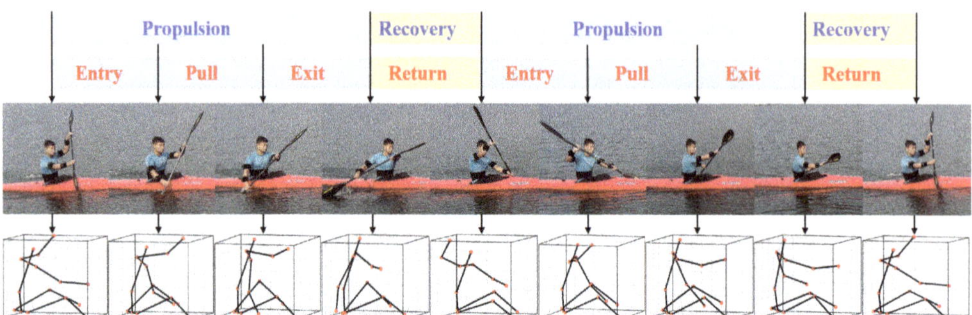

Figure 10. The result of motion reconstruction using the proposed method.

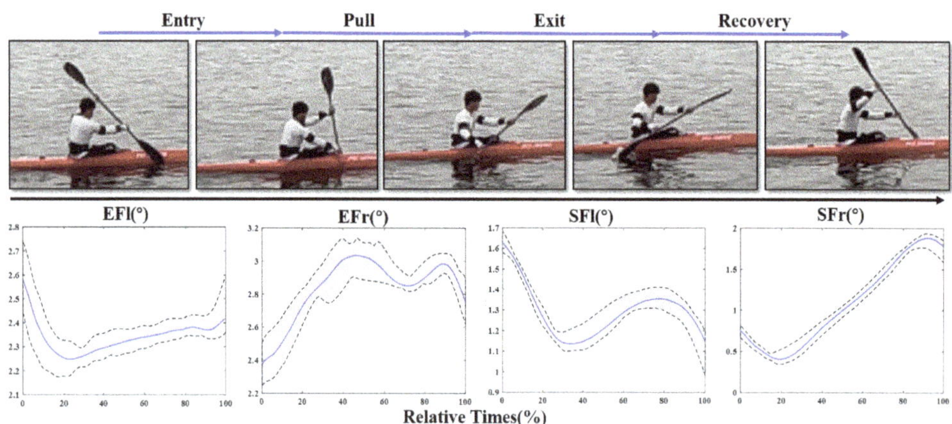

Figure 11. Joint angle transform curves of elbow and shoulder during right-side stroke cycle.

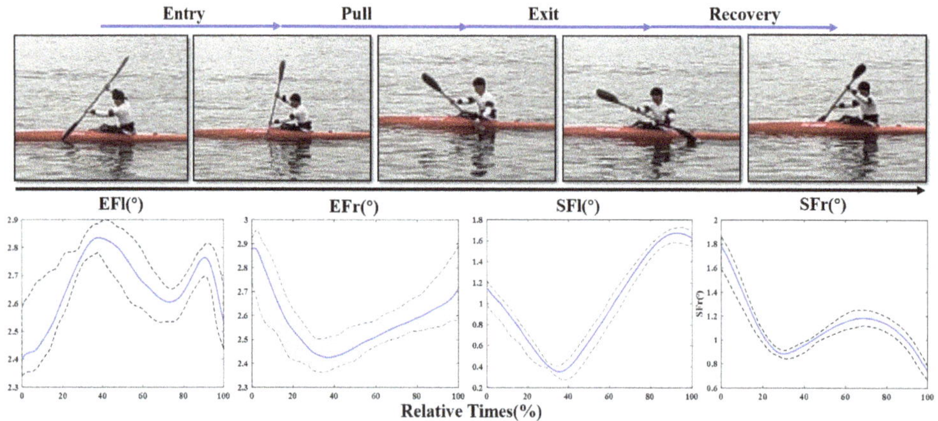

Figure 12. Joint angle transform curves of elbow and shoulder during left-side stroke cycle.

During the stroke, the draw elbow is extended, while the draw shoulder is flexed in an attempt to place the blade paddle in the water as far as toward the boat as possible. To maintain the optimal stroke performance, the kayakers alter shoulder and arm musculature to resist the external forces. Because the upper limbs movements accounted for approximately half of the stroke cycle, shoulder injuries are always observed in elite kayakers [31]. Besides the kinematic asymmetries of the whole-body during paddling, the MAX, MIN and ROM of elbow and shoulder flexion are important reasons for muscle strain [32]. As shown in Figures 11 and 12 and Table 2, the MEAN, MAX and MIN values of elbow are larger than those of shoulder. However, the ROM of shoulder is much larger than that of elbow. The maximal shoulder angle of dominant limb appears at around 80% into the cycle. The peak of shoulder extension occurs during the abductor movement of the recovery phase in a stroke cycle. These observed results are similar to previous studies [31,33]. This illustrates that the power output for rowing is mainly completed by the upper arm and shoulder muscles. According to the above analysis, the potential for shoulder and scapular injury is monitored and avoided as much as possible.

The proficient stroke would ideally be symmetrical on both sides in order to propel the kayak straight and distribute forces equally on the body. The elite kayakers always use a similar, consistent rowing technique. The body motion and posture do not change during the whole process. In this study, all athletes could be regard as junior rowers. Based on the time-normalized group joint angles series, the correlations on opposite curves of both sides are 0.8045, 0.7607, 0.9326 and 0.8548, respectively. These parameters represent the degree of proficiency, which is the kinematic difference among elite, junior and beginner kayakers, where elites have a perfect pattern of symmetry between the left to right, which are also the inadequacies for beginners to improve themselves.

Obviously, this above kinematic statistical analysis is comprehensible and intuitive, and it can assist coaches to draw up the training program and avoid sport injury. Different from the in-house analysis based on the training equipment, the real on-water movement analysis is more responsive and makes the results more pertinent.

Table 2. Summary statistics of joint angle sequencing data.

MEAN ± SD (deg)		ROM	MAX	MIN	MEAN
Left Side	EFl	21.9 ± 3.1	167.0 ± 2.1	145.2 ± 1.6	156.4 ± 1.5
	EFr	37.6 ± 6.4	158.3 ± 5.9	120.6 ± 1.6	129.0 ± 1.6
	SFl	74.1 ± 2.0	99.9 ± 1.9	25.8 ± 1.6	58.9 ± 1.3
	SFr	60.2 ± 4.5	103.1 ± 3.4	41.0 ± 1.7	63.9 ± 2.0
Right Side	EFl	20.3 ± 1.9	147.7 ± 3.8	127.5 ± 1.9	132.5 ± 1.9
	EFr	36.9 ± 4.9	174.9 ± 3.4	138.0 ± 2.9	155.8 ± 2.2
	SFl	29.6 ± 2.4	94.6 ± 2.3	65.0 ± 1.4	75.5 ± 1.4
	SFr	88.6 ± 2.6	110.0 ± 2.0	21.5 ± 2.2	62.3 ± 1.5

4.3. Phase Partition Based on the Joint Angles Series

Phases are always described in the stroke quality analysis [34]. The two-phase model including propulsion and recovery phase is a widely used indicator. For more detailed analysis on the premise of not disrupting the two-phase model, the sub-phase model including entry, pull, exit and return can be used. In this study, to predict the phase of motion process automatically, machine learning algorithms were used to classify the phase based on the wavelet scattering feature matrix of four joint angles series, and the durations of each phase were obtained.

High speed motion camera was used to capture the kayaker's movement at the recording rate of 240 frames per second, and, then, the four joint angle series were annotated with video images as the comparator groups. Next, the sliding window of 20 data points with 50% overlap was applied for the different phase series. We ended up with a dataset of 11,851 pieces of labeled sequences. Because the energy of wavelet scattering decreases rapidly as the layer level increases, and almost 99% of the energy is contained in the first two layers, we focused on the first two layer scattering coefficients [35], and then the features extracted from the labeled records using wavelet scattering were combined to predict the class membership of data. Movement recognition consists of identifying special actions and the stage at which the object is currently at. Fine-grained movement recognition is focuses on distinguishing different actions between subtle changes. We try to propose sensor-based recognition and evaluation models based on joint angle series. Because the number of scattering features is large, several simple classifiers including Support Vector Machine (SVM), Logistic Regression, Decision Tree, K-Nearest Neighbor (KNN) and Random Forest were considered, and we compared their performance. The whole feature dataset was divided into training (80%) and test sets (20%). The training dataset as used to train the models and the remaining dataset was used for predicting. Standard measurements were used to evaluate the performance of each selected model: classification accuracy, precision recall and F1-score. The receiver operating characteristic curve (ROC) was obtained from the false positive rate against the true positive rate under the different thresholds. The precision–recall curve (PRC) was derived from the relationship between precision and recall with the different thresholds. The corresponding areas under the curves (AUC) of ROC and PRC were also calculated to assess the prediction performance. The model with a higher area (between 0 and 1) value gives a better predictive performance. The specific details for recognition performance are listed in Table 3.

Table 3. Prediction performance comparison.

Evaluation Metrics	Machine Learning Algorithm				
	Decision Tree	**KNN**	**SVM**	**Logistic Regression**	**Random Forest**
Accuracy	**0.9827**	**0.9898**	**0.9898**	**0.9789**	**0.9856**
Precision	0.9826	0.9900	0.9898	0.9787	0.9856
Recall	0.9827	0.9899	0.9899	0.9789	0.9857
F1-score	0.9826	0.9899	0.9898	0.9788	0.9855
AUC of ROC	0.9834	0.9993	0.9999	0.9990	0.9992
AUC of PRC	0.9679	0.9984	0.9996	0.9976	0.9981

The trained model was used to predict the phase position of the new motion sequences collected from the participants. Next, we calculated the duration of each sub-phase of all participants. The result is shown in Table 4. As described in a previous paper [36], experienced kayakers have lower standard deviation (SD) values than novices, and SD can be regarded as an indicator of the joint angle consistency of the kayakers. Table 4 shows that the third and fourth athletes present relatively small standard deviations. The results reflect that they have achieved better performance over the experiment time.

Table 4. The duration of sub-phases during one round trip.

MEAN ± SD (ms)	Left Side				Right Side			
	Entry	**Pull**	**Exit**	**Recovery**	**Entry**	**Pull**	**Exit**	**Recovery**
Subject 1	124 ± 13	396 ± 28	116 ± 15	388 ± 22	144 ± 15	299 ± 23	95 ± 16	371 ± 28
Subject 2	120 ± 14	414 ± 14	118 ± 17	385 ± 24	152 ± 14	301 ± 22	100 ± 14	385 ± 32
Subject 3	121 ± 6	354 ± 19	86 ± 12	319 ± 18	129 ± 10	297 ± 16	96 ± 12	337 ± 14
Subject 4	132 ± 8	369 ± 17	96 ± 11	328 ± 22	131 ± 9	300 ± 16	96 ± 9	343 ± 15
Subject 5	131 ± 10	435 ± 32	105 ± 13	370 ± 19	133 ± 7	333 ± 20	101 ± 12	405 ± 22
Subject 6	127 ± 9	428 ± 27	103 ± 14	361 ± 21	127 ± 9	340 ± 23	91 ± 18	393 ± 26

The duration ratio of different phase on both sides towards all participant was also calculated, as plotted in Figure 13. The acceleration procedure of kayak is mainly completed during the propulsion phase. More precisely, the entry and pull phase complete most of the work. As shown in Figure 13, the ratio between propulsion and recovery phase of all experimenters in this study is close to 60%. For many elite athletes, efforts are made to minimize the duration of recovery process and increase the duration of the propulsion phase, and this is more effective than enhancing the stroke rate for increasing the boat velocity [37,38]. On the other hand, it can be seen that the symmetry between left side and right side is imperfect. Asymmetry has been related to injuries in kayakers during the paddle stroke [33].

By the above quantitative analysis of the stroke cycle, the stroke quality of kayakers can be evaluated from different perspectives. The coach will be able to provide specific recommendations to the athletes, and the performance of kayakers can be improved.

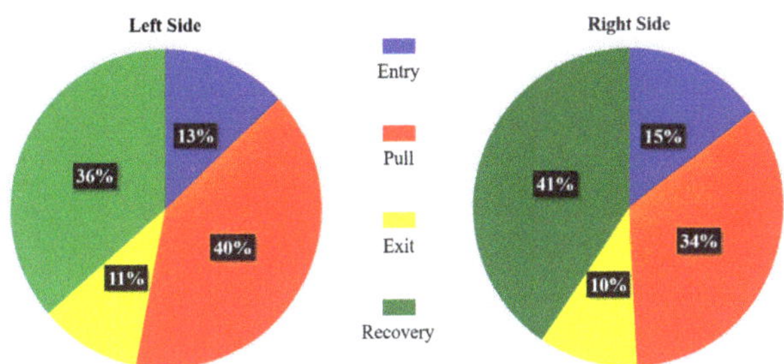

Figure 13. The ratio of different phase duration to total time in a stroke cycle.

5. Discussion

The whole-body posture is important to the paddling technique [39], and rowing technique is largely composed of posture and timing. This work tries to explore these aspects further. We used inertial sensors to acquire kayakers' motion kinematic information during rowing, which is powerful and permits more precise and specific correlations between performance and proficient level. All sequence data of whole-body activity obtained in this study are available under real on-water environment. The experimental results, including duration of stroke cycle, stroke frequency, ROM of limbs movement, similarity of both sides, rate of propulsion/recovery phase, etc. are consistent with previous study [2,3,31,36,40].

Although the satisfactory results are also seen in the comparison and prediction of the follow-up experimental arrangement, indeed, the performance of the athletes could be influenced by many factors. The distinct differences are also show in only short time period even for the same individual. Consequently, the timeliness of provided information obtained by the proposed method is important for communication between coaches and athletes. On the other hand, wearing multiple sensors have resulted in bodily discomfort and further it would affect the coordination between the limbs and the spine. Therefore, it is necessary to minimize the number of wearable devices or significantly decrease the size and weight of sensors nodes. Besides, we next set out to optimize the rigid model, including decreasing the number of degrees of freedom, and so on. After adjusting the above possible factors, a more comprehensive water sport athlete monitoring system will be established in the future.

6. Conclusions

In this work, we present a systematic method for athlete's motion capture and kinematic analysis. The customizable rigid model is used to demonstrate the kayaker's posture, and each segment attitude of the whole-body is iteratively calculated by the quaternization vector multiplication. The contrast test indicates that the proposed approach has comparable accuracy to the standard commercial optical motion capture system. This paper highlights the range of motion of the extremities, which is essential for preventing sports injury, and the duration of motion phases, which is the import index for competitive level. The detailed kinematic analysis based on the field on-water experiments could be provided, and this enable coaches to give targeted feedback and guidance based on the participant's activities in real scenario.

Author Contributions: Conceptualization, L.L., H.-H.W. and S.Q.; methodology, L.L., S.Q. and H.-H.W.; software, L.L.; validation, Z.-D.H.; visualization, Y.-C.Z.; formal analysis, L.L., H.-H.W. and S.Q.; supervision, Z.-D.H.; project administration, L.L. and Y.-C.Z.; writing—original draft, L.L. and S.Q.; writing—review and editing, S.Q. and L.L. All authors have read and agreed to the published version of the manuscript.

Funding: This work was supported by National Natural Science Foundation of China under Grant Nos. 61873044, 61473058 and 61803072; Dalian Science and Technology Innovation fund (2018J12SN077); China Postdoctoral Science Foundation under Grant Nos. 2017M621131 and 2017M621132; and the Liaoning Key R&D Guidance Project under Grant ZX2018KJ002. Natural Fund Guidance Project of Liaoning Science and Technology Department (2019-ZD-0133)

Acknowledgments: The authors would like to express their thanks to the members of laboratory of Intelligent system, Dalian University of Technology.

Conflicts of Interest: The authors declare no conflict of interest.

Abbreviations

The following abbreviations are used in this manuscript:

IMUs	Inertial Measurement Units
ECS	Earth Coordinate System
SCS	Sensor Coordinate System
BCS	Body Coordinate System
SF	Shoulder Flexion Angle
EF	Elbow Flexion Angle
KF	Knee flexion Angle
FF	Foot Flexion Angle
SFl	Left Shoulder Flexion Angle
SFr	Right Shoulder Flexion Angle
EFl	Left Elbow Flexion Angle
EFr	Right Elbow Flexion Angle
CF	Complementary Filter
GD	Gradient Descent
EKF	Externed Kalman Filter
IMC	Inertial Motion Capture System
OMC	Optical Motion Capture System
SVM	Support Vector Machine
KNN	K-Nearest Neighbor
ROM	Range of Motion
MAX	Maximum Value
MIN	Minimum Value
MEAN	Mean Value
SD	Standard Deviation
ROC	Receiver Operating Characteristic
PRC	Precision Recall Curve
AUC	Area Under the Curve

References

1. Borges, T.O.; Bullock, N.; Duff, C.; Coutts, A.J. Methods for quantifying training in sprint kayak. *J. Strength Cond. Res.* **2014**, *28*, 474–482. [CrossRef] [PubMed]
2. Lee, C.H.; Nam, K.J. Analysis of the kayak forward stroke according to skill level and knee flexion angle. *Int. J. Bio-Sci. Bio-Technol.* **2012**, *4*, 41–48.
3. Brown, B.M.; Lauder, M.; Dyson, R. Notational analysis of sprint kayaking: Differentiating between ability levels. *Int. J. Perform. Anal. Sport* **2011**, *11*, 171–183. [CrossRef]

4. Tay, C.S.; Kong, P.W. A Video-based Method to Quantify Stroke Synchronisation in Crew Boat Sprint Kayaking. *J. Hum. Kinet.* **2018**, *65*, 45–56. [CrossRef] [PubMed]
5. Gomes, B.B.; Ramos, N.V.; Conceiç ao, F.A.; Sanders, R.H.; Vaz, M.A.; Vilas-Boas, J.P. Paddling force profiles at different stroke rates in elite sprint kayaking. *J. Appl. Biomech.* **2015**, *31*, 258–263. [CrossRef]
6. Gravenhorst, F.; Tessendorf, B.; Adelsberger, R.; Arnrich, B.; Tröster, G.; Thiem, C. SonicSeat: A seat position tracker based on ultrasonic sound measurements for rowing technique analysis. In *BODYNETS*; 2012; pp. 76–81. Available online: https://www.google.com/url?sa=t&rct=j&q=&esrc=s&source=web&cd=&ved=2ahUKEwiX987v2bbuAhUF_aQKHTCoBt4QFjAAegQIBhAC&url=https%3A%2F%2Fwww.research-collection.ethz.ch%2Fbitstream%2Fhandle%2F20.500.11850%2F95501%2Feth-47161-02.pdf&usg=AOvVaw0_2M2ZEQUAOu3zSYn8tbwR (accessed on 25 January 2021).
7. Henry, J.C.; Clark, R.R.; McCabe, R.P.; Vanderby, R., Jr. An evaluation of instrumented tank rowing for objective assessment of rowing performance. *J. Sport. Sci.* **1995**, *13*, 199–206. [CrossRef]
8. Ridge, B.R.; Broad, E.; Kerr, D.A.; Ackland, T.R. Morphological characteristics of Olympic slalom canoe and kayak paddlers. *Eur. J. Sport Sci.* **2007**, *7*, 107–113. [CrossRef]
9. Plagenhoef, S. Biomechanical analysis of Olympic flatwater kayaking and canoeing. *Res. Quarterly. Am. Alliance Heal. Phys. Educ. Recreat. Danc.* **1979**, *50*, 443–459. [CrossRef]
10. Sturm, D.; Yousaf, K.; Eriksson, M. A wireless, unobtrusive kayak sensor network enabling feedback solutions. In Proceedings of the 2010 International Conference on Body Sensor Networks, Singapore, 7–9 June 2010; pp. 159–163.
11. Qiu, S.; Wang, Z.; Zhao, H.; Hu, H. Using distributed wearable sensors to measure and evaluate human lower limb motions. *IEEE Trans. Instrum. Meas.* **2016**, *65*, 939–950. [CrossRef]
12. Tesconi, M.; Tognetti, A.; Scilingo, E.P.; Zupone, G.; Carbonaro, N.; De Rossi, D.; Castellini, E.; Marella, M. Wearable sensorized system for analyzing the lower limb movement during rowing activity. In Proceedings of the 2007 IEEE International Symposium on Industrial Electronics, Vigo, Spain, 4–7 June 2007; pp. 2793–2796.
13. King, R.C.; McIlwraith, D.G.; Lo, B.; Pansiot, J.; McGregor, A.H.; Yang, G.Z. Body sensor networks for monitoring rowing technique. In Proceedings of the 2009 Sixth International Workshop on Wearable and Implantable Body Sensor Networks, Berkeley, CA, USA, 3–5 June 2009; pp. 251–255.
14. Ruffaldi, E.; Peppoloni, L.; Filippeschi, A. Sensor fusion for complex articulated body tracking applied in rowing. *Proc. Inst. Mech. Eng. Part P J. Sport. Eng. Technol.* **2015**, *229*, 92–102. [CrossRef]
15. Hamano, S.; Ochi, E.; Tsuchiya, Y.; Muramatsu, E.; Suzukawa, K.; Igawa, S. Relationship between performance test and body composition/physical strength characteristic in sprint canoe and kayak paddlers. *Open Access J. Sport. Med.* **2015**, *6*, 191.
16. Wang, Z.; Li, J.; Wang, J.; Zhao, H.; Qiu, S.; Yang, N.; Shi, X. Inertial sensor-based analysis of equestrian sports between beginner and professional riders under different horse gaits. *IEEE Trans. Instrum. Meas.* **2018**, *67*, 2692–2704. [CrossRef]
17. McKean, M.R.; Burkett, B. The relationship between joint range of motion, muscular strength, and race time for sub-elite flat water kayakers. *J. Sci. Med. Sport* **2010**, *13*, 537–542. [CrossRef]
18. Hemerly, E.M.; Coelho, F.A.A. Explicit Solution for Magnetometer Calibration. *IEEE Trans. Instrum. Meas.* **2014**, *63*, 2093–2095. [CrossRef]
19. Falbriard, M.; Mohr, M.; Aminian, K. Hurdle Clearance Detection and Spatiotemporal Analysis in 400 Meters Hurdles Races Using Shoe-Mounted Magnetic and Inertial Sensors. *Sensors* **2020**, *20*, 354. [CrossRef] [PubMed]
20. Choukroun, D.; Bar-Itzhack, I.Y.; Oshman, Y. Novel quaternion Kalman filter. *IEEE Trans. Aerosp. Electron. Syst.* **2006**, *42*, 174–190. [CrossRef]
21. Salchow-Hömmen, C.; Callies, L.; Laidig, D.; Valtin, M.; Schauer, T.; Seel, T. A tangible solution for hand motion tracking in clinical applications. *Sensors* **2019**, *19*, 208. [CrossRef]
22. Pio, R. Euler angle transformations. *IEEE Trans. Autom. Control* **1966**, *11*, 707–715. [CrossRef]
23. Wu, Z.; Jiang, X.; Zhong, M.; Shen, B.; Zhang, L. Wearable Sensors Measure Ankle Joint Changes of Patients with Parkinson's Disease before and after Acute Levodopa Challenge. *Park. Dis.* **2020**, *2020*, 1–7. [CrossRef]
24. Liu, L.; Qiu, S.; Wang, Z.L.; Li, J.; Wang, J.X. Canoeing Motion Tracking and Analysis via Multi-Sensors Fusion. *Sensors* **2020**, *20*, 2110. [CrossRef]
25. Li, J.; Wang, Z.; Qiu, S.; Zhao, H.; Yang, N. Study on Horse-rider Interaction Based on Body Sensor Network in Competitive Equitation. *IEEE Trans. Affect. Comput.* **2019**, 1. [CrossRef]
26. Worsey, M.T.; Espinosa, H.G.; Shepherd, J.B.; Thiel, D.V. A systematic review of performance analysis in rowing using inertial sensors. *Electronics* **2019**, *8*, 1304. [CrossRef]
27. McDonnell, L.K.; Hume, P.A.; Nolte, V. An observational model for biomechanical assessment of sprint kayaking technique. *Sport. Biomech.* **2012**, *11*, 507–523. [CrossRef]
28. Wang, J.; Wang, Z.; Qiu, S.; Xu, J.; Zhao, H.; Fortino, G.; Habib, M. A selection framework of sensor combination feature subset for human motion phase segmentation. *Inf. Fusion* **2020**, *70*, 1–11. [CrossRef]
29. Singh, A.; Kingsbury, N. Dual-tree wavelet scattering network with parametric log transformation for object classification. In Proceedings of the 2017 IEEE International Conference on Acoustics, Speech and Signal Processing (ICASSP), New Orleans, LA, USA, 5–9 March 2017; pp. 2622–2626.

30. Fleron, M.K.; Ubbesen, N.C.H.; Battistella, F.; Dejtiar, D.L.; Oliveira, A.S. Accuracy between optical and inertial motion capture systems for assessing trunk speed during preferred gait and transition periods. *Sport. Biomech.* **2018**, *18*, 366–377. [CrossRef] [PubMed]
31. Fleming, N.; Donne, B.; Fletcher, D. Effect of kayak ergometer elastic tension on upper limb EMG activity and 3D kinematics. *J. Sport. Sci. Med.* **2012**, *11*, 430.
32. Limonta, E.; Squadrone, R.; Rodano, R.; Marzegan, A.; Veicsteinas, A.; Merati, G.; Sacchi, M. Tridimensional kinematic analysis on a kayaking simulator: Key factors to successful performance. *Sport Sci. Health* **2010**, *6*, 27–34. [CrossRef]
33. Wassinger, C.A.; Myers, J.B.; Sell, T.C.; Oyama, S.; Rubenstein, E.N.; Lephart, S.M. Scapulohumeral kinematic assessment of the forward kayak stroke in experienced whitewater kayakers. *Sport. Biomech.* **2011**, *10*, 98–109. [CrossRef] [PubMed]
34. McDonnell, L.K.; Hume, P.A.; Nolte, V. A deterministic model based on evidence for the associations between kinematic variables and sprint kayak performance. *Sport. Biomech.* **2013**, *12*, 205–220. [CrossRef] [PubMed]
35. Bruna, J.; Mallat, S. Invariant scattering convolution networks. *IEEE Trans. Pattern Anal. Mach. Intell.* **2013**, *35*, 1872–1886. [CrossRef] [PubMed]
36. Bosch, S.; Shoaib, M.; Geerlings, S.; Buit, L.; Meratnia, N.; Havinga, P. Analysis of indoor rowing motion using wearable inertial sensors. In Proceedings of the 10th EAI International Conference on Body Area Networks, Sydney, Australia, 28–30 September 2015; pp. 233–239.
37. Trevithick, B.A.; Ginn, K.A.; Halaki, M.; Balnave, R. Shoulder muscle recruitment patterns during a kayak stroke performed on a paddling ergometer. *J. Electromyogr. Kinesiol.* **2007**, *17*, 74–79. [CrossRef] [PubMed]
38. Černe, T.; Kamnik, R.; Vesnicer, B.; Gros, J.Ž.; Munih, M. Differences between elite, junior and non-rowers in kinematic and kinetic parameters during ergometer rowing. *Hum. Mov. Sci.* **2013**, *32*, 691–707. [CrossRef] [PubMed]
39. Kemecsey, I. (British Canoe Union, Holme Pierrepont, Nottingham, UK). Personal communication, 1986.
40. Bjerkefors, A.; Tarassova, O.; Rosén, J.S.; Zakaria, P.; Arndt, A. Three-dimensional kinematic analysis and power output of elite flat-water kayakers. *Sport. Biomech.* **2018**, *17*, 414–427. [CrossRef] [PubMed]

Article

Ambulatory Human Gait Phase Detection Using Wearable Inertial Sensors and Hidden Markov Model

Long Liu [1,2,†], Huihui Wang [3,†], Haorui Li [2,†], Jiayi Liu [2], Sen Qiu [2,*], Hongyu Zhao [2] and Xiangyang Guo [2]

1 Department of Electrical & Information Engineering, Dalian Neusoft University of Information, Dalian 116023, China; liulong@neusoft.edu.cn
2 School of Control Science and Engineering, Dalian University of Technology, Dalian 116024, China; 17303779346@mail.dlut.edu.cn (H.L.); Liujiayi@mail.dlut.edu.cn (J.L.); zhaohy@dlut.edu.cn (H.Z.); 15736873491@163.com (X.G.)
3 School of Fundamental Education, Dalian Neusoft University of Information, Dalian 116023, China; wanghuihui@neusoft.edu.cn
* Correspondence: qiu@dlut.edu.cn
† Long Liu, Huihui Wang and Haorui Li contributed equally to this work and are co-first author.

Abstract: Gait analysis, as a common inspection method for human gait, can provide a series of kinematics, dynamics and other parameters through instrumental measurement. In recent years, gait analysis has been gradually applied to the diagnosis of diseases, the evaluation of orthopedic surgery and rehabilitation progress, especially, gait phase abnormality can be used as a clinical diagnostic indicator of Alzheimer Disease and Parkinson Disease, which usually show varying degrees of gait phase abnormality. This research proposed an inertial sensor based gait analysis method. Smoothed and filtered angular velocity signal was chosen as the input data of the 15-dimensional temporal characteristic feature. Hidden Markov Model and parameter adaptive model are used to segment gait phases. Experimental results show that the proposed model based on HMM and parameter adaptation achieves good recognition rate in gait phases segmentation compared to other classification models, and the recognition results of gait phase are consistent with ground truth. The proposed wearable device used for data collection can be embedded on the shoe, which can not only collect patients' gait data stably and reliably, ensuring the integrity and objectivity of gait data, but also collect data in daily scene and ambulatory outdoor environment.

Keywords: body sensor network; gait analysis; gyroscope; information fusion; hidden Markov model

Citation: Liu, L.; Wang, H.; Li, H.; Liu, J.; Qiu, S.; Zhao, H.; Guo, X. Ambulatory Human Gait Phase Detection Using Wearable Inertial Sensors and Hidden Markov Model. *Sensors* **2021**, *21*, 1347. https://doi.org/10.3390/s21041347

Academic Editor: Maria de Fátima Domingues, Ayman Radwan, Andrea Sciarrone and Massimo Sacchetti

Received: 29 December 2020
Accepted: 8 February 2021
Published: 14 February 2021

1. Introduction

Walking is one of the most common physical activities for humans and plays an important role in our daily activities. It can be performed in a variety of ways and directions, and is also an energy efficient method of mobility. For most people, walking is completely subconscious. In patients with neurological conditions such as stroke, this can be altered by gait abnormalities, which are usually caused by motor or sensory disorder. It is necessary to conduct specific rehabilitation exercise to deal with gait abnormalities, and the detection and tracking of gait can be of great help to patient recovery. Gait characterization and phase classification are widely used in the field of medical diagnosis to assess and detect the balance ability, which can be used for gait-based identification, robot control for artificial limbs and humanoid robots [1–8]. Researches about Micro-Electro-Mechanical Systems (MEMS) have developed rapidly over the past decade, enabling the development of computer communication devices, high-performance physical sensors, and especially inertial sensors. These sensors are characterized by their large memory capacity, small size and low cost, and it is due to these characteristics that they are widely used in various areas [9–22].

2. Related Works

According to the situation of foot contact with the ground, gait phase can be divided into two stages: support phase(ST) and swing phase(SW). SW and ST phases can also be divided into several subphases. The result is a model with three to eight phases. The four-stage division subdivides the support phase into four stages: Heel-strike (HS), Foot-flat (FF), Heel-off (HO) and Toe-off (TO). The time occupied by each stage is as follows: (1) HS to full FF, accounts for approximately 10% of the gait cycle. (2) Plantar Fascia FF to heel-to-ground (HF), accounts for approximately 35% of the gait cycle. (3) HF to toe-off (TF), accounts for approximately 15% of the gait cycle. (4) TO to HS in the next cycle, which accounts for about 40% of the gait cycle.

The hidden Markov model is a model of statistical analysis [23–25], which is used to describe Markov processes containing hidden parameters. Hidden Markov, as a typical statistical analysis model, is outstanding in solving problems in time series. Therefore, it has been widely studied in many areas of time series problem solving, and hidden Markov models are very useful in the analysis and processing of modeling problems. It has great advantages in modeling, such as modeling in the spatial direction. It also has advantages in the analysis of modeling directions, such as in solving the problem of modeling and analysis of non-stationary waveforms. The normal gait behavior of the human body is cyclical not only in time but also in space. It is because of the hidden Markov's outstanding advantage in resolving cyclic information and the cyclic and regular nature of the human gait that both the relationship is a high match between the method and the problem, so in this paper we decided to adopt this model to solve the problem of gait stage division. The stages of gait behavior can be treated as the state of a Markov chain, and the data obtained by the acquisition device is extracted to obtain the eigenvalues by feature extraction, putting these eigenvalues correspond to the output observations of the Markov model.

Since human gait has a certain regularity and periodicity, many researchers have started to use hidden Markov models for gait behavior recognition (e.g., gait used to distinguish actions under different behaviors, gait used to identify different people for identification, and determining pathological gait) [26–31].

After the success of gait behavior recognition, some researchers began to study the hidden Markov model for gait stage delineation. On the issue of gait stage classification, the different problems solved by the researchers and the different types of gait data collected (including the collection of different data locations and different acquisition devices) make many differences in the problem of gait stage classification.

In this paper, it is hoped that a single inertial sensor can be used to obtain gait phase recognition with higher accuracy. Figure 1a presents the correlation between the gait stage and the Markov chain and Figure 1b divides a complete period of gait into four stages based on the angular velocity signal from compact inertial sensor.

The division of gait phases in this paper is based on the angular velocity collected at the toe and divides a complete gait cycle into four stages. The transitions between these four stages are carried out from left to right, and there is no jump transition, the gait stage in this paper with a Markov chain representation containing four states. The states corresponding to the four stages of the locomotor gait cannot be directly observed and are directly obtained as the angular velocity measured by the sensor, so we map of states to outputs can be achieved with the help of Gaussian probability or Gaussian mixed probability models to achieve gait stage division.

Wearable device-based gait phase segmentation and adaptive recognition have become a useful tool for quantitative medical diagnosis and patient recovery evaluation [15,32–34]. Human gait behavior shows regularity and periodicity, at the same time, since different individuals, environments, and individual states result in gait diversity, which brings a great deal of uncertainty. Researchers have been committed to accurately describing gait and improving the adaptability of the proposed gait analysis model [33,35–41]. In this paper, we mainly focus on the identification of the different phases of a single gait cycle, and the various phases of human gait reflect the individual health issues, which makes it useful in the field of diagnosis and guidance for medical rehabilitation.

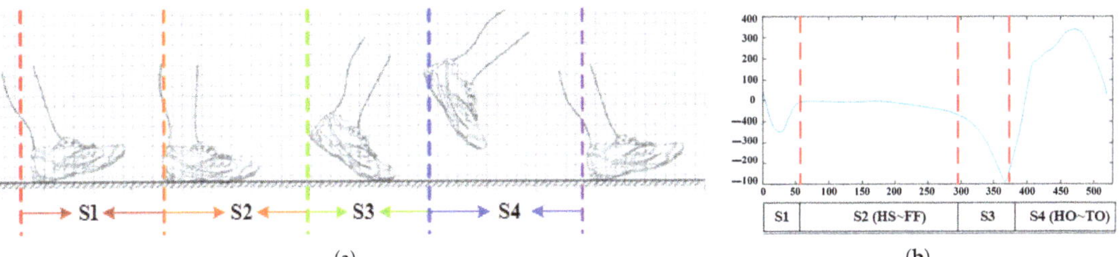

Figure 1. Traditional gait analysis methods (**a**) Correspondence between gait phases and Markov chain state (**b**) Correspondence between angular velocity and gait phases.

The rest of this paper is summarized as follows: Section 3 analyzes the hidden Markov model. Section 4 describes the proposed wearable sensor system and the gait phase segmentation methodology. Experimental results are shown in Section 5. Section 6 discusses the proposed hidden Markov model and summarizes this paper.

3. Analysis of Hidden Markov Model

3.1. Analysis of HMM Theory

HMM usually consists of five parts: hidden state, model output value, initial state probability, transition probability between States and output probability distribution. The hidden state is usually represented by S, which is the actual requirement of the model, and usually can not be obtained by direct observation; the observable output is represented by O, which is the observed output of the model, and associated with the hidden state, can be regarded as the external performance of the hidden state; the probability matrix π of the initial state represents the probability of each state at the initial time; the transition between the hidden states The probability A is the transition probability between the hidden states, and the output probability matrix B is the probability that the corresponding output of one of the hidden states is an observed output. HMM is usually represented by $\theta = (\pi, A, B)$.

The specific hidden Markov model can be described by five parts of the model:

(1) N: The number of states contained in a model is usually determined before the model is built. Suppose that the state at time t is q_t, then $q_t \in \{s_1, s_2, \cdots, s_N\}$.

(2) M: The number of observation values corresponding to each state in the model (when the output observation value is discrete value), if the observation value of the model at time t is o_t, then $o_t \in \{v_1, v_2, \cdots, v_M\}$.

(3) A: The state transition probability matrix of the model is $A = (a_{ij})_{N \times N}$. If the state of the model at time t is $q_t = s_i$, the transition probability can be expressed as $P(q_{t+1} = s_j | q_t = s_i) = a_{ij}$, and the state at time $t+1$ is $q_{t+1} = s_j$, where $1 \leqslant t \leqslant T$, T is the length of the model output observation value and $1 \leqslant i, j \leqslant N$. Meanwhile, the state transition matrix A satisfies $\sum_{j=1}^{N} a_{ij} = 1, 1 \leqslant i, j \leqslant N$.

(4) $B = (b_j(k))_{N \times M}$: If the state of the model at a certain time is $q_t = s_j$ and the output observation value is $o_t = v_k$, the relationship between the state and the observation value can be expressed as $P(o_t = v_k | q_t = s_j) = b_j(o_k), 1 \leqslant j \leqslant N, 1 \leqslant k \leqslant M$, and the observation value probability matrix needs to meet $b_j(k) = 1, 1 \leqslant k \leqslant M$.

(5) π: The probability of occurrence of each state in HMM model at the first time, $\pi = (\pi_1, \pi_2, \cdots, \pi_N)$, when the initial state is s_i, can be expressed as $p(q_1 = s_i) = \pi_i, 1 \leqslant i \leqslant N$. The initial state probability needs to satisfy $\sum_{i=1}^{N} \pi_i = 1$.

HMM contains two stochastic processes. The first process is Markov chain, which contains the initial state probability π and the state transition probability matrix A of HMM, and describes the state persistence and transition process, which is implicit. The second stochastic process describes the corresponding statistical relationship between the output state of Markov chain and the output of HMM model, which is described by the output probability matrix B.

There are two presuppositions to use HMM: homogeneous Markov hypothesis and independent observation hypothesis. In the homogeneous Markov hypothesis, the state of any time is only related to the state of the previous time of the current time, and has nothing to do with the state and output of other times. In the observation independence hypothesis, the corresponding output value at any time is only related to the state at the current time, and has nothing to do with the state and observation value at other times. The three problems of HMM and the corresponding algorithms are carried out under the premise of these two assumptions.

3.2. Implementation of EM Algorithm

In the parameter solving problem of HMM, the model parameters are unknown and the output sequence is o. the model parameters that can make the probability of output sequence o maximum are obtained. If the variables in the probability model can be directly observed, the maximum likelihood estimation (MLE) method or Bayesian estimation method is usually used to calculate from the given data. Because HMM model contains hidden variables, we can not use these two methods directly. EM algorithm keeps approaching the optimal solution through iteration. It uses maximum likelihood estimation to solve the parameters of the model with hidden variables. Each iteration of EM algorithm consists of two steps. The first step is to find the expected joint probability expectation e, and the second step is to find the model parameters when the expectation reaches the maximum value, which is called the expected maximum algorithm. Figure 2 present the flow chart of the proposed EM algorithm.

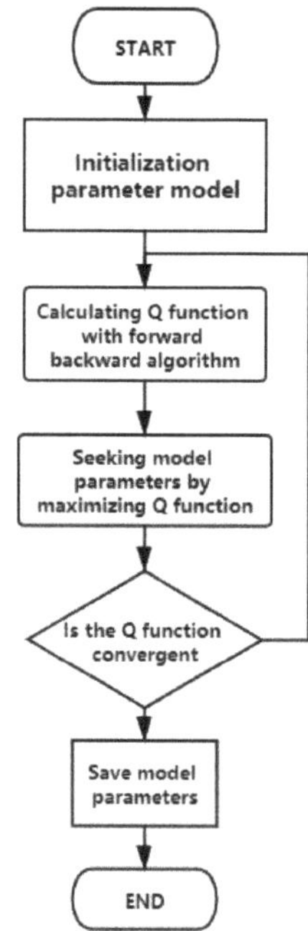

Figure 2. EM algorithm flow chart.

The first step is usually to find the Q function, and the form of Q function is different according to the distribution of output sequence. The general formula of Q is:

$$Q\left(\theta,\theta^{old}\right)=\sum_{I}p\left(I,O|\theta^{old}\right)\ln p(I,O|\theta) \tag{1}$$

When the output observations are discrete distribution $B=\{v_i\}, 1\leqslant i\leqslant M$, where M is the number of discrete observations; when the output observations obey Gaussian distribution $B=\{\mu_i,\sum i\}, i=1,\cdots,N$; when the output observations obey Gaussian mixture distribution, $B=\left\{B_{ij,}\mu_{ij},\sum ij\right\}, 1\leqslant i\leqslant N; 1\leqslant j\leqslant M$, where M is used to represent the number of Gaussian distributions corresponding to the Gaussian mixture model corresponding to each state.

If the output distribution of HMM is discrete, the Q function of the model can be decomposed into three parts, each part is only related to a single model parameter.

$$\begin{aligned}Q\left(\theta,\theta^{old}\right)&=\sum_{i=1}^{N}p\left(q_1=s_i,O|\theta^{old}\right)\ln\pi_k\\&+\sum_{t=1}^{T-1}\sum_{i=1}^{N}\sum_{j=1}^{N}p\left(q_t=s_i,q_{t+1}=s_j,O|\theta^{old}\right)\ln a_{jk}\\&+\sum_{t=1}^{T}\sum_{i=1}^{N}p\left(q_t=s_i,O|\theta^{old}\right)\ln p(o_t|q_t=s_{i,\theta})\end{aligned} \tag{2}$$

In order to simplify the calculation process, variables are introduced:

$$\gamma_t(i)=\frac{\alpha_t(i)\beta_t(i)}{P(O|\theta)}=\frac{\alpha_t(i)\beta_t(i)}{\sum_{i=1}^{N}\sum_{j=1}^{N}\alpha_t(j)\beta_t(j)} \tag{3}$$

Let

$$\xi_t(i,j)=p(q_t=s_i,q_{t+1}=s_j|O,\theta)\ (1\leqslant t\leqslant T,1\leqslant i,j\leqslant N) \tag{4}$$

The expression can also be deduced by forward backward algorithm:

$$\xi_t(i,j)=\frac{\alpha_t(i)a_{ij}b_j(o_{t+1})\beta_{t+1}(j)}{\sum_{i=1}^{N}\sum_{j=1}^{N}\alpha_t(i)a_{ij}b_j(o_{t+1})\beta_{t+1}(j)} \tag{5}$$

The second step is to find the hidden Markov model parameter θ when the Q function is maximized, and to solve the three model parameters by maximum likelihood method for the three terms of Equation (2). The parameterπ of probability model are known by solving the first term.

$$\sum_{j=1}^{K}\pi_i-1 \tag{6}$$

Then we can get the Lagrange function of the first term as follows:

$$L_1\left(\pi,\theta^{old},\lambda\right)=\sum_{i=1}^{N}p(q_1=s_i,O)\ln\pi_i+\lambda\left(\sum_{j=1}^{K}\pi_i-1\right) \tag{7}$$

The partial derivative of π_i is calculated and set to 0, we can be concluded that:

$$\pi_i=\gamma_1(i), i=1,\cdots,N \tag{8}$$

According to the second term, the parameter state transition matrix A of probability model is solved, the known state transition matrix satisfies:

$$\sum_{j=1}^{N} a_{ij} = 1, j = 1, \cdots, N \tag{9}$$

Then the Lagrange function of the second term is:

$$L_2\left(A, \theta^{old}, \lambda_1, \cdots \lambda_N\right) = \sum_{t=1}^{T-1} \sum_{i=1}^{N} \sum_{j=1}^{N} p\left(q_t = s_i, q_{t+1} = s_j, O|\theta^{old}\right) \ln\left(a_{jk}\right) + \sum_{j=1}^{N} \lambda_j \left(\sum_{k=1}^{M} b_j(k) - 1\right) \tag{10}$$

In the same way, the partial derivatives of each term of the state transition matrix are calculated and set to 0, we can get that:

$$a_{ij} = \frac{\sum_{t=1}^{T-1} \xi_t(i,j)}{\sum_{t=1}^{T-1} \gamma_t(i)} \; i, j = 1, \cdots, N \tag{11}$$

The third term of Equation (2) has different forms of solution according to the distribution of output sequence. When the output sequence obeys the discrete distribution, the Lagrange function is:

$$L_3\left(B, \theta^{old}, \lambda_1, \cdots, \lambda_N\right) = \sum_{t=1}^{T} \sum_{i=1}^{N} p\left(q_t = s_i, O|\theta^{old}\right) \ln b_j(o_t) + \sum_{j=1}^{N} \lambda_j \left(\sum_{k=1}^{M} b_j(k) - 1\right) \tag{12}$$

The partial derivative of the function to B is juxtaposed to 0, we can get that:

$$b_j(k) = \frac{\sum_{t=1, o_r = v_k}^{T} \gamma_t(j)}{\sum_{i=1}^{T} \gamma_t(j)} \tag{13}$$

If the output observations corresponding to HMM states obey Gaussian distribution $B = \{\mu_i, \sum i\}, i = 1, \cdots, N$, assume that each state corresponds to a Gaussian distribution, and n is the dimension of the output observations, then the number of output distributions and states is the same as N. The Lagrange function is written as:

$$L_3\left(B, \theta^{old}\right) = -\frac{1}{2} \sum_{t=1}^{T} \sum_{i=1}^{N} p\left(q_t = s_i, O|\theta^{old}\right) [n \; \ln(2\pi) + \ln|\Sigma_i| + (o_t - \mu_i)] \tag{14}$$

We can get the partial derivative of function to μ_i and $\sum i$ and make the partial derivative 0:

$$\mu_i = \frac{\sum_{t=1}^{T} \gamma_i(t) o_t}{\sum_{n=1}^{T} \gamma_i(t)} \; i = 1, \cdots, N \tag{15}$$

$$\sum i = \frac{\sum_{t=1}^{T} \gamma_i(t)(o_t - \mu_i)(o_t - \mu_i)^T}{\sum_{n=1}^{N} \gamma_i(t)} \; i = 1, \cdots, N \tag{16}$$

If the output sequence corresponding to the state of HMM model is represented by Gaussian mixture model distribution, $B = \{B_{km}, \mu_{km}, \sum km\}$, N are the number of States, and M is the number of Gaussian distributions contained in the Gaussian mixture model corresponding to the output sequence in each state. Then the probability of output corresponding to state j can be expressed as:

$$B = \{B_{km}, \mu_{km}, \Sigma_{km}\} \tag{17}$$

Here we introduce the intermediate variable v and satisfy the following conditions:

$$p(v_m|q_i) = B_{im}, 1 \leqslant i \leqslant N; 1 \leqslant m \leqslant M \tag{18}$$

$$v_j(o_t) = N\left(o_t|\mu_{jm}, \Sigma_{jm}\right) \tag{19}$$

The constraint condition is as follows:

$$\sum_{m=1}^{M} B_{jm} = 1, \cdots, N \tag{20}$$

Finally, the likelihood function can be written as:

$$\begin{aligned} L_3\left(B, \theta^{old}, \lambda_1, \cdots, \lambda_N\right) &= \sum_{t=1}^{T}\sum_{j=1}^{N}\sum_{m=1}^{M} p\left(v_t = v_m, q_t = s_j, O|\theta^{old}\right) \\ &\left[\ln B_{jm} - \frac{n}{2}\ln(2\pi) - \frac{1}{2}\ln|\Sigma_{jm}| - \frac{1}{2}(o_t - \mu_{jm})^T \sum_{jm}^{-1} (o_t - \mu_{jm})\right] \\ &+ \sum_{j=1}^{N}\lambda_j\left(\sum_{m=1}^{M} B_{jm} - 1\right) \end{aligned} \tag{21}$$

Because of the introduction of intermediate variables, the forward and backward algorithms including implicit variables are redefined:

$$\alpha_t(im) = p(v_t = v_m, q_t = s_i, o_1, \cdots, o_t|B) \tag{22}$$

$$\beta_t(im) = \sum_{j=1}^{N}\sum_{l=1}^{M} a_{ij}B_{jl}N\left(o_{t+1}|\mu_{jl}, \Sigma_{jl}\right)\beta_{t+1}(jl) \tag{23}$$

$$\eta_t(im) = p(v_t = v_m, q_t = s_i|O, \theta) \tag{24}$$

Similarly, let the likelihood function calculate the partial derivative for each member of $B = \{B_{km}, \mu_{km}, \Sigma_{km}\}$ and set it to 0 to obtain the extreme value, and finally get that:

$$B_{jm} = \frac{\sum_{t=1}^{T}\eta_t(jm)}{\sum_{t=1}^{T}\sum_{m=1}^{M}\eta_t(jm)} \quad j = 1, \cdots, N; m = 1, \cdots, M \tag{25}$$

$$\mu_{jm} = \frac{\sum_{t=1}^{T}\eta_t(jm)o_t}{\sum_{t=1}^{T}\eta_t(jm)} \quad j = 1, \cdots, N; m = 1, \cdots, M \tag{26}$$

$$\Sigma_{jm} = \frac{\sum_{t=1}^{T}\eta_t(jm)\left(o_t - \mu_{jm}\right)\left(o_t - \mu_{jm}\right)^T}{\sum_{t=1}^{T}\eta_t(jm)} \quad j = 1, \cdots, N; m = 1, \cdots, M \tag{27}$$

In this paper, the gait phase is divided into four phases based on the angular velocity collected at the ankle. In the process of the four stages transition, the states are from left to right in turn, and there is no jump transition. In this paper, the four gait stages are represented by the four states of Markov chain in HMM. The corresponding states of the four stages of gait can not be directly observed, but the angular velocity measured by the sensor. So in this paper, Gaussian distribution or Gaussian mixture distribution is used to realize the mapping relationship between the state and the output, and realize the division of gait stages.

4. Materials and Methods

Based on the data acquisition system, this section gives the overall scheme of gait phase recognition based on inertial sensors. The sensor data acquisition, feature extraction and gait phase division of the system are described.

4.1. Gait Data Collection

As shown in Figure 3, the data acquisition part consists of a minimum system composed of a main control chip and peripheral circuits, an inertial sensor module for gait signal sensing, a wireless signal transmission module and a data receiving device.

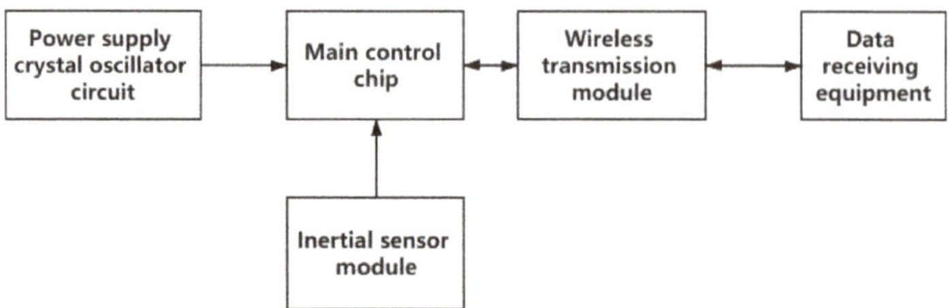

Figure 3. Data Acquisition System.

4.1.1. Data Acquisition Hardware Platform

The selection criteria of data acquisition system usually include sensor accuracy, sensor drift and sampling frequency. With regards to the transmission module, it is always necessary to consider the transmission speed and transmission reliability as well as the convenience of practical use, and the gait behavior recognition acquisition system needs to meet the characteristics of small size and low power consumption of wearable devices as far as possible. Compared with image acquisition, the acquisition of angular velocity by inertial sensors is convenient and easy to operate, and due to the development of MEMS technology, the use of inertial sensors now has a low power consumption, small size and lightweight and other superior performance. The sensor module is manufactured by Invensense (Sunnyvale, CA, USA). The accelerometer sensitivity is $\pm$4800 LSB/g, and the range of accelerometer can be set as $\pm$2, $\pm$4, $\pm$8, $\pm$16 g. According to the needs of this research, the measuring range of the gyroscope range is set at $\pm 2000^{\circ}$ and the sampling frequency is 100 Hz. The developed system hardware is shown in Figure 4.

Figure 4. System hardware.

Human gait signal collected by the system can be stored directly by adding memory in the system, or by wired and wireless communication mode. Each of these methods has its advantages and disadvantages. The advantages of wired mode are fast storage speed and low packet loss rate, the disadvantage is that the process of data storage is not observable and may be lost. Wired transmission is rarely used in practice, although it can have both fast transmission and real-time data visualization. The wireless transmission uses electromagnetic waves to send and receive signals for communication. For example researchers usually use data collected simultaneously from the waist, thighs, calves, instep toes and ankles or multiple parts to identify the gait stage. In this paper, a bandage is used to attach a wearable sensor to the human toe (as shown in Figure 5a) to capture the angular velocity generated by the human gait behavior, which is passed through Wi-Fi to the mobile phone.

4.1.2. Software Platform

The data collected by the hardware circuit will be sent to the mobile phone through wireless transmission. Therefore, Wi-Fi wireless communication software based on the Android operating system is compiled. As shown in Figure 5b, the software can receive the data collected by the Wi-Fi signal transmitter in real-time during normal operation, and save the received data to a file. The data can be read to the computer through a USB serial port to facilitate the subsequent algorithm research, and the received signal change curve is displayed on the mobile phone. The collected original 3D angular velocity signal is shown in Figure 5c.

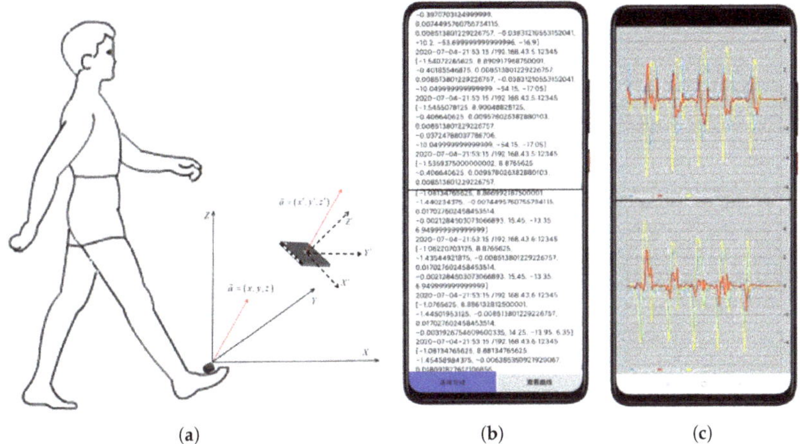

Figure 5. Data acquisition and software interface (**a**) Sensor placement (**b**) Raw data (**c**) Raw data curve.

4.2. Gait Data Preprocessing

Smoothing and de-noising the collected raw data is an indispensable step in the data processing process. Human gait behavior is low-frequency motion, in this case, it is necessary to filter the mixed high-frequency noise. The ways of generating these noises are noise caused by relative movement between acquisition equipment and the human body; there will also be noise interference in the process of digital signal conversion; electromagnetic interference introduced in the process of data transmission from acquisition equipment to receiving equipment; noise generated by power supply circuit of acquisition equipment [42–44]. Most of the noise can be eliminated by preprocessing original data. The common data preprocessing methods include signal denoising, smoothing, and normalization. Data preprocessing can be done by using a filter circuit in the data acquisition system or by the software program. Butterworth filter, FIR low-pass filter, moving average filter, median filter, Wiener filter, and wavelet filter software filtering methods are commonly used.

The common data smoothing methods include moving average filtering, median filtering, three-point linear smoothing filtering, and five point linear smoothing filtering. The selection of filtering methods is often related to the similarity of the signals before and after filtering. It is more likely to judge which filtering method to use by observing the dynamic perception and identification of subjective factors of human motion imbalance state. The filtering method selected in this way can have a better effect, but it may not be the best method. Based on information theory, this paper proposes a selective filtering method by analyzing information contained in the signal. By comparing the signal-to-noise ratio (SNR) and root mean square error (RMSE), the optimal filtering method is selected by comparing the moving average filtering, median filtering, and five point cubic filtering. The effect of sliding filtering is related to the size of the sliding window, too large windows will lead to serious signal distortion and signal delay. By comparing the window size to 15, as shown in Figure 6a,b, the changing trend of angular velocity signal after filtering is more smooth, which can reduce most of the interference.

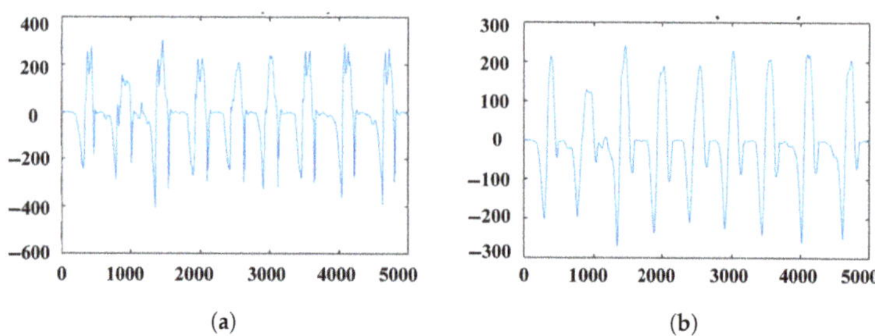

Figure 6. Angular velocity signal (**a**) Before preprocessing (**b**) After preprocessing.

4.3. Window Segmentation for Gait Data

Window segmentation is the process of cutting a set of gait data according to the actual requirements and then extracting features from the data within the window as a recognition algorithm. The window is used as the basic unit of data in the feature extraction and gait phase identification process. There are three popular window classification methods: behavior-based window, event-based window, and sliding-based window [45–49]. The behavior-based window determines the location of the window segmentation in the raw data based mainly on the behavior change in the data; the event-based window determines the location of the window segmentation in the raw data through the specific events in the data; the sliding-based window segmentation method is similar to the previous two methods of determining the location of window segmentation, however, unlike the original data, the sliding-based window is not directly related to the raw data and it uses an equidistant window to partition the data.

Among the three window segmentation methods, the sliding-based window is well adapted to the periodic, stable, and some sporadically distributed behaviors [50]. Because the threshold of behavior-based and event-based segmentation needs to be set and adjusted according to different curves. As well as the interference signal will affect the threshold judgment and the gait changes periodically, the use of behavior-based and event-based window segmentation methods is not as effective as the sliding-based window segmentation method. In this paper, the sliding-based window segmentation method is used for both feature extraction and gait event recognition, and we test the number of overlaps of adjacent window. The size of the window needs to be determined based on factors such as the actual data type and sampling frequency, the larger the window size, the more pronounced the feature differences are, and the more latency is exhibited. The window size defined in this paper does not exceed the state that occupies the smallest percentage of the gait phase, to avoid a window of data containing multiple gaits states which leads to reduced stage recognition. A window which is too small extracts features that are not representative of the gait stage. The choice of the gait stage window size is a process of balancing recognition speed and recognition accuracy. The data segmentation method based on the sliding window is shown in Figure 7.

As mentioned previously, the principles and feasibility of the hidden Markov model for gait recognition were introduced, and the model was analyzed in the context of gait stage recognition. Different gait stage divisions and different sources of gait data make the structure of the Hidden Markov model different. A complete gait cycle can be divided into multiple gait stages according to practical requirements, and the states corresponding to the gait stages have Markov property, so the hidden Markov model is widely used in the field of gait stage classification and gait recognition.

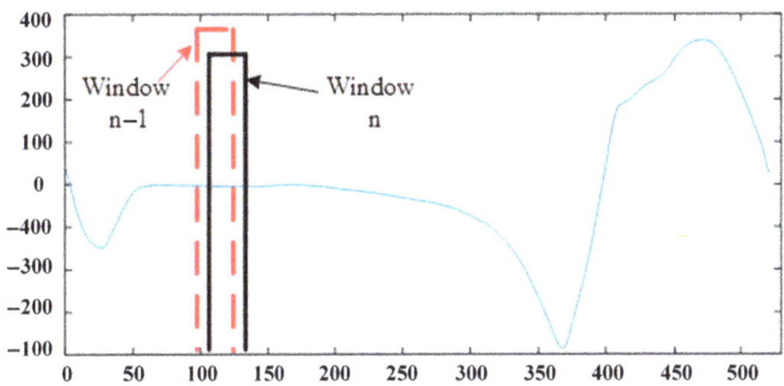

Figure 7. Using sliding window to split gait data.

Since the actual situation is far more complex than the ideal situation, there are many problems when using the hidden Markov model to solve the gait stage recognition problem, the main problem is that the three parameters of the hidden Markov model are fixed, and there is no way to adaptively handle the specific use case.

The gait motion signal is quasi-periodic, and although the motion gait is periodic, the length of the gait period, the length of a complete cycle factor such as the percentage of each stage, and the variation in data magnitude can be greatly disturbed in actual scene. At the same time, gait behavior is arbitrary and variable due to individual and environmental diversity, making gait signals (through various gait information acquisition devices to obtain kinematic, kinetic, and physiological information about gait) exhibit periodicity as well as uncertainty and nonlinearities, and complex and non-unique correlations, which add to the importance of period determination, phase delineation, similarities in gait stage analysis and identification.

Usually, the process of gait stage segmentation by hidden Markov models is to firstly estimate the parameters using certain data and then the parameters are brought into the model and the state sequence is then computed using the Vibbit algorithm, i.e., the trained model is used to identify the gait data of others. Good results can be obtained when gait data used for recognition and training are not very different. If the difference between the recognition data and the training data is large, the actual recognition accuracy will become very low. A simpler approach is to use a larger training data set to allow the model to cover a wider gait space. Although it may reduce the effect of the difference between the data used for training and the data used for recognition, it makes the gait phase recognition of the parameters widely distributed, which makes the performance of model recognition degrade. Another approach is to calculate a specific model based on a specific situation and select a specific model based on the actual situation during the use of the identification. This approach is also problematic in that gait staging based on a particular situation requires extensive analysis of movement data to do so; there is no explicit method to determine the gait for different situations, which in turn poses a more complex problem for gait stage identification.

It would be an interesting direction of research to start with the problem of gait modeling itself to determine a universally applicable method. Due to the quasi-periodic nature of human gait and the variability of human gait in different situations, a better approach is to use adaptive techniques. Adaptive approaches have been widely studied and used in solving speech recognition problems using hidden Markov models, and this research expects to put this technology applied in gait stage recognition, motion gait adaptation required correcting the model parameters with some of the gait data used for recognition. This makes the model parameters more suitable for gait recognition, to improve the accuracy and recognition rate of the hidden Markov model. The analysis from the model perspective enables the adaptive motion gait technology to realize the self-adaptive motion behavior of different individual in different environments with different movements.

Adaptive methods for models can be generally divided into two categories: feature layer-based adaptive methods and model layer-based adaptive methods. Adaptive methods based on the feature layer from the acquisition system to get the gait signal in different cases independent of the differences between the different cases of the feature extraction ensemble, i.e., the extracted features are not correlated with the factors causing the differences. The model layer-based adaptive method modifies the model parameters based on the differences between the actual measurements and the trained model so that they can be better used in the current identification and classification of actual data.

As shown in Figure 8, the model layer-based adaptive approach is based on modifying the model parameters according to the actual gait signal to be identified. The gait stage recognition hidden Markov model parameters is calculated from the training data. After the parameter training is completed, if the parameters remain unchanged, since the model parameters are collected by a specific individual under specific conditions, so that during the recognition phase the results may vary considerably for different individuals under different conditions. To counteract the effect of the differences, one way is to use a large number of gait data collected from different individuals under different conditions at the same time during the training phase. The data is used to target train multiple gait stage recognition models, and the recognition stage selects the appropriate model from these models, or firstly using a small amount of data to substitute all models to select models with high recognition rates for gait stage recognition; the other way is to take these model parameters to generate new model parameters by linear combination, an approach that simply sums up the set of parameters in various cases, and failure to consider special cases can make the model parameters become too widely distributed rather than sharply distributed, so the recognition effect is neither too high nor too low.

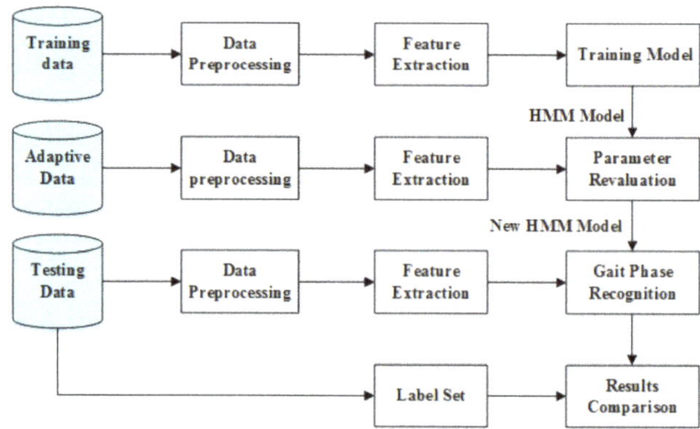

Figure 8. Flow chart of adaptive algorithm based on model layer.

The model-layer based adaptive approach is based on modifying the model parameters according to the actual gait signal to be identified. The model layer-based adaptation can be broadly divided into two categories depending on the algorithm: direct model adaptation algorithms (Direct Model Adaptation) and transform-based model adaptation methods (Transform-based Model Adaptation). Typical direct model adaptation algorithms have a Maximum a Posteriori (MAP), Minimum Classification Error (MCE), and Structural Maximum a Posteriori (SMAP), et al.

The model adaptive algorithms that directly adjust parameters can adjust those parameters for which there is a distribution of observed output in the adaptive data. More data is often required to achieve better adaptive results. Transform-based adaptive algorithms are maximum likelihood linear regression, maximum a posteriori linear regression, minimum classification error linear regression, etc. The conversion-based model adaptive algorithm obtains a series of linear transformations based on the differences between the source model and the target model to achieve a good adaptive effect on the model and can adjust all distribution parameters in the model. The adaptive algorithm can get better results with a small amount of data compared to the direct adjustment of the model parameters.

In this paper, MAP and MLLR are selected to adjust the parameters of the hidden Markov model for the human gait stage division. Bayesian theory, which combines a priori information about the data being adapted and the model parameters so that the model's posterior probability is maximized. MLLR with transformation mechanism is used to transform the parameters of the model into a feature space that is close to the adapted data.

5. Experimental Results

In the previous sections, we have analyzed the algorithm principle and feasibility of the hidden Markov model in gait stage recognition and the process of solving the model parameters, as well as two common methods of adjusting the model parameters, which will be tested and verified in this section.

5.1. Experimental Data Source

5.1.1. Data Collection Object

It is common for researchers to use, for example, the waist, thighs, calves, backs of feet, toes, ankles, or multiple sites to collect data simultaneously to perform gait phase recognition [16,51–58]. In this paper, a bandage is used to attach a wearable sensor to the human toe to capture the angular velocity generated by the human gait behavior, which is passed through Wi-fi to the mobile phone.

There were sixteen volunteers involved in the experimental data collection, eleven males and five females, ranging in age from 30 to 60 years old, height from 1.59 to 1.84 m, and weight from 49 to 88 kg. The data collection sensors were fixed on the both toes, and the angular velocity of gait during walking of normal volunteers was collected during the experiment. These volunteers were quite different in age and fluctuated in weight, which can be used to compare model accuracy before adaptation and after parameter adaptation.

5.1.2. Gait Data Collection

During the collection of data from the inertial sensors that embody gait behavior, the angular velocity signals of eight volunteers during normal walking were collected, and each of the volunteers walked on level ground with their customary walking habits, and inertial data were collected twice for each volunteer, with the distance walked each time 11 m, removing the first and the last cycle that differs significantly from normal gait data during model training and gait phase identification gait signal. The sampling frequency of the data acquisition device is 100 Hz, the normal human walking speed is around 1 m/s, and the distance between each step is about 60∼75 cm, so the gait frequency is approximately one and a half steps per second, so the set sampling frequency can capture a complete sample of each gait phase.

5.2. Model Performance Evaluation Metrics

When evaluating the effectiveness of an analytical model for gait recognition, comparing the accuracy alone is not sufficient to fully evaluate the performance of the model. The problem of gait phase recognition in this paper can be seen as a binary classification problem, including both correct and incorrect recognition. Usually, dichotomous problems are classified by Precision (P), Recall (R), Sensitivity (True Positive Rate, TPR), Specificity (False Positive Rate, FPR), and F-Measure, which are evaluated by the confusion matrix.

The horizontal coordinate of the P-R curve is the recall rate, and the vertical coordinate is the precision rate, and the ROC(Receiver Operator Characteristic) curve with specificity as the horizontal coordinate and sensitivity as the vertical coordinate. When the numbers of positive and negative samples in the data tested are similar, both ROC and P-R have good performances. However, if the negative sample number is larger, the ROC curve effect can still maintain the trend, while the effect on the PR curve is poor. The uesd Performance evaluation indicators are shown in Table 1.

Table 1. Performance evaluation indicators.

Index	Formulation	Definition
Accuracy rate (A)	$A = \frac{TP+TN}{TP+FP+TN+FN}$	The percentage of samples correctly identified as a percentage of the samples used for testing indicates the overall identification rate of the system.
Recalling rate (R)	$R = \frac{TP}{TP+FN}$	The ratio of correctly identified positive samples in all positive, indicating the identification rate ofindividual states of the HMM model.
Precision ratio (P)	$P = \frac{TP}{TP+FP}$	The proportion of correctly identified positive samples in all positive samples, showing the effect of sample distribution on recognition rate.
F1 value	$F1 = \frac{2TP}{2TP+FP+FN}$	The reconciled mean of P and R was combined to evaluate P and R.
Sensitivity (TPR)	$TPR = R$	Sensitivity is the same as recalling rate.
False accuracy (TPR)	$FPR = \frac{FP}{FP+TN}$	TPR is predicted to be the ratio of the positive sample to the negative sample.
ROC Curve	Not applicable	The larger the area enclosed by the curve, the higher the classification rate.

The classes used for identification are usually called positive classes, denoted by 1, while the others are negative classes, denoted by 0. Four scenarios will emerge from the classification model's identification of test data.

TP: The model identifies a positive sample as a positive sample.
FN: The model identifies a positive sample as a negative sample.
FP: The model identifies a negative sample as a positive sample.
TN: The model identifies a negative sample as a negative sample.

5.3. *Analysis of Results Based on Hidden Markov Models and Improved Models*

In previous sections, we have analyzed the algorithm principle and feasibility of the hidden Markov model in gait stage recognition and the process of solving the model parameters, as well as two common methods of adjusting the model parameters, which will be tested and verified in this section.

Since the information collected by the acquisition device is continuously changing, so the output signal of the hidden Markov model for human gait stage recognition is continuous, and the output corresponding distribution needs to be described by a continuous function, so this paper will verify the model performance when the corresponding output signal of the gait stage obeys two distributions respectively, i.e., the Gaussian and Gaussian mixed distribution.

5.3.1. Recognition Results of HMM with Gaussian Distribution

The appropriate size of the sliding window for each feature was first tested separately, and Figure 9 gives the values of some features. The recognition rate of the sliding window varies with the sliding window size. The size of the overlapping portion between adjacent sliding windows is taken to be half the size of the sliding window. The curve in Figure 9 lists the five eigenvalues: mean, variance, root mean square, mean gradient, waveform factor recognition rates and window size relationship, it can be seen that each feature takes the optimal value in the sliding window size range of 5 to 15 when combining multiple eigenvalues for gait stage recognition. One may use this range as a reference to find the most appropriate window within the range size which can shorten the time it takes to find a model.

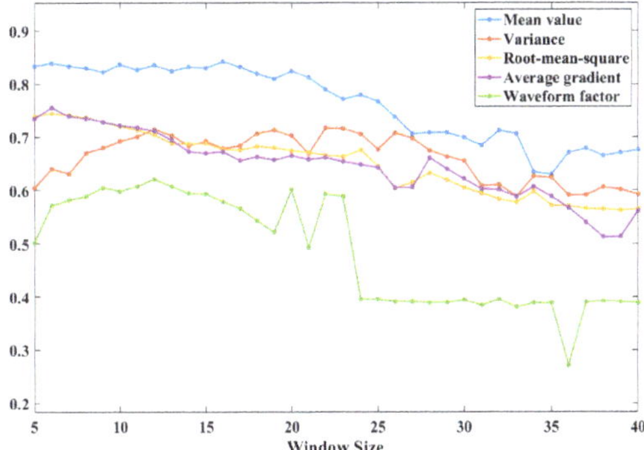

Figure 9. Change of eigenvalue recognition rate with sliding window size.

In Figure 10, the recognition rates of 15 common time-domain features are shown with the sliding window size set to 10, from which all the mean value of raw value, root mean square (RMS), and absolute mean reach the recognition rate of 70% and above. When using multidimensional features for gait stage recognition, these features should be given priority.

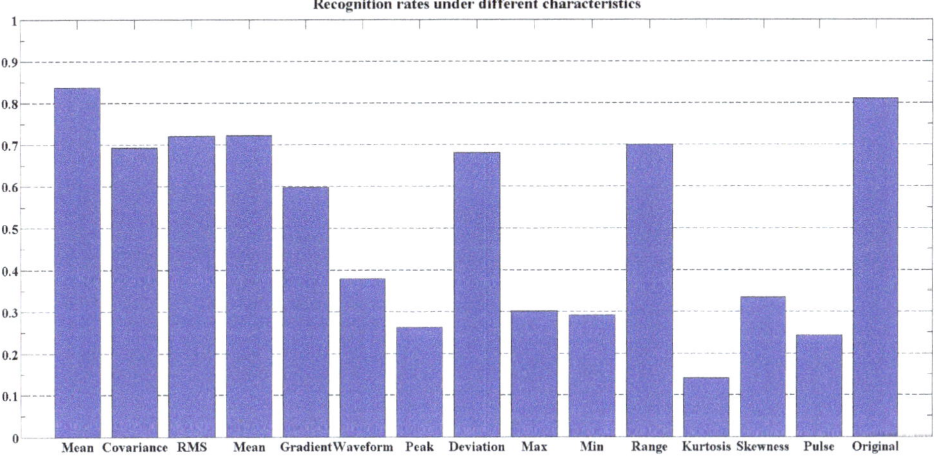

Figure 10. Recognition rate of eigenvalues with sliding window size of 10.

Previously, features of individual dimensions were analyzed for gait stage recognition rates for different size windows and different features, and Figure 11 draws the trend of recognition rate with the sliding window size after combining multiple features into a multidimensional feature. The recognition rate of the eigenvalues in Figure 9 is selected from the largest to the smallest. For example, the 3D features selected in the figure are the mean, the raw data sampling values, the average of the absolute values, the five-dimensional and seven-dimensional features are also selected according to this rule. The optimal sliding window size for gait stage recognition using 3D features is 14 and the recognition rate is 85.82%; four-dimensional features had an optimal recognition rate of 84.59% at a sliding window size of 12. Seven-dimensional features have the best recognition rate of 86.86% at a window size of 14; When the feature size is increased to 15, the best sliding window is 9, corresponding to the gait phase recognition rate of 89.54%.

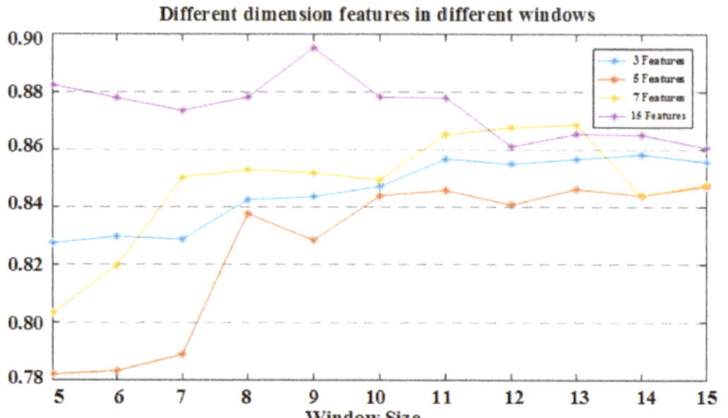

Figure 11. The recognition rate of different feature dimensions under different sliding windows.

5.3.2. Recognition Results of HMM with Gaussian Mixed Distribution

The Gaussian mixed distribution is more likely to contain the distribution of the output of the corresponding state of the gait phase than the distribution whose output is Gaussian. When the hybrid Gaussian model is used as the output distribution of the hidden Markov model identified by the gait stage, assuming that each state corresponds to the mixed Gaussian model and contains the same number of Gaussian distributions, assuming it was found previously that when the feature dimension is 15, the highest gait stage recognition rate can be obtained, and therefore no longer uses unidimensional features as observations for the gait stage recognition model in this section.

Figure 12 shows the recognition rate of the gait stage recognition model with a sliding window size in the range of 5 to 15 when M is taken at different values. Each curve in the figure corresponds to the recognition rate trend, which firstly goes up and then goes down, the appropriate window range is roughly between 9 and 15. It can be seen from the figure that the trend of the recognition rate increases gradually as the value of M increases to the highest recognition rate. Considering the influence of model parameter complexity on the time complexity of the subsequent state sequence calculations, the Gaussian hybrid model is used when the output distribution of the stage is distributed, we selected the recognition effect of the model when M is 5 and the sliding window size is 12 for comparison.

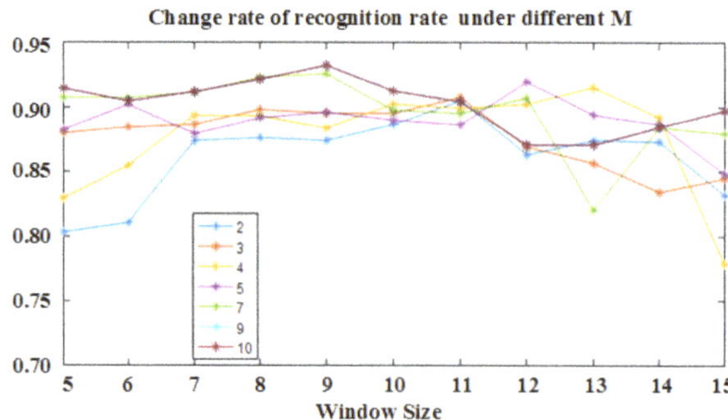

Figure 12. Change rate of recognition rate with sliding window under different M.

5.4. *Identification Results after Model Parameter Adaptation*

5.4.1. Performance Analysis of Gaussian HMM after Parameter Adaptation

It can be seen from Figure 10 that the highest recognition rate is onbtained when the unidimensional feature is the mean. Taking the mean as an example, it is meaningful to analyze maximum likelihood linearity regression estimation and maximum a posteriori probability estimation of the effect of parameter adaptation of the model on recognition rates. Table 2 presents the parameters before and after model adaptation. Since this paper assumes sequential transitions between gait stages, the initial state probabilities will no longer be included in the table, and the parameters used in this paper are obtained from the adaptation algorithm by adjusting the mean and variance, and the table shows the mean and variance of the parameters before and after adaptation.

Table 2. Comparison of performance indicators.

Parameters	Mean Value	Variance
HMM	[−2.6518 3.7528 −1.0578 −0.1528]	[3.9164 2.4664 0.7880 0.0146]
MLLR	[−2.4168 3.6096 −1.3194 −0.2117]	[3.0880 2.2895 0.8294 0.0184]
MAP	[−2.3472 3.5249 −1.5249 −0.2398]	[2.7750 2.2851 0.7171 0.0192]

The state transfer matrix A is a 4 by 4 square with the same state transfer matrix for all three models as shown in the following equation.

$$A = \begin{bmatrix} 0.8460 & 0.1540 & 0 & 0 \\ 0 & 0.8253 & 0.1747 & 0 \\ 0 & 0 & 0.7395 & 0.2605 \\ 0.1649 & 0 & 0 & 0.8351 \end{bmatrix}$$

The experiments in this paper all based on the assumption that each experiment starts from a state where the heel is off the ground, so the initial probability matrix is shown as:

$$\pi = \begin{bmatrix} 1 & 0 & 0 & 0 \end{bmatrix}$$

From Figure 13, it can be seen that the number of correctly recognized state sequences for states S1, S2, and S4 has increased considerably. There is a smaller decrease in the number of correct identifications for state S3, but the overall identification rate has increased.

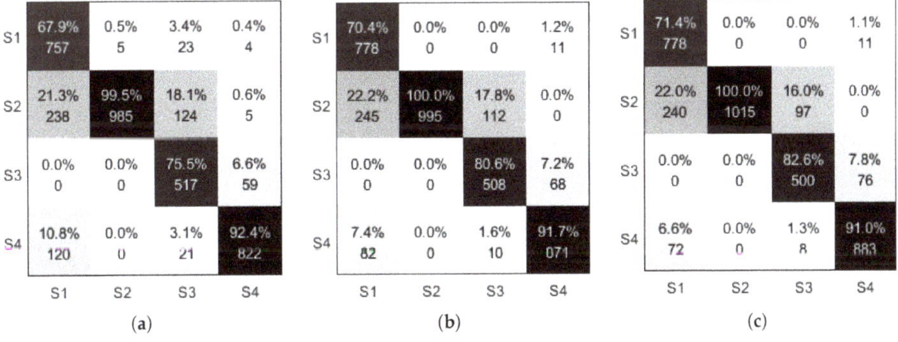

Figure 13. Confusion matrix for model recognition (**a**) HMM (**b**) MLLR (**c**) MAP.

5.4.2. Performance Analysis of Gaussian Mixed HMM with Adaptive Parameters

After obtaining the effect of the number of Gaussian distributions M and the size of the sliding window on the recognition rate in the Gaussian mixed model. The Gaussian distribution can be viewed as a special case of mixed Gaussian distributions, the discussion of parameter adaptation model in this section is based on the output of the gait stage recognition model which obeys the basis of the hybrid Gaussian model.

This section presents the MAP and MLLR based algorithm. The results after parameter adaptation are compared with the identification model of the gait stage with unadjusted parameters merely using the correct identification state. The ratio of the number of states to the total number of states used for identification is not comprehensive enough to evaluate the model, so this section evaluates the gait stage model by using the confusion matrix The model before and after parameter adaptation is compared, as shown in Figure 14, with the horizontal axis in the confusion matrix indicating the original state of the data and the vertical coordinate indicates the identified state.

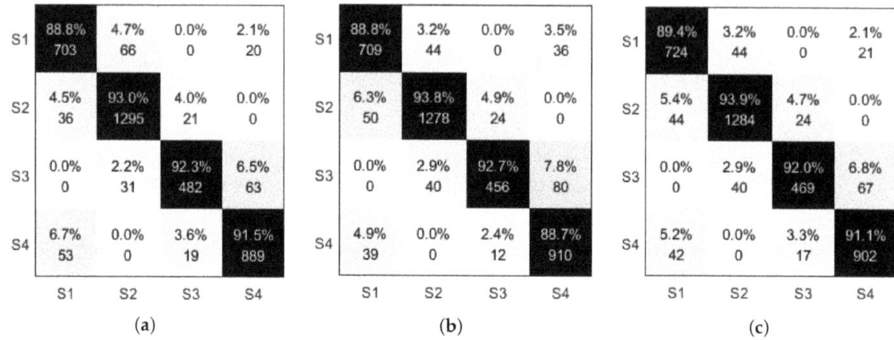

Figure 14. Confusion matrix of the model before and after parameter adaptation (**a**) HMM (**b**) MLLR (**c**) MAP.

As shown in the data in Figure 14a,b, the number of correct identifications of states S1 and S4 by maximum likelihood linear regression is greater than the number of states corresponding to correct recognition when the model is not adapted, and the opposite is true for states S2 and S3, where the number of states corresponding to correct recognition is less than the number of states corresponding to correct recognition when the model is not adapted. The confusion matrix of Figure 14c results from parameter revaluation using the maximum posterior probability estimated a priori information about the model parameters, so this paper adjusts the maximum a posteriori estimation by the maximum likelihood linear regression method of the mean and variance obtained which are used as a priori information. The maximum a posteriori probability estimate combined with the model identification information obtained from the maximum likelihood linear regression can be seen summarized as follows: the number of correct identifications for S1 is higher than the first two methods; the number of correct identifications for states S2 to S4 is intermediate between the first two methods position, as the maximum, a posteriori estimate combines the original model with the parameters of the adapted maximum likelihood linear regression in the form of weights. It is theoretically normal that the iterative result does not exceed the highest of the two.

From the performance metrics in Table 3, it can be seen that when the output of the gait stage recognition model is Gaussian mixed distribution, relative to the single Gaussian distribution univariate and the recognition rate is improved when multidimensional features are present. When the output signal features of the model obey a Gaussian mixed distribution, maximum likelihood linearity when comparing the accuracy of the model before and after parameter adaptation and the parameter adjustment scheme combining regression and maximum a posteriori probability estimation results in an increase in the overall identification rate, and for the lower correct identification in the original model state, the recognition rate increases after parameter adjustment; meanwhile, the recognition rate is close to unadjusted after parameter adjustment with higher correct identification in the original model state.

Table 3. Comparison of performance using different methods.

Performance Index	Without Parameter Adaptive	MLLR	MAP
Accuracy rate (A)	91.59%	91.16%	91.88%
Precision ratio (P)	0.8876, 0.9303, 0.9234, 0.9146	0.8885, 0.9383, 0.9268, 0.8869	0.8938, 0.9386, 0.9196, 0.9111
Recalling rate (R)	91.59%	91.16%	91.88%
Precision ratio (P)	0.8910, 0.9578, 0.8368, 0.9251	0.8986, 0.9453, 0.7917, 0.9469	0.9176, 0.9497, 0.8142, 0.9386
F1 value	0.8893, 0.9439, 0.8780, 0.9198	0.8935, 0.9418, 0.8539, 0.9160	0.9056, 0.9441, 0.8637, 0.9247
Sensitivity (TPR)	0.0308, 0.0424, 0.0127, 0.0307	0.0309, 0.0363, 0.0113, 0.0437	0.0300, 0.0364, 0.0129, 0.0327

The ROC curves in Figure 15 for the three models show that the parameters of the maximum likelihood linear regression are used as the maximum posterior probability estimates to obtain the overall performance that is higher than when no parameter adaptation and maximum likelihood linear regression estimation are performed. As shown in the data in Table 3, the highest recognition accuracy of the gait stage model based on the angular velocity signal is around 92%, and the performance is not significantly higher after the adaptation of the model parameters.

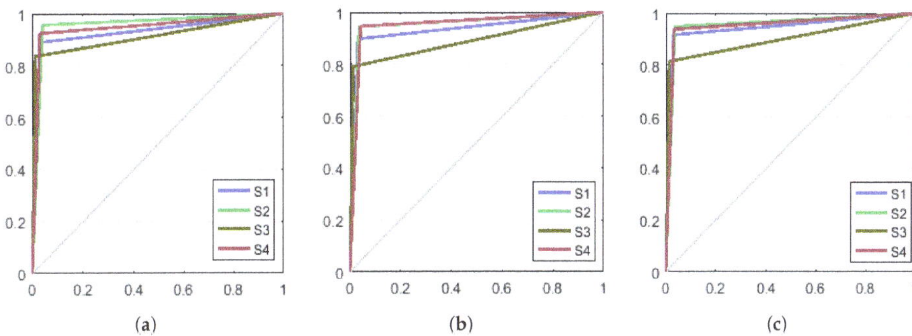

Figure 15. ROC curve of HMM parameters before and after adaptation (**a**) HMM (**b**) MLLR (**c**) MAP.

The possible reasons for the lift of the gait phase are: first, overlapping data between adjacent windows at the time of feature extraction, close to the moment of gait phase transition stage judgments will be subject to errors; second, the hidden Markov model may be used for parameter adjustment schemes and there are differences in the speech recognition domain and gait stage recognition.

5.5. Comparison with Other Algorithms

In order to test the advantages of HMM and the improved algorithm, this paper uses other six commonly used algorithms as a comparison: K-Nearest Neighbor(KNN), Logistic Regression(LR), Linear Discriminant Analysis(LDA), Random Forest(RF), Naive Bayesian Model(NBM) and SVM. The experimental results are shown in Figure 16. The resolution of the classification method is between 80% and 92%, which is close to the recognition rate of the HMM model. If only from the perspective of recognition rate, KNN, LR, and SVM can replace HMM as a model of gait cycle stage division. However, the gait phase division not only identifies specific stages but also needs to consider the transition relationship between states on time series. As a comparison, the six algorithms do not consider the transition between states, which may lead to state transition sequence errors, and then judge the number of cycles and the proportion of each state in each cycle proportion. The recognition rates of these six algorithms are shown in Figure 16.

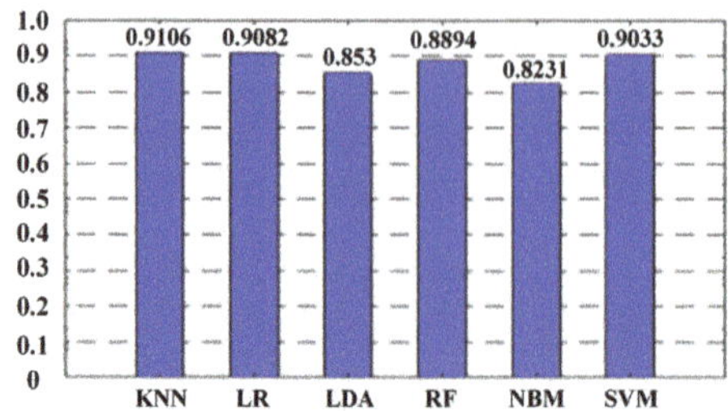

Figure 16. Comparison of recognition rates of different classification algorithms.

Figure 17a,b show the state sequence recognition diagram of the improved HMM model and KNN respectively. The red solid line in the figures is the state sequence as the control, and the blue dotted line is the time series actually recognized by the model. It can be seen from the figures that the recognition error state mainly occurs near state transition time. In Figure 17a, the recognition error of the improved HMM model is to identify the state as the previous or the next state of the current state, which is determined by the state transition matrix. Therefore, there is almost no error in judging the number of cycles contained in a complete gait cycle. The length of the recognition error is about five time points, which is close to the coincidence length of two adjacent windows. The HMM result is acceptable because the minimum unit to distinguish the state of the HMM model is the length of a sliding window and adjacent windows. In addition, there will be errors in the state sequence compared with the state recognition, so the recognition error of the HMM model is inevitable. In Figure 17b, state S1 occurs in the process of state S3 and S4 transition, which has an impact on the number of cycles, cycle length, and the proportion of each stage. The same phenomenon also occurs in the other five algorithms during algorithm testing.

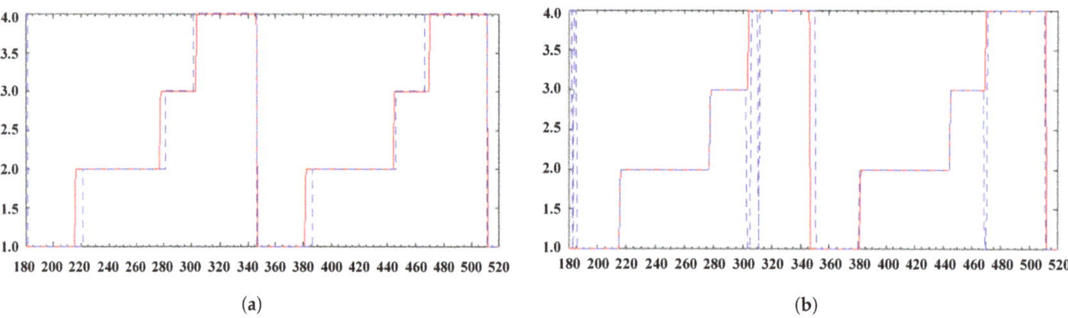

Figure 17. State sequence (**a**) Recognized by HMM (**b**) Recognition sequence of KNN.

5.6. Analysis of Human Balance Ability by Gait Parameters

The swing phase of normal people accounts for about 40% of the whole gait cycle. Patients with abnormal gait usually have shorter support phase and longer swing phase. The swing phase in the gait cycle corresponds to the state S2 of the HMM model in this paper. The red and blue histograms in Figure 18 show the proportion of the swing phase recognized by the control group and HMM models in the complete cycle, and the green curve represents the error between them. The maximum error accounts for 3.97% of the complete cycle and the average value is 2.02%, which proves the feasibility of HMM when dealing with gait cycle analysis.

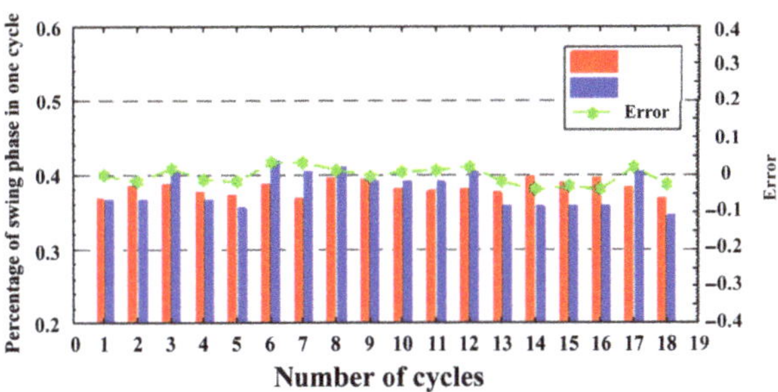

Figure 18. The ratio of the swing phase to the complete cycle.

Table 4 shows the spatiotemporal parameters related to human balance of the left and right lower limbs(LLL,RLL) of a normal person. The recognition results of HMM are very close to the control group results, and the left-right ratio of swing time, stance time to gait cycle reach 0.9453, 0.9241 and 0.9936 and 1.0000, respectively, indicating that the subject maintained good balance, which is similar to the control group results.

Table 4. Comparison of gait spatiotemporal parameters.

	Reference Value of LLL	Reference Value of RLL	Recognition Value of LLL	Recognition Value of RLL
Swing time (s)	0.5879 ± 0.0321	0.5954 ± 0.0446	0.5864 ± 0.0536	0.5543 ± 0.0857
Standing time (s)	0.4093 ± 0.0807	0.3977 ± 0.0423	0.4079 ± 0.0182	0.4414 ± 0.0129
Gait cycle (s)	1.5636 ± 0.1064	1.5746 ± 0.3254	1.5607 ± 0.1993	1.5707 ± 0.3293

5.7. Experimental Conclusions

In this section, the actual effects of gait phase recognition models are compared. Firstly, the data source and data acquisition scheme are introduced; secondly, the indexes and curves for evaluating the algorithm are introduced; then, the overall recognition rate of single dimensional and multi-dimensional features before and after the model parameters are adapted and the recognition rate when the output is subject to different distribution are analyzed; compared with the recognition rate of the model before and after the parameter adjustment, the model parameters result in better perforamance. The recognition rate of the model is 91.88%, even though the age, height, and weight of the participants are different. At the end of this chapter, the gait parameters related to the balance ability of the human body are analyzed. The gait comparison parameters calculated according to the gait stages are consistent with the reference values. The limitations of the previous proposed Hidden Markov Model-based method mainly lie in that HMM depends only on each state and its corresponding observer. HMM models are memoryless and cannot take advantage of contextual information. Because it's only related to the previous state, if we want to take advantage of more known information, we must build a high-level HMM model. We have been trying to address this recognized issue in this research, though much work remains.

6. Discussions and Conclusions

Gait analysis has been widely used in disease diagnosis, orthopedic surgery, and rehabilitation training in recent years, so it is more and more important to establish an accurate and effective walking model. One of the technical challenges in our previous work is the overflow problem of algorithm recursion, so we normalized the eigenvalues. Secondly, we have screened to extract the characteristic signal, and marked the phase division effect on some data. When the angular velocity signal has interference, the effect of maintaining regularity is better. In this paper, the data source and data acquisition scheme based on the inertial sensors are proposed, and the index of the evaluation algorithm is given. The segmentation of gait with the hidden Markov Model(HMM) is used to analyze the global recognition rate of single dimension and multi-dimension features before and after the model parameters are adaptive. Through analysis and comparison, the method has a high recognition rate, ensures the integrity and objectivity of gait data, and provides a new theoretical basis for medical diagnosis. This research tested different methods and the experimental results showed that MAP and MLLR achieved best performance with regard to parameter adaptation. In future work, we plan to conduct more comparative trials and try to find the deep mechanism of what kind of factors affect the performance of parameter adaptation. Deep learning models including CNN and RNN have been applied to clinical gait analysis applications, however, these methods require a lot of computation and the Real-time monitoring cannot be guaranteed. We will try to address this issue in future work.

Author Contributions: Conceptualization, L.L., H.W. and H.L.; Methodology, L.L., S.Q. and H.W.; Software, L.L. and H.Z.; Validation, H.L. and J.L.; Data analysis, L.L., H.W. and X.G.; Project administration, L.L. and S.Q.; Supervision, S.Q. and H.Z.; Writing original draft, X.G. and H.L.; Writing review & editing, H.L. and J.L. All authors have read and agreed to the published version of the manuscript.

Funding: This work was jointly supported by the National Natural Science Foundation of China under Grant 61803072 and 61903062, and China Postdoctoral Science Foundation under Grants 2017M621132 and 2017M621131, and the Dalian Science and Technology Innovation Fund under Grant 2019J13SN99 and 2020JJ27SN067, and in part by the Fundamental Research Funds for the Central Universities under Grant DUT20JC03 and DUT20JC44. The authors would like to express their thanks to these funding bodies.

Institutional Review Board Statement: Not applicable.

Informed Consent Statement: Not applicable.

Data Availability Statement: Data available in a publicly accessible repository that does not issue DOIs. Publicly available datasets were analyzed in this study. This data can be found here: [http://lis.dlut.edu.cn/database_en.htm].

Acknowledgments: The authors would like to express their thanks to the members of LIS (laboratory of Intelligent system, Dalian University of Technology).

Conflicts of Interest: The authors declare no conflict of interest.

Abbreviations

The following abbreviations are used in this manuscript:

MEMS	Micro-electro-mechanical Sensor
BSN	Body Sensor Network
IMU	Inertial Measurement Unit
ZUPT	Zero Velocity Updating
HMM	Hidden Markov Model

References

1. Barth, J.; Oberndorfer, C.; Pasluosta, C.; Schülein, S.; Gassner, H.; Reinfelder, S.; Kugler, P.; Schuldhaus, D.; Winkler, J.; Klucken, J.; et al. Stride segmentation during free walk movements using multi-dimensional subsequence dynamic time warping on inertial sensor data. *Sensors* **2015**, *15*, 6419–6440. [CrossRef]
2. Bhargava, N.; Taro, N.; Amanda, R.; Jerry, T.; Jerry, L.; Tam, N.; Steven, Z.; Donald, L. Real-Time Classification of Patients with Balance Disorders vs. Normal Subjects Using a Low-Cost Small Wireless Wearable Gait Sensor. *Biosensors* **2016**, *6*, 58.
3. Bao, S.D.; Meng, X.L.; Xiao, W.; Zhang, Z.Q. Fusion of inertial/magnetic sensor measurements and map information for pedestrian tracking. *Sensors* **2017**, *17*, 340. [CrossRef] [PubMed]
4. Majumder, S.; Mondal, T.; Deen, M.J. A Simple, Low-cost and Efficient Gait Analyzer for Wearable Healthcare Applications. *IEEE Sens. J.* **2018**, *19*, 2320–2329. [CrossRef]
5. Lahmiri, S. Gait Nonlinear Patterns Related to Parkinson's Disease and Age. *IEEE Trans. Instrum. Meas.* **2019**, *68*, 2545–2551. [CrossRef]
6. Chiang, C.Y.; Chen, K.H.; Liu, K.C.; Hsu, S.; Chan, C.T. Data Collection and Analysis Using Wearable Sensors for Monitoring Knee Range of Motion after Total Knee Arthroplasty. *Sensors* **2017**, *17*, 418. [CrossRef] [PubMed]

7. Hayati, H.; Mahdavi, F.; Eager, D. Analysis of agile canine gait characteristics using accelerometry. *Sensors* **2019**, *19*, 4379. [CrossRef]
8. Wang, H.; Li, L.; Chen, H.; Li, Y.; Qiu, S. Motion Recognition for Smart Sports Based on Wearable Inertial Sensors. In Proceedings of the 14th EAI International Conference on BodyNets 2019, Florence, Italy, 2–3 October 2019; pp. 114–124.
9. Wang, Z.; Liu, R.; Zhao, H.; Qiu, S.; Shi, X.; Wang, J.; Li, J. Motion Analysis of Deadlift for Trainers with Different Levels based on Body Sensor Network. *IEEE Trans. Instrum. Meas.* **2021**, *71*, 1–12.
10. Wang, Y.; Wang, H.; Xuan, J.; Leung, D.Y.C. Powering future body sensor network systems: A review of power sources. *Biosens. Bioelectron.* **2020**, *166*, 112410. [CrossRef]
11. Qiu, S.; Wang, Z.; Zhao, H.; Hu, H. Using Distributed Wearable Sensors to Measure and Evaluate Human Lower Limb Motions. *IEEE Trans. Instrum. Meas.* **2016**, *65*, 939–950. [CrossRef]
12. Gravina, R.; Li, Q. Emotion-relevant activity recognition based on smart cushion using multi-sensor fusion. *Inf. Fusion* **2019**, *48*, 1–10. [CrossRef]
13. Li, Q.; Raffaele, G.; Qiu, S.; Wang, Z.; Zang, W.; Li, Y. Group Walking Recognition Based on Smartphone Sensors. In Proceedings of the 14th EAI International Conference on BodyNets 2019, Florence, Italy, 2–3 October 2019; Volume 1, pp. 91–102.
14. Ma, C.; Li, W.; Cao, J.; Du, J.; Li, Q.; Gravina, R. Adaptive sliding window based activity recognition for assisted livings. *Inf. Fusion* **2020**, *53*, 55–65. [CrossRef]
15. Li, J.; Wang, Z.; Qiu, S.; Zhao, H. Using Body Sensor Network to Measure the Effect of Rehabilitation Therapy on Improvement of Lower Limb Motor Function in Children with Spastic Diplegia. *IEEE Trans. Instrum. Meas.* **2020**, *69*, 9215–9227. [CrossRef]
16. Choe, N.; Zhao, H.; Qiu, S.; So, Y. A sensor-to-segment calibration method for motion capture system based on low cost MIMU. *Meas. J. Int. Meas. Confed.* **2019**, *131*, 490–500. [CrossRef]
17. Liu, L.; Wang, Z.; Qiu, S. Driving Behavior Tracking and Recognition Based on Multisensors Data Fusion. *IEEE Sens. J.* **2020**, *20*, 10811–10823. [CrossRef]
18. Gravina, R.; Ma, C.; Pace, P.; Aloi, G.; Russo, W.; Li, W.; Fortino, G. Cloud-based Activity-aaService cyber–physical framework for human activity monitoring in mobility. *Future Gener. Comput. Syst.* **2017**, *75*, 158–171. [CrossRef]
19. Qiu, S.; Wang, Z.; Zhao, H.; Liu, L.; Li, J.; Jiang, Y.; Fortino, G. Body Sensor Network based Robust Gait Analysis: Toward Clinical and at Home Use. *IEEE Sens. J.* **2019**, *19*, 8393–8401. [CrossRef]
20. Liu, L.; Wang, H.; Qiu, S.; Zhang, Y.; Hao, Z. Paddle Stroke Analysis for Kayakers Using Wearable. *Sensors* **2021**, *21*, 914. [CrossRef] [PubMed]
21. Fortino, G.; Di Fatta, G.; Pathan, M.; Vasilakos, A.V. Cloud-assisted body area networks: State-of-the-art and future challenges. *Wirel. Netw.* **2014**, *20*, 1925–1938. [CrossRef]
22. Li, Q.; Gravina, R.; Li, Y.; Alsamhi, S.H.; Sun, F.; Fortino, G. Multi-user activity recognition: Challenges and opportunities. *Inf. Fusion* **2020**, *63*, 121–135. [CrossRef]
23. Mannini, A.; Sabatini, A.M. Machine learning methods for classifying human physical activity from on-body accelerometers. *Sensors* **2010**, *10*, 1154–1175. [CrossRef]
24. Nickel, C.; Busch, C. Classifying accelerometer data via hidden Markov models to authenticate people by the way they walk. *IEEE Aerosp. Electron. Syst. Mag.* **2013**, *28*, 29–35. [CrossRef]
25. Huang, J.; Yi, Z.; Wang, X.; Wu, H. Gait recognition system based on (2D)2 PCA and HMM. In Proceedings of the Eighth International Conference on Digital Image Processing, Chengu, China, 20–22 May 2016.
26. Cavallo, F.; Sabatini, A.M.; Genovese, V. A step toward GPS/INS personal navigation systems: Real-time assessment of gait by foot inertial sensing. In Proceedings of the 2005 IEEE/RSJ International Conference on Intelligent Robots and Systems, Edmonton, AB, Canada, 2–6 August 2005; pp. 1187–1191.
27. Qiu, S.; Wang, Z.; Zhao, H. Heterogeneous data fusion for three-dimensional gait analysis using wearable MARG sensors. *Int. J. Comput. Sci. Eng.* **2017**, *14*, 222–233. [CrossRef]
28. Wang, Z.; Qiu, S.; Cao, Z.; Jiang, M. Quantitative assessment of dual gait analysis based on inertial sensors with body sensor network. *Sens. Rev.* **2013**, *33*, 48–56. [CrossRef]
29. Stalin, M.; Bennett, C.L. Development of an Insole System for Real-time Capture of Ground Reaction Forces in Lower-limb Amputees. In Proceedings of the 2013 29th Southern Biomedical Engineering Conference, Seville, Spain, 25–28 September 2013; pp. 137–138.
30. Schepers, H.M.; Roetenberg, D.; Veltink, P.H. Ambulatory human motion tracking by fusion of inertial and magnetic sensing with adaptive actuation. *Med. Biol. Eng. Comput.* **2010**, *48*, 27–37. [CrossRef] [PubMed]
31. Qiu, S.; Wang, H.; Li, J.; Zhao, H.; Wang, Z.; Wang, J.; Wang, Q.; Plettemeier, D.; Bärhold, M.; Bauer, T.; et al. Towards Wearable-Inertial-Sensor-Based Gait Posture Evaluation for Subjects with Unbalanced Gaits. *Sensors* **2020**, *20*, 1193. [CrossRef] [PubMed]
32. Mikos, V.; Heng, C.H.; Tay, A.; Yen, S.C.; Chia, N.S.Y.; Koh, K.M.L.; Tan, D.M.L.; Au, W.L. A Wearable, Patient-Adaptive Freezing of Gait Detection System for Biofeedback Cueing in Parkinson's Disease. *IEEE Trans. Biomed. Circuits Syst.* **2019**, *13*, 503–515. [CrossRef]
33. Sun, F.; Zang, W.; Gravina, R.; Fortino, G.; Li, Y. Gait-based identification for elderly users in wearable healthcare systems. *Inf. Fusion* **2020**, *53*, 134–144. [CrossRef]
34. Ahmad, Z.; Khan, N. Human Action Recognition Using Deep Multilevel Multimodal (M^2) Fusion of Depth and Inertial Sensors. *IEEE Sens. J.* **2020**, *20*, 1445–1455. [CrossRef]

35. Anwary, A.R.; Yu, H.; Vassallo, M. Optimal Foot Location for Placing Wearable IMU Sensors and Automatic Feature Extraction for Gait Analysis. *IEEE Sens. J.* **2018**, *18*, 2555–2567. [CrossRef]
36. Zhao, H.; Wang, Z.; Qiu, S. Adaptive gait detection based on foot-mounted inertial sensors and multi-sensor fusion. *Inf. Fusion* **2019**, *52*, 157–166. [CrossRef]
37. Brzostowski, K. Toward the Unaided Estimation of Human Walking Speed Based on Sparse Modeling. *IEEE Trans. Instrum. Meas.* **2018**, *67*, 1389–1398. [CrossRef]
38. Qiu, S.; Wang, Z.; Zhao, H.; Liu, L.; Jiang, Y. Using Body-Worn Sensors for Preliminary Rehabilitation Assessment in Stroke Victims with Gait Impairment. *IEEE Access* **2018**, *6*, 31249–31258. [CrossRef]
39. An, W.W.; Ting, K.H.; Au, I.P.; Zhang, J.H.; Chan, Z.Y.; Davis, I.S.; So, W.K.; Chan, R.H.; Cheung, R.T. Neurophysiological Correlates of Gait Retraining With Real-Time Visual and Auditory Feedback. *IEEE Trans. Neural Syst. Rehabil. Eng.* **2019**, *27*, 1341–1349. [CrossRef] [PubMed]
40. Ahmed, M.; Naude, J.; Birkholtz, F.; Glatt, V.; Tetsworth, K. Gait & Posture The relationship between gait and functional outcomes in patients treated with circular external fi xation for malunited tibial fractures. *Gait Posture* **2019**, *68*, 569–574.
41. Qiu, S.; Yang, Y.; Hou, J.; Ji, R. Ambulatory estimation of 3D walking trajectory and knee joint angle using MARG Sensors. In Proceedings of the Fourth International Conference on Innovative Computing Technology (INTECH), Luton, UK, 13–15 August 2014; pp. 191–196.
42. Teknomo, K.; Estuar, M.R. Visualizing Gait Patterns of Able bodied Individuals and Transtibial Amputees with the Use of Accelerometry in Smart Phones. *Rev. Colomb. Estad.* **2014**, *37*, 471–488. [CrossRef]
43. Wang, J.; Wang, Z.; Qiu, S.; Xu, J.; Zhao, H.; Fortino, G.; Habib, M. A selection framework of sensor combination feature subset for human motion phase segmentation. *Inf. Fusion* **2021**, *70*, 1–11. [CrossRef]
44. Zhao, H.; Wang, Z.; Qiu, S.; Shen, Y.; Zhang, L.; Tang, K. Heading Drift Reduction for Foot-Mounted Inertial Navigation System via Multi-Sensor Fusion and Dual-Gait Analysis. *IEEE Sens. J.* **2019**, *19*, 8514–8521. [CrossRef]
45. Noor, M.H.M.; Salcic, Z.; Wang, I.K. Adaptive sliding window segmentation for physical activity recognition using a single tri-axial accelerometer. *Pervasive Mob. Comput.* **2016**, *38*, 41–59. [CrossRef]
46. Caderby, T.; Yiou, E.; Peyrot, N.; Bonazzi, B.; Dalleau, G. Detection of swing heel-off event in gait initiation using force-plate data. *Gait Posture* **2013**, *37*, 463–466. [CrossRef]
47. Farah, J.D.; Baddour, N.; Lemaire, E.D. Gait phase detection from thigh kinematics using machine learning techniques. In Proceedings of the 2017 IEEE International Symposium on Medical Measurements and Applications (MeMeA), Rochester, MN, USA, 7–10 May 2017.
48. Wang, T.; Wang, Z.; Zhang, D.; Tao, G.U.; Hongbo, N.I.; Jia, J.; Zhou, X.; Jing, L.V. Recognizing Parkinsonian Gait Pattern by Exploiting Fine-Grained Movement Function Features. *ACM Trans. Intell. Syst. Technol.* **2016**, *8*, 1–22. [CrossRef]
49. Qiu, S.; Liu, L.; Zhao, H.; Wang, Z.; Jiang, Y. MEMS Inertial Sensors Based Gait Analysis for Rehabilitation Assessment via Multi-Sensor Fusion. *Micromachines* **2018**, *9*, 442. [CrossRef] [PubMed]
50. Liu, Y.; Yue, L.; Hou, J. Gait recognition based on MEMS accelerometer. In Proceedings of the IEEE 10th International Conference on Signal Processing Proceedings, Beijing, China, 24–28 October 2010.
51. Leal-Junior, A.G.; Frizera, A.; Avellar, L.M.; Marques, C.; Pontes, M.J. Polymer Optical Fiber for In-Shoe Monitoring of Ground Reaction Forces during the Gait. *IEEE Sens. J.* **2018**, *18*, 2362–2368. [CrossRef]
52. Martinez-Hernandez, U.; Mahmood, I.; Dehghani-Sanij, A.A. Simultaneous Bayesian Recognition of Locomotion and Gait Phases with Wearable Sensors. *IEEE Sens. J.* **2018**, *18*, 1282–1290. [CrossRef]
53. Qiu, S.; Wang, Z.; Zhao, H.; Qin, K.; Li, Z.; Hu, H. Inertial/magnetic sensors based pedestrian dead reckoning by means of multi-sensor fusion. *Inf. Fusion* **2018**, *39*, 108–119. [CrossRef]
54. Meng, X.; Zhang, Z.Q.; Wu, J.K.; Wong, W.C.; Yu, H. Self-contained pedestrian tracking during normal walking using an inertial/magnetic sensor module. *IEEE Trans. Biomed. Eng.* **2014**, *61*, 892–899. [CrossRef]
55. Young, S.S.; Sangkyung, P. Pedestrian inertial navigation with gait phase detection assisted zero velocity updating. In Proceedings of the ICARA 2009—4th International Conference on Autonomous Robots and Agents, Wellington, New Zealand, 10–12 February 2009; pp. 336–341.
56. Suh, Y.S. A smoother for attitude and position estimation using inertial sensors with zero velocity intervals. *IEEE Sens. J.* **2012**, *12*, 1255–1262. [CrossRef]
57. Feliz, R.; Zalama, E.; Garcia-Bermejo, J.G. Pedestrian tracking using inertial sensors. *J. Phys. Agents* **2009**, *3*, 35–43.
58. Skog, I.; Händel, P.; Nilsson, J.O.; Rantakokko, J. Zero-velocity detection—An algorithm evaluation. *IEEE Trans. Biomed. Eng.* **2010**, *57*, 2657–2666. [CrossRef]

Article

Out-of-Distribution Detection of Human Activity Recognition with Smartwatch Inertial Sensors

Philip Boyer [1,2,*], David Burns [3] and Cari Whyne [1,2,3]

1 Orthopaedic Biomechanics Lab, Holland Bone and Joint Program, Sunnybrook Research Institute, Toronto, ON M4N 3M5, Canada; cwhyne@sri.utoronto.ca
2 Institute of Biomedical Engineering, University of Toronto, Toronto, ON M5S 1A1, Canada
3 Division of Orthopaedic Surgery, Department of Surgery, University of Toronto, Toronto, ON M5S 1A1, Canada; d.burns@utoronto.ca
* Correspondence: philip.boyer@mail.utoronto.ca

Abstract: Out-of-distribution (OOD) in the context of Human Activity Recognition (HAR) refers to data from activity classes that are not represented in the training data of a Machine Learning (ML) algorithm. OOD data are a challenge to classify accurately for most ML algorithms, especially deep learning models that are prone to overconfident predictions based on in-distribution (IIN) classes. To simulate the OOD problem in physiotherapy, our team collected a new dataset (SPARS9x) consisting of inertial data captured by smartwatches worn by 20 healthy subjects as they performed supervised physiotherapy exercises (IIN), followed by a minimum 3 h of data captured for each subject as they engaged in unrelated and unstructured activities (OOD). In this paper, we experiment with three traditional algorithms for OOD-detection using engineered statistical features, deep learning-generated features, and several popular deep learning approaches on SPARS9x and two other publicly-available human activity datasets (MHEALTH and SPARS). We demonstrate that, while deep learning algorithms perform better than simple traditional algorithms such as KNN with engineered features for in-distribution classification, traditional algorithms outperform deep learning approaches for OOD detection for these HAR time series datasets.

Keywords: human activity recognition; out of distribution; anomaly detection; open set classification; physiotherapy; inertial sensors; smart watch; rehabilitation; machine learning

Citation: Boyer, P.; Burns, D.; Whyne, C. Out-of-Distribution Detection of Human Activity Recognition with Smartwatch Inertial Sensors. *Sensors* **2021**, *21*, 1669. https://doi.org/10.3390/s21051669

Academic Editors: Maria de Fátima Domingues, Ayman Radwan and Andrea Sciarrone

Received: 28 January 2021
Accepted: 18 February 2021
Published: 1 March 2021

Publisher's Note: MDPI stays neutral with regard to jurisdictional claims in published maps and institutional affiliations.

1. Introduction

Human activity recognition (HAR) constitutes automatic characterization of activity and movement through intelligent "learning" algorithms. Under the HAR umbrella lies the detection and identification of such varied tasks as hand gestures [1], walking up stairs [2], commuting [3], detecting falls [4], and even smoking [5], with applications in a growing number of fields, including physiotherapy [6]. HAR can be accomplished via a variety of strategies, including machine vision technology [7–9] or analysis of inertial data. Inertial measurement units (IMUs) are embedded in many widely available commercial devices, including smart phones [10,11], and smartwatches [6,12]. IMUs enable data capture in natural (home or varied) settings without any need to perform actions in front of a fixed camera or other apparatus. The analysis of inertial data via machine learning (ML) has been demonstrated to yield robust HAR [6].

In attempting to track specific human activities in unsupervised environments, subjects may perform unexpected, unknown, or unrelated activities. This presents a challenge to the use of ML based classifiers as training a ML algorithm on all possible human actions for activity recognition is impractical, and supervised ML algorithms may not accurately classify such out-of-distribution (OOD) activities. In the context of identifying at-home physiotherapy exercises with a smartwatch, subjects may be instructed to only wear a watch while performing their exercises. However, even in this case, subjects may perform other activities between exercises—such as getting a drink of water—or may forget to remove their smartwatch entirely after completing their exercise routines and go about other daily activities. A classification process for at-home physiotherapy exercises that does not include OOD detection risks detecting these activities as exercises, and thereby reporting incorrect estimations of physiotherapy participation and adherence.

Methods to address the OOD problem exist for image classification but have been less widely applied in the context of time series activity recognition. In this paper, we experiment with several methods commonly used in the image domain to address the OOD problem in the context of shoulder physiotherapy activity recognition. We test three traditional algorithms for OOD-detection on engineered statistical features (One-Class State Vector Machine (OCSVM) [13,14], K-Nearest Neighbor (KNN) [15], and Kmeans [16]), KNN with deep feature embeddings generated by two neural network models, and well-known methods in the image domain based on deep learning: thresholding Softmax confidence [17], confidence calibration via entropy regularization [18], confidence calibration via temperature scaling and input perturbations (ODIN) [19], and extending the Softmax layer to allow prediction of an unknown class (OpenMax) [20]. We evaluate these techniques on a novel physiotherapy exercise dataset we collected (SPARS9x) as well as on two publicly-available activity datasets (MHEALTH [21] and SPARS [6]).

Contributions of this study include:

- A new physiotherapy activity dataset SPARS9x (DOI: 10.21227/cx5v-vw46), with additional inertial data captured from the smartwatch of each subject while they performed activities of daily living. We believe this study is unique in its approach of capturing a dataset that explicitly simulates the distinction between known target human activities and unknown a priori OOD activities.
- Evaluation of methods of OOD detection from the image domain as applied to physiotherapy inertial data captured by smartwatches, in comparison to traditional algorithms using both hand-crafted engineered statistical features and deep learning model-derived features.

This paper is organized as follows. In Section 2, we include a brief synopsis of OOD detection methods and other background important to this study. Section 3 provides an overview of the OOD detection methods that are explored and the datasets used as well as details related to the setup of the analyses, including model architectures and evaluation methods used. Section 4 presents the results of the analyses and Section 5 the discussion. Section 6 proposes potential future work, Section 7 details the study limitations, and Section 8 provides a summary and concluding remarks.

2. Background and Related Work

2.1. Human Activity Recognition with Machine Learning of Inertial Data

There is a large body of research on the use of IMUs and the inertial data they collect for HAR [22–24], with a considerable subset evaluating the practical applications of these technologies in the health domain [25]. In particular, gait assessment [26,27] and fall prediction [28,29] have garnered significant attention, likely due in large part to their immediate clinical need. More recently, researchers have also begun applying these same methodologies to physiotherapy and rehabilitation [6,30–32] where unique challenges may be encountered. The focus of these studies has typically been the exploration of techniques for improving the accuracy of correctly identifying physiotherapy exercises. Less effort has been spent on identifying when subjects are performing the exercises that are within the scope of activities represented in the training dataset.

In a closed-set classification problem, an ML model is tasked with predicting the class that a sample belongs to from within a set of classes that the ML model was trained on. However, input samples may deviate from these classes, often into classes that are both previously unknown and unforeseen. The challenge of accurately identifying such samples is known as the out-of-distribution (OOD) or open set classification problem (for a review of OOD detection methodologies, see [33]). It is worth noting that OOD detection is often conflated with novelty, anomaly, or outlier detection in the literature [34]. However, predictions in the case of novelty or outlier detection in time series data often implies identifying deviation from an expected input based on a chosen error metric [35–40].

OOD detection methods are commonly employed in the image domain, but many of these solutions have largely remained untested in the HAR time series domain. The following sections provide a short summary of OOD detection methods of interest to this study.

2.2. Out-of-Distribution Detection Techniques

Scheirer et al. formalized the definition of the open set classification problem and identified a binary SVM method to separate open set inputs from the closed set with a linear kernel [41]. This method was then extended by Scheirer et al. with Compact Ablating Probability (CAP) to identify unknown class samples and Extreme Value Theory (EVT) to calibrate prediction results using a Weibull distribution [42]. Many techniques have been proposed for OOD detection based on traditional classification algorithms, such as OCSVM [41,43–45], KNN [15], and Kmeans [16]. Recently, deep learning-based OOD-detection techniques have been proposed [17,19,20,26–35,46–55], some of which we now highlight.

In deep neural network classifiers with multi-class outputs, the final layer commonly performs a Softmax operation on the output from the previous layers (the "core"), which transforms input values into probabilities estimates between 0 and 1, the total of which sum to 1. The Softmax layer is typically used in deep learning classifier models to assign probability estimates or activations for input samples to known classes. One weakness of the Softmax layer is that all predictions must be assigned to one of a range of predetermined classes. To detect OOD samples, the activation, or "confidence", of the Softmax prediction may be evaluated against a threshold, whereby input data that elicit a maximum Softmax output below the threshold may be designated as out-of-distribution [17]. However, Softmax outputs are prone to overconfident predictions, which may impact the accuracy of this approach [56]. Other methods have been proposed to ameliorate the overconfidence issue, including temperature scaling [19,56] and entropy regularization [18]. Alternatively, one may reformulate the Softmax layer to predict an unknown class, such as with the OpenMax technique [20]. Traditional and deep learning algorithms may be used together, such as using an OCSVM to detect anomalous motion (e.g., jumping and walking) with features extracted from accelerometer and gyroscopic data by a convolutional neural network [57].

Generative adversarial networks (GANs) have also been explored for the OOD problem [58]. GANs are generally incorporated into an open set discriminative framework by training a GAN on the known input dataset and converting the trained discriminator into an OOD classifier [46,59]. Alternatively, the generator used to synthesize samples may designate them as from an OOD class, and then add them to the full training set used to train a classifier [47–50]. However, such GAN augmentation techniques may produce samples that do not adequately cover the entirety of in-distribution (IIN) decision boundaries, resulting in non-optimal OOD classifiers [60]. Searching a trained generator's latent space for a close match to an input is another approach [61]. Autoencoders or similar compact representation models may also be used to detect OOD-samples based on high input reconstruction error [51–54], or in adversarial frameworks with clustering methods [55]. Recently, generative likelihood ratio methods for OOD detection have been proposed, including that of Ren et al. for genomics data with application to the image domain [62].

2.3. OOD Detection with Inertial Data

Studies on OOD detection in the context of HAR inertial data are less common than those in the image domain. In the recent HAR time series outlier-detection work by Organero, a model is trained on a particular activity for each user and the Pearson correlation coefficient is used to identify deviation of real data from predicted data [36].

Omae et al. deployed a convolutional neural network to perform feature extraction of accelerometer and gyroscopic data from the swimming activity, and an OCSVM is used to detect anomalous motion (e.g., jumping and walking) based on extracted features in consideration of an individualized optimization algorithm [57]. It is notable that each of these studies either approach the problem in terms of anomaly detection, and/or use "traditional" algorithms such as OCSVM to identify OOD data. To our knowledge, the selection of solutions in the image domain for OOD detection presented in this study (i.e., Softmax Thresholding [17], ODIN [19], and OpenMax [20]) have not yet been applied to the domain of HAR time series inertial data, nor has a direct comparison of these to methodologies using traditional algorithms using engineered or deep learning features been performed.

3. Materials and Methods

3.1. *Out-of-Distribution Detection*

In this section, an overview of methods considered for OOD detection on HAR time series datasets is presented. The deep learning approaches are well-known in the image domain.

3.1.1. One Class State Vector Machines (OCSVMs)

One class state vector machines (OCSVMs) test whether an input sample is a member of a single class of interest. For the OOD problem, this one class would correspond to all classes of IIN data. Using the approach described by Schölkopf et al. [13], data are cordoned off from the origin with a hyperplane, and data beyond this plane are classified as OOD.

3.1.2. K-Nearest Neighbor (KNN)

The K-Nearest Neighbor (KNN) algorithm finds the (Euclidean) distance between inputs. For classification, KNN typically uses a voting mechanism based on the closest N-neighbors to the location in space of an input to be tested. OOD classes can be identified when the distance to the closest N-neighbors exceeds a given threshold.

3.1.3. Kmeans

Kmeans is an unsupervised algorithm that uses an iterative updating process to assign similar data to clusters. OOD data can be detected by comparing the distance to the center of the cluster assigned to an input against a threshold value.

3.1.4. Deep Feature Embedding

Deep features refer to the outputs of layers immediately preceding the dense/Softmax layers of fully trained deep neural network models, effectively embedding test data into the feature space of each core model. This deep feature representation may be used with traditional algorithms that require vectorized input data (e.g., KNN). We use an L2 normalization layer in our deep models to produce an embedding with consistent scale in the feature space.

3.1.5. Softmax Thresholding

It is common practice in deep learning classification algorithms to base predictions on the maximum output of a Softmax layer [17]. The output of the Softmax layer is often interpreted as the confidence of a model in its prediction that the sample is from a particular class [56]. If the Softmax output is used as a confidence metric, samples that fail to meet that threshold may be classified as OOD.

3.1.6. Entropy Regularization

Softmax predictions in deep learning models are prone to overconfidence [63] and may be adjusted through confidence calibration techniques. Entropy-regularization accomplishes this by penalizing the negative log-loss with the addition of a negative entropy term that effectively regularizes the loss to prevent overconfident predictions. This has been shown to be an effective technique on several diverse datasets with models incorporating dense, convolutional (CNN), and long-short term memory (LSTM) layers [18]. The intuition behind this technique for OOD detection is that reducing overconfidence may allow for better discrimination of the more confident Softmax predictions of IIN activities from the less confident predictions of OOD activities.

3.1.7. ODIN

The out-of-Distribution detector for neural networks (ODIN) is a two-part method for OOD detection, performed post-process during prediction by a trained network [19]. In temperature scaling, the logits (the output of the final output layer of a model) of the classifier are scaled by a scalar value called the "temperature" (T) prior to performing Softmax.

The input data are then perturbed by a fixed amount, which has a greater effect on the IIN data than the OOD data, increasing their separation. The input perturbations are based on the directionality of input gradients, obtained by backpropagating the inputs through the network once. The absolute magnitudes of the inputs are perturbed by an epsilon value opposite the gradient direction. The perturbed and temperature-scaled inputs are run through the classifier, and OOD samples are detected as those falling beneath a learned threshold of Softmax.

3.1.8. OpenMax

OpenMax [20] is a model layer that assigns a probability to an unknown class. It is a replacement for the Softmax layer. Its core concept is to reject inputs that are "far" from the training data distribution in feature space. Each class is represented as a point with the mean calculated from the correctly classified training examples. A Weibull distribution is fitted on the largest distances between all correctly identified training samples of a class and the mean. The result is a probability that an input is from this class, and a set threshold is used to determine OOD.

Unlike Softmax, the OpenMax layer does not restrict probabilities to sum to 1. Instead, a pseudo-activation (of the logits, also termed the activation vector) is calculated for an OOD class by using the Weibull distribution to redistribute activations to include an unknown class. This can be used to identify OOD data if this class has the highest probability output from the OpenMax layer.

3.2. Experimental Setup

3.2.1. Experimental Datasets

This study considers two publicly available inertial sensor datasets, MHEALTH (Mobile Health) [21] and SPAR (Supervised Physiotherapy Activity Recognition) [6], and a novel dataset (SPARS9x) we collected specifically to assess OOD detection. The properties of these datasets are summarized in Table 1. Only the data from inertial devices attached to one wrist from each study subject in MHEALTH are used, effectively simulating smartwatch IMUs.

The new inertial dataset, SPARS9x (Smart Physiotherapy Activity Recognition System 9-axis), was captured from 20 healthy subjects (8 male, 12 female, median age 25). Supervised physiotherapy exercises, followed by a minimum 3 h of data from unrelated and unstructured activities (OOD data), were captured from a smartwatch with a 9-axis IMU (accelerometer, gyroscope and magnetometer). This study was approved by the Research Ethics Board of Sunnybrook Health Sciences Centre.

Table 1. Dataset Characteristics.

Dataset	Ns	Ne	Type	Sensors	fz [Hz]	Time [h]
SPARS	20	7	Shoulder Physiotherapy	Wrist 6-axis IMU	50	3.4
SPARS9x [a]	20	6	Shoulder Physiotherapy	Wrist 9-axis IMU	50	95.4
MHealth [b]	10	12	General Fitness	Wrist 9-axis IMU	50	1.9

Ns, number of subjects; Ne, number of exercises; f_s, sampling frequency. [a] Four isometric exercises are included in the SPARS9x dataset, but these are excluded for this study. [b] Only the data corresponding to the wrist-IMU of MHEALTH are used.

Shoulder physiotherapy exercises in SPARS9x were selected from the Basic Shoulder Rehabilitation Program provided by the Sunnybrook Holland Orthopaedic & Arthritic Centre. Exercises were selected to include both concentric (muscle activation with shortening – shown in Figure 1) and isometric (muscle activation without length change) types. Exercises were performed with a resistance band when needed. Concentric exercises included: active flexion, cross chest adduction, shoulder girdle stabilization with elevation, biceps muscle strengthening, triceps pull downs, and external rotation in 90-degree abduction in the scapular plane. Isometric exercises included external rotation, internal rotation, abduction, and extension.

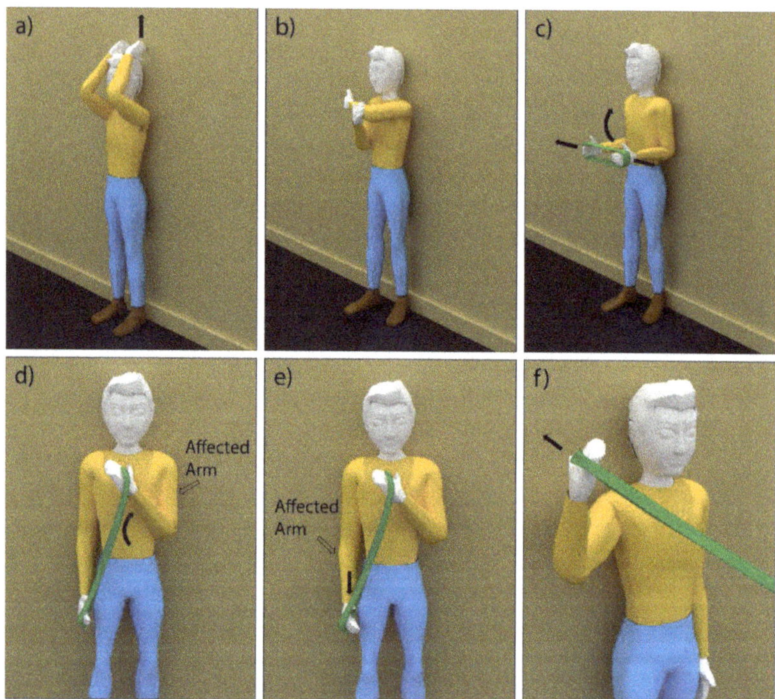

Figure 1. SPARS9x concentric exercises: (**a**) active flexion; (**b**) cross chest adduction; (**c**) shoulder girdle stabilization with elevation; (**d**) biceps muscle strengthening; (**e**) triceps pull downs; and (**f**) external rotation in 90-degree abduction in the scapular plane. Black arrows indicate direction of motion or tension.

3.2.2. Data Transformation Pipeline

The raw time series inertial data were split into equally-sized windows, with constant step size of 1 s, using the Seglearn package for time series machine learning [64]. Samples counts at 10s window size for each dataset were 24,004 for SPARS9x, 9527 for SPARS, and

5676 for MHealth. Artifacts in SPARS were located and removed based on sampling rate and duplicate time stamp checks. For SPARS9x, each record was plotted and visually inspected for artifacts immediately after each subject's session, and records that contained artifacts were either truncated or sessions were re-recorded.

Inputs to traditional models are often generated based on statistics describing the raw data, such as the mean and kurtosis of each input channel [1]. The statistical features engineered for input to the traditional algorithms in this analysis are median, absolute energy (root mean squared), standard deviation, variance, minimum, maximum, skewness, kurtosis, mean spectral energy, and mean crossings. These are also generated by the Seglearn Python package.

For the purposes of this study, we define OOD activities as any activity performed by a subject that is not in the list of labeled activities available to a classifier during training. The activities performed in SPAR [6] and MHEALTH [21] are described elsewhere. However, it is especially important to note which classes of these datasets are to be designated as OOD for reproducibility of these experiments since activities similar to IIN classes may be more difficult to detect as OOD. Highly confused classes were those identified through in-distribution classification, corresponding to activities of similar motion that produce patterns in the data that are difficult to discriminate between by machine learning classifiers (e.g., jogging and running in MHEALTH). In general, highly confused classes in supervised classification were kept together in IIN for these experiments. Beyond that, OOD classes were simply selected as the final two and three activities in the order of activities for SPARS and MHEALTH respectively. Lower trapezius row with resistance band and bent over row with 3 lb. dumbbell are selected as OOD classes for SPARS. Frontal elevation of arms, cycling, and jump front and back are selected as OOD activities for MHEALTH. MHEALTH null class (Class 0) is not used for this analysis as it is unclear how much overlap there is between those data and the in-distribution activities (e.g., walking or standing), potentially skewing results. Only concentric exercises from SPARS9x are used for this analysis.

Since the recording time of OOD data for each subject significantly exceeds that of their physiotherapy exercises, for this study, the length of each OOD record in SPARS9x was cut to match the length of time spent performing target exercises. Magnetometer data are excluded for these experiments from SPARS9x.

3.2.3. Model Architecture

Two separate deep neural networks were tested as core architectures in the analysis.

The first core is a convolutional recurrent neural network (CRNN), which was found in our previous study to perform well on classification tasks [6]. In the particular case of time series data, temporal relationships may be captured by recurrent neural network (RNN) layers [65]. This model consists of two convolutional layers with 128 filters and kernel size of 2, followed by RELU and Maxpooling layers. Next, the outputs of the convolutional layers are passed into two LSTM layers of 100 units each and dropout of 0.1. This is consistent with findings in literature that suggest that two sequential LSTM layers may be beneficial when processing time series data [66].

The second core is a fully convolutional neural network (FCN) used for time series classification [67], which we have previously shown to be effective in a personalized activity recognition approach [68].

A final dense layer outputs the logits for each class (prior to Softmax, method depending). Hyperparameters of each core were grid-searched for their optimal values in our previous works [6]. The deep learning models were developed in PyTorch. The traditional models were implemented using packages from the scikit-learn Python library.

3.2.4. In-Distribution Classification Experiments

Supervised classification of in-distribution activities is performed to validate the performance of the core models with each of the three datasets.

We compare performance of:

- KNN with engineered statistical features
- KNN with deep features (CRNN/FCN)
- CRNN and FCN Cores

3.2.5. Out-of-Distribution Prediction Experiments

We compare the following methods for OOD detection:

- Traditional algorithms: KNN, OCSVM, and Kmeans with engineered features
- KNN with deep features (CRNN/FCN)
- Deep learning methods: SoftMax threshold, entropy regularization, ODIN, and OpenMa

3.2.6. Class Removal Experiments

The lowest confidence classes (or those with the lowest mean activations or greates overlap with OOD) in the OOD detection experiments according to SoftMax prediction confidence (i.e., potentially the most likely to be confused with OOD classes) are identified in the SPARS9x and MHEALTH datasets and either removed from the experiments (SPARS9x or moved into the IIN training set (MHEALTH). OOD experiments are then repeated for these two datasets at a segment length of 10s to evaluate impact on OOD detection. These experiments are performed to analyze the effect that removing or re-designating highly confused classes (in terms of OOD detection) has on OOD prediction performance.

3.2.7. Training and Validation

A grid search is performed on segment length for each experiment, with segment sizes from 2 to 10 s at 2 s intervals. A 5-fold cross-validation strategy is used with equal-sized folds for each experiment. Each deep learning model uses a batch size of 256 and is run for 100 epochs. Testing was performed on a system with an NVidia GTX1080 GPU with 8 GB on-board memory, an Intel Core i7-6700 CPU (@3.4 GHz), and 16.0 GB of RAM.

3.2.8. Evaluation Metrics

For the in-distribution classification tasks, an accuracy metric was used, as these datasets demonstrated approximate equal class balance. In the OOD problem, we are concerned with how well a method is able to differentiate between distributions, which can be measured by the area under the receiver operating curve (AUROC) metric. The AUROC can be interpreted as a measure of how separable the IIN and OOD data are under the model.

4. Results

4.1. In-Distribution Classification

Table 2 presents the in-distribution supervised classification experiment results. Deep learning methods or KNN with deep features consistently outperformed KNN with engineered features.

Figure 2 shows the confusion matrices for each of the three datasets for classification using a KNN with deep learning features as input generated by the FCN architecture at 10s segment size. As reflected in Table 2, the KNN algorithm is able to classify SPARS9x with near perfect accuracy, whereas there are pairs of highly confused classes for both SPARS (Internal Rotation/External Rotation) and MHEALTH (Jogging/Running).

Table 2. In-Distribution Classification Results.

	Accuracy % (Confidence Interval)														
	MHEALTH Segment Length					**SPARS Segment Length**					**SPARS9x Segment Length**				
Method	2.0s	4.0s	6. 0s	8.0s	10.0s	2.0s	4.0s	6. 0s	8.0s	10.0s	2.0s	4.0s	6. 0s	8.0s	10.0s
KNN	88.0 (3.0)	87.7 (2.3)	86.5 (2.2)	87.7 (2.4)	87.1 (1.5)	81.0 (1.7)	82.3 (2.6)	82.9 (0.83)	81.4 (2.8)	82.2 (1.8)	96.4 (0.67)	97.6 (0.78)	97.8 (0.71)	97.7 (0.80)	97.7 (0.77)
KNN (CRNN Deep Features)	95.2 (1.4)	94.1 (2.3)	94.8 (1.3)	**96.0 (1.1)**	92.7 (1.4)	87.7 (0.86)	67.5 (4.7)	81.2 (1.0)	78.7 (1.7)	75.6 (1.2)	97.6 (0.41)	96.6 (0.76)	96.8 (0.92)	95.5 (0.57)	94.8 (0.61)
KNN (FCN Deep Features)	93.1 (0.8)	93.7 (2.3)	93.3 (1.7)	92.8 (1.7)	93.1 (1.8)	89.9 (0.98)	92.5 (1.2)	92.2 (1.1)	**93.1 (1.4)**	92.1 (1.5)	98.9 (0.19)	99.8 (0.12)	99.7 (0.14)	**99.9 (0.066)**	99.4 (0.43)
CRNN	94.8 (1.3)	93.8 (2.3)	94.6 (1.4)	95.9 (1.2)	90.0 (2.2)	87.6 (0.87)	67.4 (5.1)	80.8 (1.3)	76.2 (1.7)	72.9 (0.76)	97.7 (0.46)	96.6 (0.76)	96.7 (0.94)	95.4 (0.53)	94.7 (0.64)
FCN	94.2 (1.6)	94.8 (1.6)	93.0 (0.79)	93.5 (1.4)	95.3 (1.3)	87.9 (1.4)	86.4 (1.6)	87.1 (1.6)	87.9 (1.2)	88.0 (2.4)	98.5 (0.38)	99.7 (0.19)	99.6 (0.20)	**99.9 (0.066)**	98.1 (1.7)

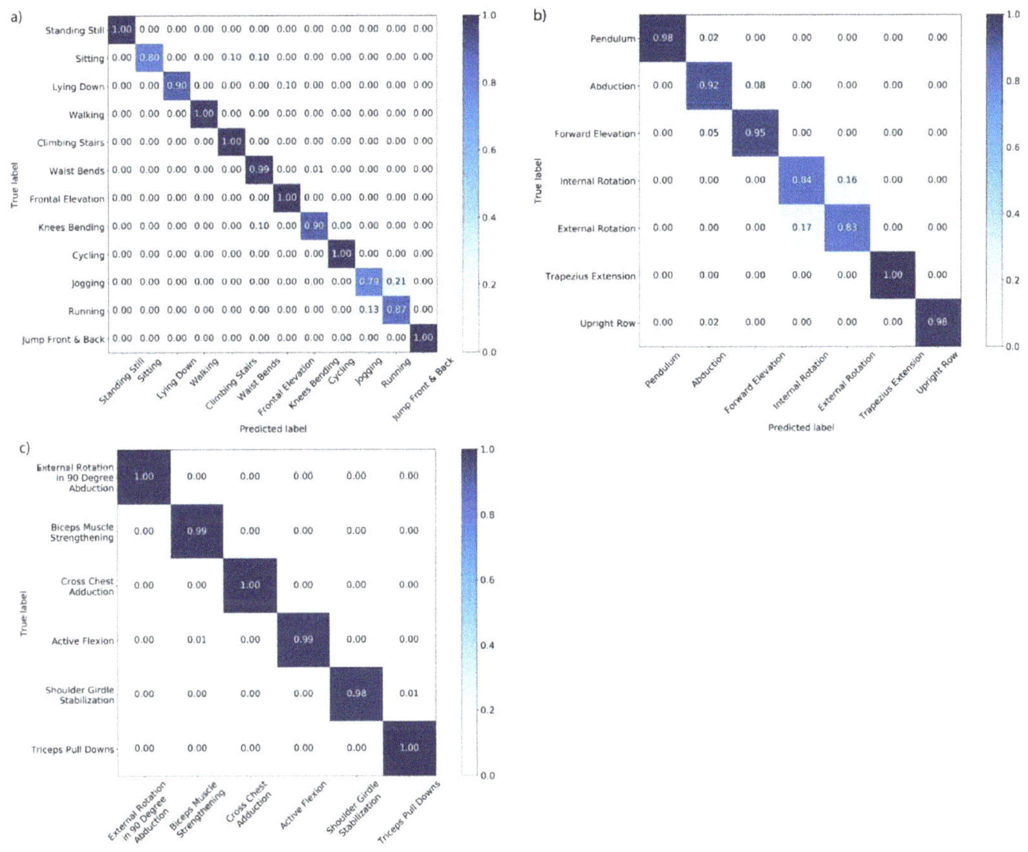

Figure 2. In-distribution classification confusion matrices with KNN with FCN Deep Features at 10 s segment size for: (**a**) MHEALTH; (**b**) SPARS; and (**c**) SPARS9x.

4.2. Out-of-Distribution Detection

Table 3 presents the results of the OOD experiments. Each reported accuracy results includes the standard error of five folds for the grid search of optimal hyperparameter settings.

Table 3. Out-of-Distribution Detection Results.

	AUROC (Confidence Interval)														
	MHEALTH Segment Length					SPARS Segment Length					SPARS9x Segment Length				
Method	2.0s	4.0s	6. 0s	8.0s	10.0s	2.0s	4.0s	6. 0s	8.0s	10.0s	2.0s	4.0s	6. 0s	8.0s	10.0s
Traditional Methods – Engineered Statistical Features															
KNN	0.903 (0.026)	0.902 (0.019)	**0.905 (0.018)**	0.904 (0.017)	0.898 (0.014)	0.865 (0.027)	0.912 (0.021)	0.920 (0.019)	0.927 (0.018)	0.934 (0.017)	0.918 (0.0028)	0.963 (0.0033)	0.975 (0.0029)	0.980 (0.0019)	**0.982 (0.0021)**
Kmeans	0.881 (0.020)	0.887 (0.015)	0.886 (0.010)	0.884 (0.014)	0.878 (0.0064)	0.842 (0.021)	0.892 (0.024)	0.903 (0.019)	0.913 (0.018)	0.917 (0.017)	0.872 (0.011)	0.937 (0.0059)	0.955 (0.0046)	0.964 (0.0043)	0.969 (0.0042)
OCSVM	0.796 (0.019)	0.796 (0.027)	0.784 (0.022)	0.776 (0.023)	0.759 (0.015)	0.802 (0.027)	0.863 (0.029)	0.871 (0.028)	0.883 (0.027)	0.887 (0.027)	0.804 (0.017)	0.896 (0.010)	0.927 (0.0080)	0.940 (0.00079)	0.949 (0.0074)
KNN – Deep Feature Embedding															
CRNN Features	0.852 (0.017)	0.839 (0.024)	0.791 (0.055)	0.839 (0.031)	0.837 (0.047)	0.903 (0.023)	0.858 (0.020)	0.859 (0.029)	0.829 (0.052)	0.877 (0.029)	0.754 (0.017)	0.819 (0.0074)	0.794 (0.011)	0.788 (0.0053)	0.774 (0.020)
FCN Features	0.854 (0.023)	0.883 (0.024)	0.874 (0.014)	0.877 (0.017)	0.891 (0.028)	0.969 (0.0080)	0.976 (0.0073)	0.971 (0.0064)	0.974 (0.0082)	**0.978 (0.0058)**	0.921 (0.011)	0.960 (0.0053)	0.967 (0.0076)	0.965 (0.0062)	0.965 (0.0060)
CRNN Core Model															
Softmax Threshold	0.611 (0.048)	0.609 (0.047)	0.553 (0.057)	0.656 (0.038)	0.578 (0.068)	0.777 (0.036)	0.819 (0.014)	0.840 (0.020)	0.788 (0.026)	0.839 (0.024)	0.700 (0.016)	0.718 (0.013)	0.680 (0.014)	0.669 (0.025)	0.634 (0.031)
Entropy Regularization	0.736 (0.015)	0.731 (0.032)	0.680 (0.064)	0.705 (0.035)	0.644 (0.067)	0.911 (0.0080)	0.917 (0.0091)	0.928 (0.012)	0.903 (0.019)	0.930 (0.0087)	0.691 (0.021)	0.689 (0.024)	0.708 (0.015)	0.690 (0.023)	0.636 (0.039)
ODIN	0.633 (0.036)	0.664 (0.050)	0.597 (0.060)	0.546 (0.082)	0.512 (0.036)	0.860 (0.0086)	0.860 (0.020)	0.846 (0.029)	0.855 (0.010)	0.815 (0.032)	0.692 (0.017)	0.698 (0.017)	0.636 (0.037)	0.668 (0.029)	0.595 (0.030)
OpenMax	0.661 (0.050)	0.614 (0.076)	0.693 (0.053)	0.707 (0.055)	0.589 (0.059)	0.840 (0.034)	0.863 (0.0036)	0.873 (0.024)	0.851 (0.037)	0.838 (0.036)	0.776 (0.019)	0.694 (0.058)	0.689 (0.035)	0.682 (0.066)	0.758 (0.036)
FCN Core Model															
Softmax Threshold	0.731 (0.016)	0.622 (0.039)	0.680 (0.055)	0.600 (0.031)	0.597 (0.017)	0.779 (0.024)	0.812 (0.018)	0.783 (0.042)	0.799 (0.034)	0.795 (0.034)	0.756 (0.014)	0.752 (0.012)	0.759 (0.017)	0.764 (0.0082)	0.771 (0.0058)
Entropy Regularization	0.752 (0.018)	0.637 (0.039)	0.666 (0.062)	0.610 (0.039)	0.672 (0.023)	0.792 (0.027)	0.815 (0.024)	0.791 (0.037)	0.781 (0.037)	0.807 (0.034)	0.747 (0.015)	0.752 (0.011)	0.762 (0.021)	0.743 (0.017)	0.772 (0.014)
ODIN	0.699 (0.022)	0.601 (0.030)	0.649 (0.026)	0.645 (0.035)	0.673 (0.046)	0.820 (0.022)	0.816 (0.023)	0.826 (0.031)	0.849 (0.020)	0.821 (0.016)	0.743 (0.017)	0.752 (0.012)	0.747 (0.025)	0.735 (0.020)	0.749 (0.0095)
OpenMax	0.794 (0.036)	0.645 (0.043)	0.734 (0.054)	0.714 (0.069)	0.708 (0.014)	0.845 (0.021)	0.856 (0.013)	0.855 (0.017)	0.875 (0.021)	0.850 (0.030)	0.910 (0.011)	0.897 (0.020)	0.916 (0.011)	0.915 (0.020)	0.922 (0.019)

Traditional algorithms performed better than the deep learning-based models. The KNN method in particular achieved superior results. Among the deep learning algorithms, Softmax thresholding without any other intervention was particularly ineffective, while OpenMax generally performed best. Deep feature embedding with KNN yielded competitive results compared to the deep learning methods.

Figure 3 illustrates the results of a segment length grid search for each of the algorithms on SPARS9x. Increasing segment length consistently improved accuracy for the traditional algorithms but not for the deep learning algorithms.

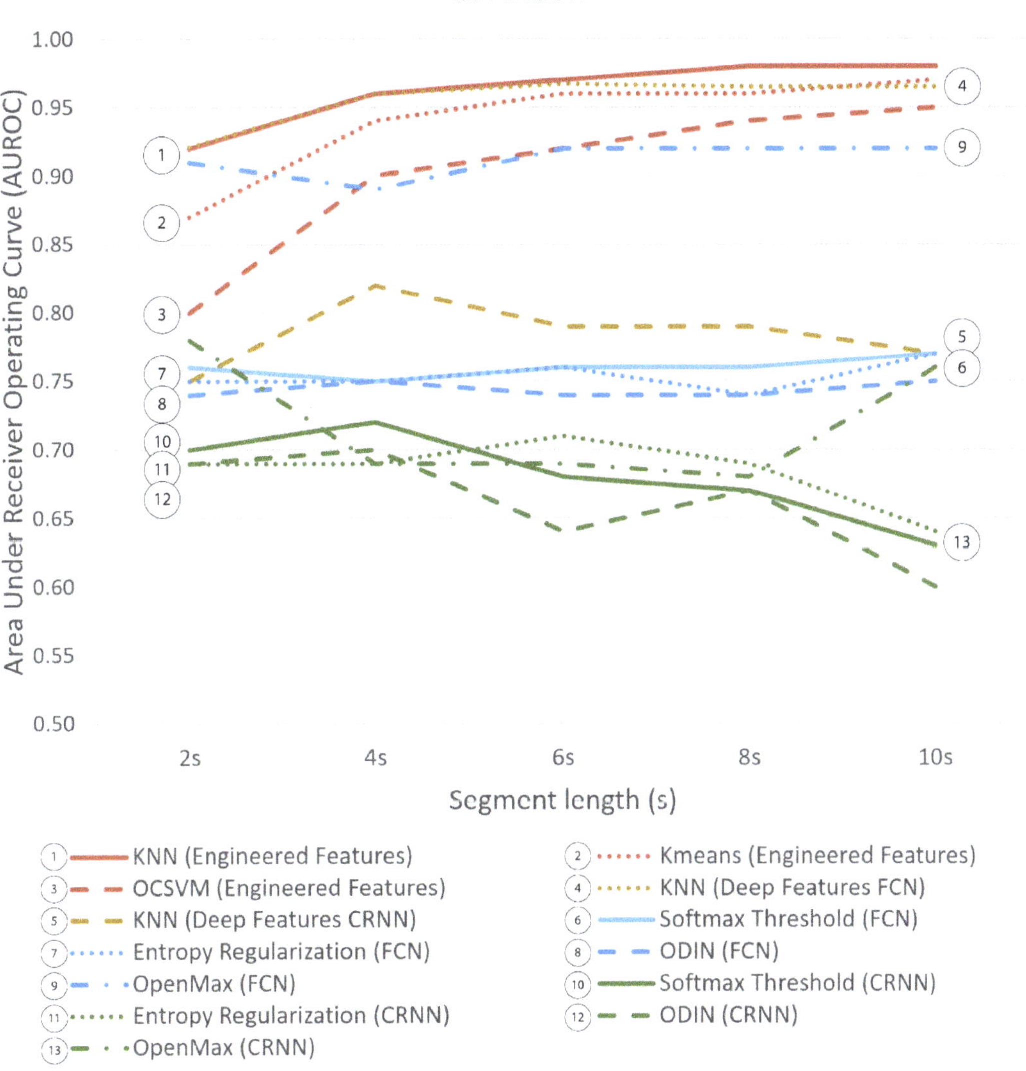

Figure 3. OOD-Detection AUROC for SPARS9x for each method by segment length. Traditional algorithms with engineered features are shown in red (1–3), deep features in yellow (4 and 5), deep learning approaches with FCN in blue (6–9), and deep learning approaches with CRNN in green (10–13).

4.3. Class Removal Experiments

The distributions of activations for the four deep learning algorithms for SPARS9x with the CRNN core are shown in Figure 4. Distributions in vanilla Softmax in Figure 4a are clustered near to the maximum of 1.0, whereas the other algorithms appear to be effective at reducing overconfidence judging by their improved spread. Mean activations with the FCN core of the four deep learning algorithms for each activity of SPARS9x are shown in Figure 4. This figure illustrates a similar reduction in overconfidence for the FCN, but it is important to note that there is still significant overlap in activations between the OOD and IIN classes for each.

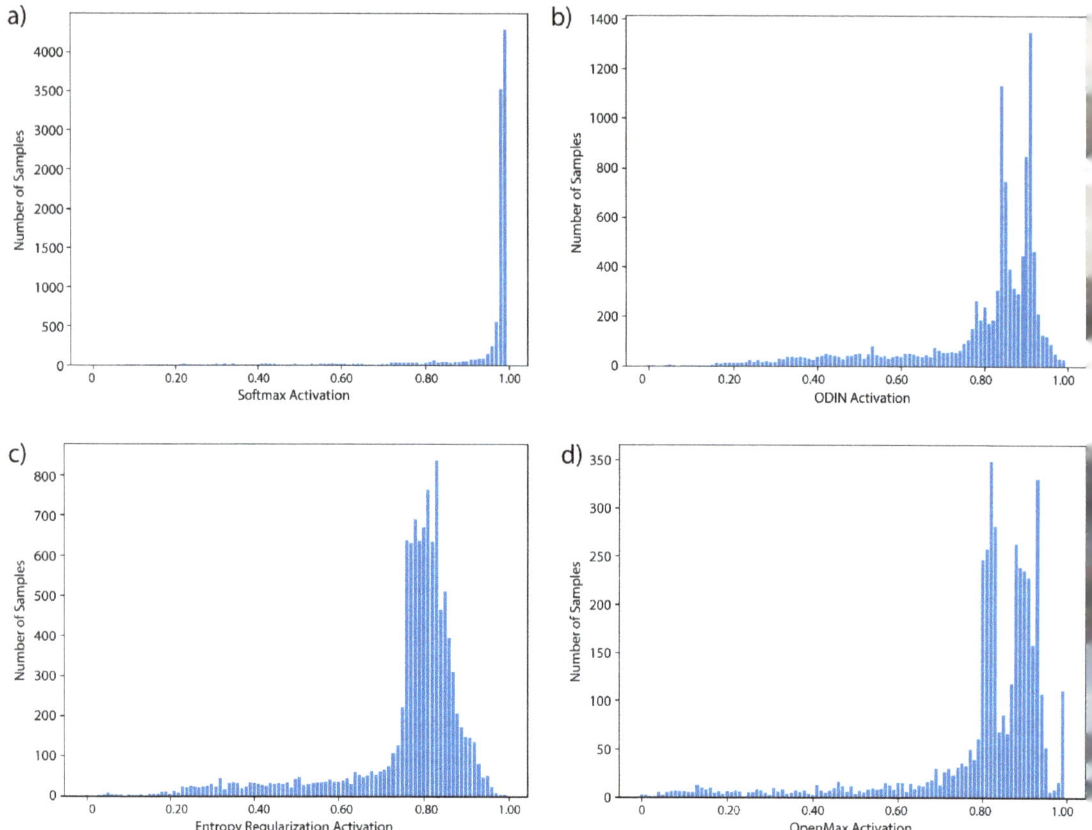

Figure 4. Distributions of activations of SPARS9x prediction with CRNN core: (**a**) Softmax activation; and (**b**–**d**) activation post-processing with ODIN, entropy regularization, and OpenMax respectively. Activations are scaled to between 0 and 1 for each method for illustrative purposes.

The boxplots in Figure 5a demonstrate the lower range of mean Softmax activations for the shoulder girdle stabilization activity with the FCN model for SPARS9x. Removal of this activity class from the experiments results in less overlap of Softmax confidence with the OOD data, but also results in altering the confidence in predicting every remaining class.

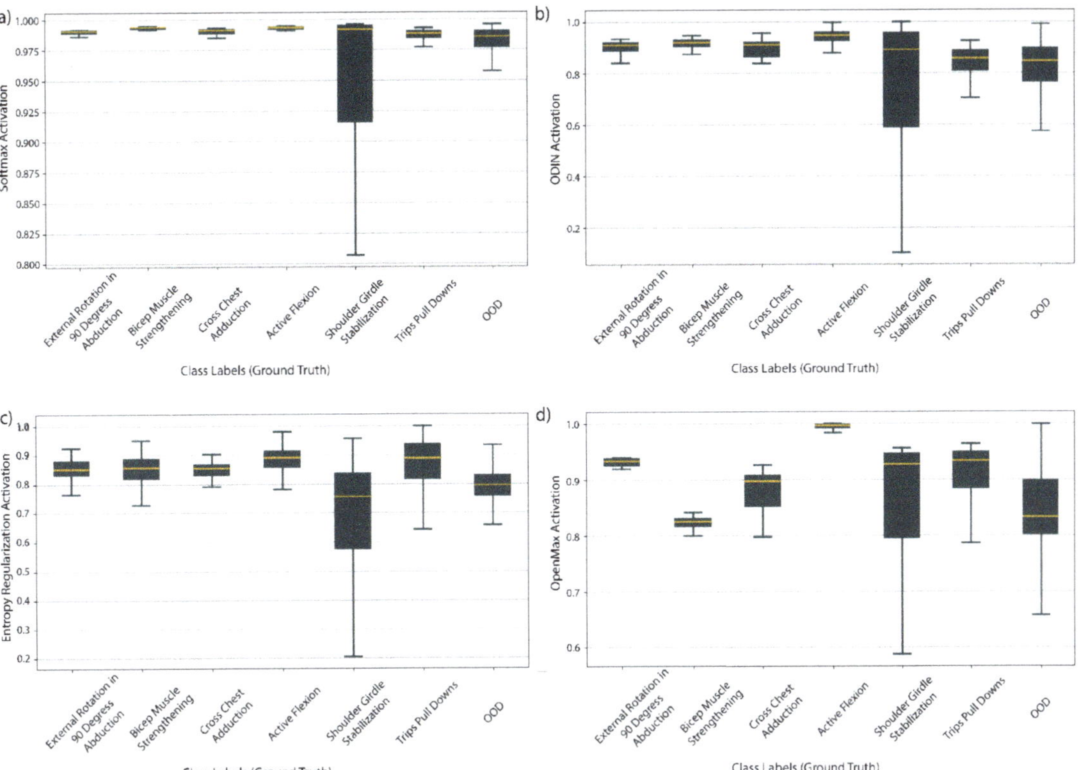

Figure 5. Mean activations for each method by ground truth class labels for SPARS9x prediction with FCN core: (**a**) Softmax activation; and (**b–d**) activation post-processing with ODIN, entropy regularization, and OpenMax respectively. Activations are scaled to between 0 and 1 for each method for illustrative purposes.

The KNN nearest neighbor distance predictions for SPARS9x are shown in Figure 6b, where the OOD class samples would appear to already be easily separable from the IIN samples with an appropriately chosen distance threshold. Indeed, it appears that not much is gained in terms of separability with removal of the Shoulder Girdle Stabilization class, and this is reflected in the results shown in Figure 7b, where AUROC of the KNN algorithm remains stable.

Figure 8 illustrates the effect on Softmax activation from re-designating the cycling activity of MHealth from OOD to IIN. As expected, there is a significant increase in the prediction confidence for cycling (because there are now samples of this activity in the training data), but the important takeaway from this figure is that the activations of each of the other activities are affected as well.

The effects of class removal or re-designation on each dataset for a selection of methods are shown in Figure 7. Figure 7a illustrates that moving the cycling activity of MHEALTH from OOD to IIN improves OOD detection for each algorithm, especially in the case of the deep learning algorithms. KNN using FCN features for cycling IIN now performs better than the KNN algorithm using engineered features. For SPARS9x, removal of the shoulder girdle stabilization activity, as shown in Figure 7b, yields little improvement, and the KNN algorithm with engineered statistical features still performs better than the other algorithms.

4.4. Train and Prediction Time

Mean prediction times of the models are shown in Table 4. Softmax thresholding and entropy regularization had the shortest prediction times. OpenMax was found to require the longest computation times for OOD prediction. The training times of the base models shown in Table 5 illustrate the longer training times required for deep learning methods.

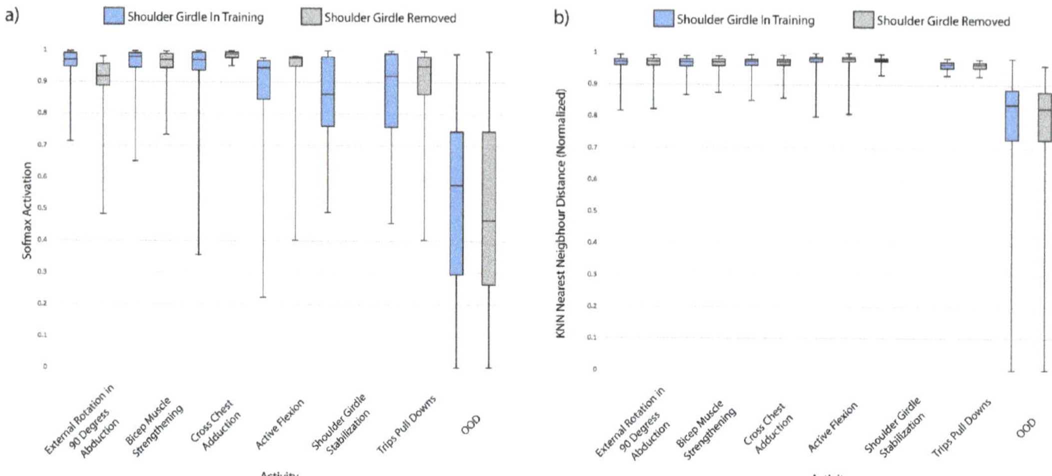

Figure 6. Effect on shoulder girdle stabilization exercise inclusion/exclusion from training data for OOD detection of the SPARS9x dataset on: (**a**) FCN Softmax activation; and (**b**) normalized KNN nearest neighbor distance. This figure highlights the resiliency to changes in training class inclusion/exclusion of the KNN algorithm compared to deep learning algorithms for detecting OOD data with this dataset. Note that, even though the mean Softmax activation decreases for the OOD data and increases for the in-distribution classes with shoulder girdle stabilization removed, there is still a slight decrease in AUROC (see Softmax Threshold of Figure 7b).

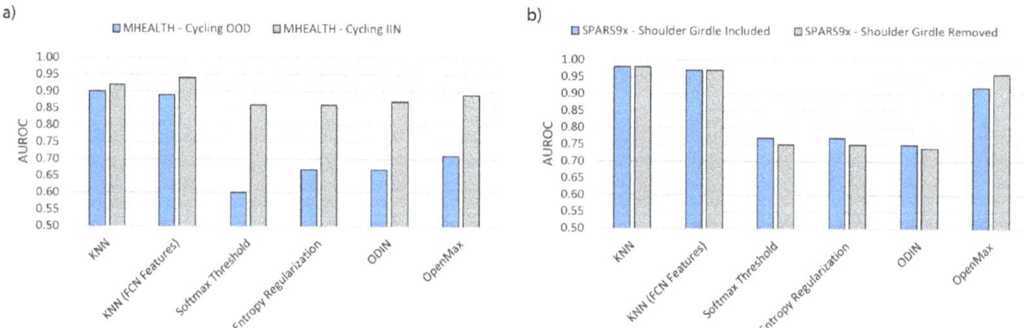

Figure 7. Effect on OOD Detection AUROC of selected algorithms by: (**a**) moving MHEALTH "cycling" class from OOD to IIN; and (**b**) removing the "Shoulder Girdle Stabilization" exercise from SPARS9x. FCN core is used for the deep learning methods.

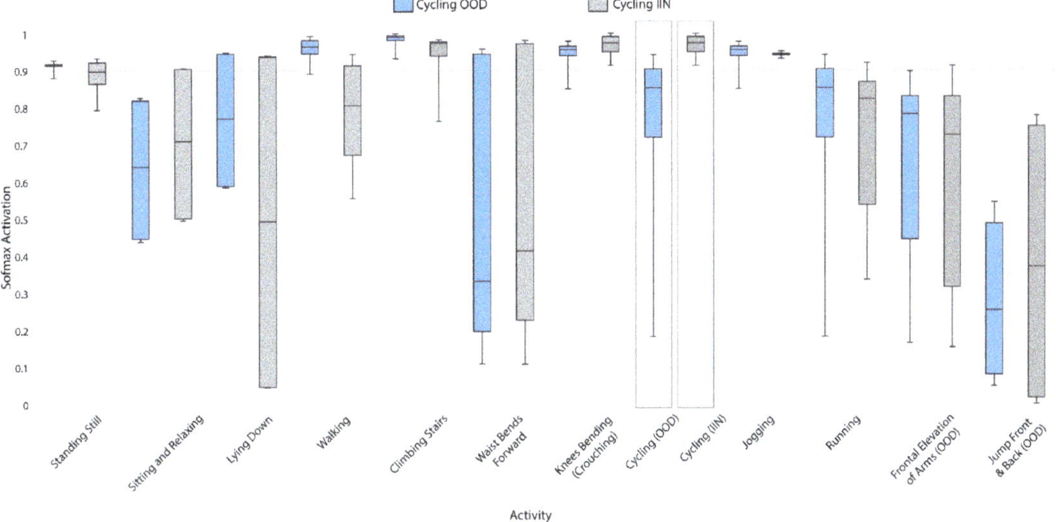

Figure 8. Effect on FCN Softmax activation for each activity in MHEALTH when cycling activity is in OOD and when cycling activity is IIN.

Table 4. Out-of-Distribution Prediction Time.

Method	Prediction Time (s)		
	MHEALTH	**SPAR**	**SPARS9x**
Traditional Methods—Engineered Statistical Features			
KNN	0.66 (0.002)	0.37 (0.005)	2.47 (0.03)
Kmeans	0.47 (0.001)	0.32 (0.007)	1.46 (0.02)
OCSVM	0.52 (0.002)	0.35 (0.005)	2.11 (0.03)
KNN—Deep Feature Embedding			
CRNN Features	0.10 (0.0008)	0.15 (0.006)	0.93 (0.05)
FCN Features	0.16 (0.0009)	0.30 (0.01)	2.00 (0.03)
CRNN Core Model			
Softmax Threshold	0.066 (0.002)	**0.082 (0.004)**	**0.41 (0.005)**
Entropy Regularization	**0.059 (0.0003)**	0.084 (0.004)	**0.41 (0.005)**
ODIN	0.22 (0.0003)	0.34 (0.02)	1.66 (0.02)
OpenMax	3.39 (0.08)	0.93 (0.02)	5.41 (0.06)
FCN Core Model			
Softmax Threshold	0.099 (0.0004)	0.22 (0.007)	1.09 (0.01)
Entropy Regularization	0.14 (0.0008)	0.21 (0.003)	0.84 (0.03)
ODIN	0.43 (0.001)	0.69 (0.002)	3.16 (0.10)
OpenMax	3.64 (0.09)	1.04 (0.02)	6.04 (0.06)

Table 5. Training Time.

Model	Training Time (s)		
	MHEALTH	**SPAR**	**SPARS9x**
KNN	3.08 (0.009)	4.69 (0.02)	5.58 (0.02)
Kmeans	1.98 (0.06)	3.03 (0.03)	3.50 (0.07)
OCSVM	2.43 (0.06)	4.40 (0.03)	5.74 (0.05)
CRNN core	81.2 (0.05)	124 (0.7)	144 (1.0)
FCN core	164 (6.0)	261 (7.4)	317 (1.0)

5. Discussion

In this paper, we address the OOD problem in the context of shoulder physiotherapy activity recognition using traditional algorithms on engineered statistical features, deep feature embeddings generated by two neural network models, as well as deep learning approaches. We evaluated these techniques on a novel physiotherapy exercise dataset (SPARS9x) that may best reflect the clinical use case for OOD of physiotherapy data as well as on two publicly-available activity datasets (MHEALTH [21] and SPARS [6]). Deep learning performance was superior to traditional algorithms (i.e., KNN) with engineered features for in-distribution classification; however, surprisingly, the opposite is true for OOD detection of SPARS9x in particular. Since the KNN algorithm also performs worse with deep features for SPARS9x, this would suggest the answer lies in the features generated by the deep learning models themselves. This may be an indication that the deep learning models are learning representations of the data that do not generalize as well to data distributions outside of their training experience for human activity inertial datasets such as SPARS9x. These deep learning models may be learning complex relationships in the data in order to discriminate between in-distribution classes. Due to the inscrutable nature of the models and even the inertial data themselves, it is difficult to describe or understand exactly what the learned deep features represent. In future work, it would be very interesting to identify what aspects of the data the neural network models in this study are using to discriminate classes, and why these are not as effective as using simple engineered features for OOD detection with a dataset such as SPARS9x that explicitly simulates OOD activity data.

HAR classification research involving IMU sensors embedded in wearables is fairly common given the ubiquity of these devices, including those using smartphones [2,10,22,69,70] and smartwatches [5,6,12,71]. Studies examining feature extraction techniques have previously been performed for time series [72–74]. However, in this paper, we are interested in the impact of engineered versus deep learning-generated features on HAR inertial data for OOD detection in particular. Numerous human activity inertial datasets exist with notable examples including MHEALTH [21], PAMAP2 [75], and the recent WISDM gesture-recognition dataset [76]. While these datasets often have an intermediary "null" class between exercises (e.g., MHEALTH), this is often restricted. For example, in the gesture recognition study by Bulling et al. [1], participants were asked not to engage in any other activities between target activities. Classification of repetitive exercises, as in the case of physiotherapy inertial data, add an additional nuance to the analysis. OOD activities may resemble IIN activities, but IIN activities may be distinct in their repetitive nature (e.g., reaching for a mug versus lifting an arm to perform a physiotherapy exercise for several repetitions in a short time frame). We believe this study is unique in its approach of capturing a dataset that explicitly simulates the distinction between known target human activities and unknown a priori OOD activities.

Table 4 where samples in the test set are bucketed according to activation for SPARS9x with a CRNN core. The other deep learning methods, ODIN, entropy regularization, and OpenMax, appear to be effective at reducing this overconfidence for HAR inertial data, judging by the improved spread in the distributions. However, when using confidence to threshold between IIN and OOD classes, relative confidence between classes is more important than absolute confidence. The activations in Figure 5 have been normalized to illustrate this point. Of particular interest in this figure is that some degree of overlap is present for each of the algorithms between the OOD classes and the IIN activities. For every model other than OpenMax (which does not use this information directly for OOD class prediction), this means that thresholding based on activation will be little more accurate than thresholding Softmax, an observation that is borne out in the results of Table 3.

As shown in Figure 7, moving the cycling class of MHEALTH from OOD to INN dramatically increases OOD detection accuracy for deep learning approaches. The algorithms are able to accurately discriminate this class in supervised IIN classification with near perfect accuracy, just as with the other two chosen OOD classes, however, the cycling class in particular is wrongly overconfident. Moving a class that causes confusion as measured by Softmax confidence increases AUROC for Softmax Thresholding, but it also increases KNN prediction AUROC. This is seen as designating the cycling activity as IIN increased the AUROC of Softmax Thresholding from 0.60 to 0.86, but also increased KNN from 0.90 to 0.92. Conversely, in Figure 7b, the KNN algorithm has no such issue with the shoulder girdle stabilization activity, so the advantage that the traditional algorithms have over Softmax cannot be explained by class confusion alone.

Removing or adding a class to a deep learning model's training set impacts both the difficulty of the OOD task and also alters the learned feature representation. As shown in Figure 6a, removing a low-confidence class did not greatly improve OOD prediction accuracy. Removing this class alters the model's parameters which may have decreased generalizability to the test and OOD set. With the case of shoulder girdle stabilization of SPARS9x and cycling in MHEALTH, other classes may become less or more confident as a result. This may explain why removing the shoulder girdle stabilization activity from SPARS9x did not result in a dramatic change as in the case of moving the cycling class to IIN in the MHEALTH dataset.

This is illustrative of some major pitfalls in using activation threshold-based techniques for OOD-detection, including entropy regularization and ODIN: some classes may have more variance than others in how they are performed, and in-distribution classes may share similarities to one another. Either of these issues would reduce confidence in predicting these classes. This spread of confidence explains why Softmax thresholding without any other intervention was found to be a particularly ineffective method, as there is activation overlap between samples from less confident in-distribution classes and OOD samples. This is in contrast to the traditional algorithms such as KNN that did not exhibit this overlap—while discriminating between in-distribution classes were generally less accurate than deep learning methods, the activation (i.e., distance) spreads were still far removed from those of the OOD data. Further investigation reveals that while each of the deep learning methods appear effective at reducing Softmax overconfidence (i.e., entropy regularization, ODIN, and OpenMax), the relative overlap of confidence spread between the IIN and OOD classes appears to differ little from vanilla Softmax. For every model other than OpenMax (which does not use this information directly for OOD class prediction), this means that thresholding based on activation will be little more accurate than thresholding Softmax, an observation that is borne out in the results of Table 3.

Table 3 demonstrates that choice of segment length may cause large variance in AUROC. This is another advantage that traditional algorithms have over deep learning methods in these experiments—one can set a relatively high segment length of 10s and be confident that this is a reasonable choice that will result in near-optimal prediction accuracy. For deep learning algorithms, optimization of segment length is important to achieve optimal performance of the classification pipeline, and this likely reflects a tradeoff between the model size and number of training samples versus the amount of information available in each individual segment.

While several methods exist to perform activity segmentation or windowing for time series data, for this analysis, we focus on the sliding window technique due to the periodicity of physiotherapy exercises. Analyses have been performed on the impact of sliding window size for HAR inertial data [77], including adaptive sliding window segmentation techniques [78–80]. To our knowledge, an analysis of impact of window size for OOD-detection with HAR inertial data has yet to be performed.

This analysis evaluated the effect of fixed window size on OOD detection. While setting a fixed window size or including that parameter in a hyperparameter grid search is not optimal, to our knowledge existing adaptive window algorithms that have been proposed for inertial data are not easily extensible to the OOD problem as they rely on first predicting the class the segment belongs to. As an example, the adaptive window size algorithms as proposed by Noor et al. [78] increases window size based on increasing probability that a segment is from a particular predicted activity, activity information which would not be available in the OOD case. While an adaptive sliding window algorithm such as this may help differentiate between IIN activities, it is unclear if it would be beneficial in identifying OOD activities. This point is made clearer when examining the OOD detection in Table 3, where larger window sizes generally increased detection accuracy, versus IIN classification in Table 2, where highest detection accuracies are obtained in the mid-sized windows. This implies that the optimal window size for OOD detection accuracy is dependent on the entirety of the dataset, not just the in-distribution ones.

While confidence trends are similar overall, the CRNN model exhibited much higher Softmax confidence in its predictions than the FCN model (i.e., exhibiting much more of the overconfidence that neural network models are known for). This may be an indication of why the FCN model performs better in most supervised learning scenarios.

The ODIN method was developed for OOD detection on image datasets (as were many of the other methods tested here), rather than time series datasets. There are manipulations that make sense in the image domain but may be less applicable in the time domain For example, while it might be reasonable to flip an image as an augmentation method, performing an activity in reverse is not a valid example of that same activity. This argument also applies to input perturbations of ODIN. In the original study by Liang et al. [19], increasing perturbations improved DenseNet results, but significantly negatively impacted Wide ResNet results, where AUROC decreased from 0.93–0.95 to 0.80–0.85. This suggests that the success of this approach at improving OOD detection depends heavily on the model architecture, but the type of dataset is also likely a factor.

The OpenMax results presented in the work of Bendale and Boult are generated by models trained on ImageNet [20] using AlexNet as a feature extractor [81]. To our knowledge, no such large dataset exists for physiotherapy time series data. In this research, we trained with a significantly smaller dataset in a completely separate domain (i.e., inertial time series data). There are also some conceptual concerns with the OpenMax approach. Specifically, as indicated by the range of OpenMax activations we found, this approach may be impacted in a similar fashion to the other methods by the wide variation in activations for some IIN classes. Since the OOD decision in OpenMax is based on the distance from the mean activation vector, and the Weibull distribution is fit to the largest deviations in the correct training samples per class from each mean activation, this method will still make inaccurate predictions if there is significant activation overlap between IIN and OOD activations. Distance from the MAV may be too simplified of an approach in this case. A different distance measure and/or loss function may be needed, such as the il-loss proposed by Hassen and Chan [82], while retaining the remainder of the unknown class approach of OpenMax. A GAN-augmented approach may also be used to ameliorate these issues, but as noted previously, generated samples are unlikely to fully encapsulate the IIN decision-boundary, so such a classifier is unlikely to be an optimal detector [60]. However, as seen in the work by Ge et al., GAN augmentation may still improve results [47].

Traditional algorithms tended to have the lengthiest prediction time requirements, but OpenMax required the longest time to run. As implemented in these experiments, OpenMax uses a costly sample-by-sample approach in generating the final activation vector. Unlike ODIN and OpenMax, entropy regularization only requires a change to the model loss function compared to standard supervised learning and does not require additional costly processing steps in the prediction phase, shortening its prediction time. The increase in prediction times for MHEALTH versus SPARS for traditional algorithms and OpenMax may be partly explained by the greater number of IIN classes in the MHEALTH dataset.

6. Future Work

Regardless of approach, we suggest that future work should attempt to resolve and explain the apparent decrease in generalizability of deep learning features compared to engineered features towards OOD detection with time series human activity inertial datasets similar to SPARS9x. Identifying patterns in the data that are key to discriminating between in-distribution classes by these deep learning models may assist in this process.

The present work focuses on purely discriminative methods rather than delving into the myriad of generative options available. These generative models may also be combined with traditional algorithms. An adversarial autoencoder method, such as that presented by Pidhorskyi et al. [83], may also be a promising future approach to explore, or one based on likelihood ratios [62].

This work evaluated IIN classification and OOD detection accuracy using a range of fixed window sizes. While this study found that larger window sizes increased OOD detection accuracy, this may not always be the case. A new algorithm for adaptive sliding window segmentation for this kind of periodic HAR data specifically accounting for the potential for either periodic or non-periodic OOD data would be ideal.

7. Limitations

The SPARS9x dataset was captured in laboratory conditions with healthy subjects. We expect that performance of exercises by injured patients in the home setting to exhibit more variance in motion than those present in the SPARS9x dataset.

Only the lower arm sensor in the MHEALTH dataset is used, essentially simulating inertial data captured by a smartwatch. Including data from other sensors in this dataset would likely improve classification and detection accuracy, but we limited the data to align with the proposed clinical solution to tracking shoulder physiotherapy with a smartwatch alone.

Only concentric exercises from SPARS9x were used in this analysis. Isometric exercises in SPARS9x were found to be too highly confused with OOD data in preliminary classification validation experiments to be viable for use in the OOD detection analysis.

Magnetometer data were excluded from SPARS9x for these analyses as their inclusion was not found to improve classification accuracy when combined with data from the accelerometer and gyroscope in preliminary experiments.

This paper applies a selection of well-known OOD detection techniques from the image domain to HAR time series inertial datasets. There are many other OOD detection methods in use in the image domain and elsewhere, and those tested in this paper represent only a selection of those most widely cited. However, these results are likely reflective of most discriminative models that base prediction on Softmax activation, whether explicitly or implicitly.

8. Conclusions

In this paper, we present a novel physiotherapy inertial dataset captured specifically for the analysis of the out-of-distribution problem in HAR and tested a range of OOD detection methods common in the image domain. Our results indicate that simple and rapid OOD detection techniques based on traditional algorithms such as KNN using engineered statistical features outperform sophisticated deep learning techniques on some HAR time series datasets.

Author Contributions: Conceptualization, P.B., C.W. and D.B.; methodology, P.B. and D.B.; software, P.B. and D.B.; validation, P.B. and D.B.; formal analysis, P.B. and D.B.; investigation, P.B. and D.B.; resources, C.W.; data curation, P.B.; writing—original draft preparation, P.B.; writing—review and editing, P.B., C.W. and D.B.; visualization, P.B.; supervision, C.W.; project administration, C.W.; funding acquisition, C.W. and D.B. All authors have read and agreed to the published version of the manuscript.

Funding: This work was supported in part by the Workplace Safety and Insurance Board of Ontario under Grant RICH2018, the Natural Sciences and Engineering Research Council of Canada under grant CHRP-538866, and the Canadian Institutes of Health Research under grant CPG-163963, and the Ontario Graduate Scholarship.

Institutional Review Board Statement: The study was conducted according to the guidelines of the Declaration of Helsinki, and approved by the Research Ethics Board of Sunnybrook Health Sciences Centre (Project Identification Number 2765).

Informed Consent Statement: Informed consent was obtained from all subjects involved in the stud

Data Availability Statement: The SPARS9x dataset presented in this study is publicly availabl in the open access repository "Shoulder Physiotherapy Activity Recognition 9-Axis Dataset" DO 10.21227/cx5v-vw46.

Conflicts of Interest: David Burns is a cofounder and David Burns and Cari Whyne hold equity i Halterix Corporation, a digital physiotherapy company.

References

1. Bulling, A.; Blanke, U.; Schiele, B. A tutorial on human activity recognition using body-worn inertial sensors. *ACM Comput. Sur* **2014**, *46*, 1–33. [CrossRef]
2. Anguita, D.; Ghio, A.; Oneto, L.; Parra, X.; Reyes-Ortiz, J.L. A Public Domain Dataset for Human Activity Recognition Usin Smartphones. In Proceedings of the European Symposium on Artificial Neural Networks, Computational Intelligence an Machine Learning, Bruge, Belgium, 24–26 April 2013.
3. Garcia-Ceja, E.; Brena, R.F.; Carrasco-Jimenez, J.C.; Garrido, L. Long-Term Activity Recognition from Wristwatch Acceleromete Data. *Sensors* **2014**, *14*, 22500–22524. [CrossRef] [PubMed]
4. Özdemir, A.T.; Barshan, B. Detecting Falls with Wearable Sensors Using Machine Learning Techniques. *Sensors* **2014**, *14* 10691–10708. [CrossRef] [PubMed]
5. Shoaib, M.; Incel, O.D.; Scholten, H.; Havinga, P. SmokeSense: Online Activity Recognition Framework on Smartwatches. I *Mobile Computing, Applications, and Services*; Springer: Berlin/Heidelberg, Germany, 2018; pp. 106–124. [CrossRef]
6. Burns, D.M.; Leung, N.E.; Hardisty, M.; Whyne, C.M.; Henry, P.; McLachlin, S. Shoulder physiotherapy exercise recognition Machine learning the inertial signals from a smartwatch. *Physiol. Meas.* **2018**, *39*, 075007. [CrossRef]
7. Vakanski, A.; Ferguson, J.M.; Lee, S. Mathematical Modeling and Evaluation of Human Motions in Physical Therapy Usin Mixture Density Neural Networks. *J. Physiother. Phys. Rehabil.* **2016**, *1*, 118.
8. Han, J.; Shao, L.; Xu, D.; Shotton, J. Enhanced Computer Vision with Microsoft Kinect Sensor: A Review. *IEEE Trans. Cyber* **2013**, *43*, 1318–1334. [CrossRef] [PubMed]
9. Dolatabadi, E.; Taati, B.; Mihailidis, A. An Automated Classification of Pathological Gait Using Unobtrusive Sensing Technolog *IEEE Trans. Neural Syst. Rehabil. Eng.* **2017**, *25*, 2336–2346. [CrossRef] [PubMed]
10. Almaslukh, B.; Artoli, A.M.; Al-Muhtadi, J. A Robust Deep Learning Approach for Position-Independent Smartphone-Base Human Activity Recognition. *Sensors* **2018**, *18*, 3726. [CrossRef]
11. Kwapisz, J.R.; Weiss, G.M.; Moore, S.A. Activity recognition using cell phone accelerometers. *ACM SIGKDD Explor. Newsl.* **2011** *12*, 74–82. [CrossRef]
12. Shen, S.; Wang, H.; Choudhury, R.R. I am a Smartwatch and I can Track my User's Arm. In Proceedings of the 14th Annua International Conference on Mobile Systems, Applications, and Services, Singapore, 25–30 June 2016; pp. 85–96.
13. Schölkopf, B.; Platt, J.C.; Shawe-Taylor, J.C.; Smola, A.J.; Williamson, R.C. Estimating the Support of a High-Dimensiona Distribution. *Neural Comput.* **2001**, *13*, 1443–1471. [CrossRef] [PubMed]
14. Mourão-Miranda, J.; Hardoon, D.R.; Hahn, T.; Marquand, A.F.; Williams, S.C.; Shawe-Taylor, J.; Brammer, M. Patient classificatio as an outlier detection problem: An application of the One-Class Support Vector Machine. *NeuroImage* **2011**, *58*, 793–804 [CrossRef]
15. Dudani, S.A. The Distance-Weighted k-Nearest-Neighbor Rule. *IEEE Trans. Syst. Man Cybern.* **1976**, *6*, 325–327. [CrossRef]
16. Chawla, S.; Gionis, A. k-means–: A unified approach to clustering and outlier detection. In Proceedings of the 2013 SIAM International Conference on Data Mining, Austin, TX, USA, 2–4 May 2013; pp. 189–197.
17. Hendrycks, D.; Gimpel, K. A Baseline for Detecting Misclassified and Out-of-Distribution Examples in Neural Networks. In Proceedings of the International Conference on Learning Representations, Toulon, France, 24–26 April 2017.
18. Pereyra, G.; Tucker, G.; Chorowski, J.; Kaiser, Ł.; Hinton, G. Regularizing Neural Networks by Penalizing Confident Output Distributions. In Proceedings of the International Conference on Learning Representations, Toulon, France, 24–26 April 2017.
19. Liang, S.; Li, Y.; Srikant, R. Enhancing the Reliability of Out-of-distribution Image Detection in Neural Networks. In Proceedings of the International Conference on Learning Representations, Vancouver, BC, Canada, 30 April–3 May 2018.
20. Bendale, A.; Boult, T.E. Towards Open Set Deep Networks. In Proceedings of the 2016 IEEE Conference on Computer Vision and Pattern Recognition (CVPR), Las Vegas, NV, USA, 27–30 June 2016; pp. 1563–1572. [CrossRef]
21. Banos, O.; Garcia, R.; Holgado-Terriza, J.A.; Damas, M.; Pomares, H.; Rojas, I.; Saez, A.; Villalonga, C. mHealthDroid: A Novel Framework for Agile Development of Mobile Health Applications. In *Ambient Assisted Living and Daily Activities*; Springer Berlin/Heidelberg, Germany, 2014; pp. 91–98.
22. Ronao, C.A.; Cho, S.-B. Human activity recognition with smartphone sensors using deep learning neural networks. *Expert Syst Appl.* **2016**, *59*, 235–244. [CrossRef]
23. Köping, L.; Shirahama, K.; Grzegorzek, M. A general framework for sensor-based human activity recognition. *Comput. Biol. Med* **2018**, *95*, 248–260. [CrossRef] [PubMed]
24. Hassan, M.M.; Huda, S.; Uddin, Z.; Almogren, A.; Alrubaian, M.A. Human Activity Recognition from Body Sensor Data using Deep Learning. *J. Med. Syst.* **2018**, *42*, 99. [CrossRef] [PubMed]

25. Wang, Y.; Cang, S.; Yu, H. A survey on wearable sensor modality centred human activity recognition in health care. *Expert Syst. Appl.* **2019**, *137*, 167–190. [CrossRef]
26. Kluge, F.; Gaßner, H.; Hannink, J.; Pasluosta, C.; Klucken, J.; Eskofier, B.M. Towards Mobile Gait Analysis: Concurrent Validity and Test-Retest Reliability of an Inertial Measurement System for the Assessment of Spatio-Temporal Gait Parameters. *Sensors* **2017**, *17*, 1522. [CrossRef] [PubMed]
27. Qiu, S.; Liu, L.; Zhao, H.; Wang, Z.; Jiang, Y. MEMS Inertial Sensors Based Gait Analysis for Rehabilitation Assessment via Multi-Sensor Fusion. *Micromachines* **2018**, *9*, 442. [CrossRef]
28. Tedesco, S.; Barton, J.; O'Flynn, B. A Review of Activity Trackers for Senior Citizens: Research Perspectives, Commercial Landscape and the Role of the Insurance Industry. *Sensors* **2017**, *17*, 1277. [CrossRef]
29. Godfrey, A. Wearables for independent living in older adults: Gait and falls. *Maturitas* **2017**, *100*, 16–26. [CrossRef]
30. Liao, Y.; Vakanski, A.; Xian, M.; Paul, D.; Baker, R. A review of computational approaches for evaluation of rehabilitation exercises. *Comput. Biol. Med.* **2020**, *119*, 103687. [CrossRef]
31. Houmanfar, R.; Karg, M.; Kulic, D. Movement Analysis of Rehabilitation Exercises: Distance Metrics for Measuring Patient Progress. *IEEE Syst. J.* **2014**, *10*, 1014–1025. [CrossRef]
32. Liao, Y.; Vakanski, A.; Xian, M. A Deep Learning Framework for Assessing Physical Rehabilitation Exercises. *IEEE Trans. Neural Syst. Rehabil. Eng.* **2020**, *28*, 468–477. [CrossRef] [PubMed]
33. Geng, C.; Huang, S.-J.; Chen, S. Recent Advances in Open Set Recognition: A Survey. *IEEE Trans. Pattern Anal. Mach. Intell.* **2020**, *1*, 1. [CrossRef] [PubMed]
34. Pimentel, M.A.; Clifton, D.A.; Clifton, L.; Tarassenko, L. A review of novelty detection. *Signal Process.* **2014**, *99*, 215–249. [CrossRef]
35. Li, D.; Chen, D.; Goh, J.; Ng, S. Anomaly Detection with Generative Adversarial Networks for Multivariate Time Series. *arXiv* **2019**, arXiv:1809.04758.
36. Munoz-Organero, M. Outlier Detection in Wearable Sensor Data for Human Activity Recognition (HAR) Based on DRNNs. *IEEE Access* **2019**, *7*, 74422–74436. [CrossRef]
37. Meng, C.; Jiang, X.S.; Wei, X.M.; Wei, T. A Time Convolutional Network Based Outlier Detection for Multidimensional Time Series in Cyber-Physical-Social Systems. *IEEE Access* **2020**, *8*, 74933–74942. [CrossRef]
38. Lu, H.; Liu, Y.; Fei, Z.; Guan, C. An Outlier Detection Algorithm Based on Cross-Correlation Analysis for Time Series Dataset. *IEEE Access* **2018**, *6*, 53593–53610. [CrossRef]
39. Lin, S.; Clark, R.; Birke, R.; Schonborn, S.; Trigoni, N.; Roberts, S. Anomaly Detection for Time Series Using VAE-LSTM Hybrid Model. In Proceedings of the ICASSP 2020—2020 IEEE International Conference on Acoustics, Speech and Signal Processing (ICASSP), Barcelona, Spain, 4–8 May 2020; pp. 4322–4326.
40. Laptev, N.; Amizadeh, S.; Flint, I. Generic and Scalable Framework for Automated Time-series Anomaly Detection. In Proceedings of the 21st ACM SIGKDD International Conference on Knowledge Discovery and Data Mining, Sydney, Australia, 10–13 August 2015; pp. 1939–1947.
41. Scheirer, W.J.; Rocha, A.D.R.; Sapkota, A.; Boult, T.E. Toward Open Set Recognition. *IEEE Trans. Pattern Anal. Mach. Intell.* **2013**, *35*, 1757–1772. [CrossRef]
42. Bendale, A.; Boult, T. Towards Open World Recognition. In Proceedings of the 2015 IEEE Conference on Computer Vision and Pattern Recognition (CVPR), Boston, MA, USA, 7–12 June 2015; pp. 1893–1902.
43. Scheirer, W.J.; Jain, L.P.; Boult, T.E. Probability Models for Open Set Recognition. *IEEE Trans. Pattern Anal. Mach. Intell.* **2014**, *36*, 2317–2324. [CrossRef]
44. Erfani, S.M.; Rajasegarar, S.; Karunasekera, S.; Leckie, C. High-dimensional and large-scale anomaly detection using a linear one-class SVM with deep learning. *Pattern Recognit.* **2016**, *58*, 121–134. [CrossRef]
45. Yang, K.; Ahn, C.R.; Vuran, M.C.; Aria, S.S. Semi-supervised near-miss fall detection for ironworkers with a wearable inertial measurement unit. *Autom. Constr.* **2016**, *68*, 194–202. [CrossRef]
46. Ryu, S.; Koo, S.; Yu, H.; Lee, G.G. Out-of-domain Detection based on Generative Adversarial Network. In Proceedings of the 2018 Conference on Empirical Methods in Natural Language Processing, Brussels, Belgium, 31 October–4 November 2018; pp. 714–718.
47. Ge, Z.; Demyanov, S.; Garnavi, R. Generative OpenMax for Multi-Class Open Set Classification. In Proceedings of the British Machine Vision Conference, London, UK, 4–7 September 2017. [CrossRef]
48. Neal, L.; Olson, M.; Fern, X.; Wong, W.-K.; Li, F. Open Set Learning with Counterfactual Images. In Proceedings of the European Conference on Computer Vision—ECCV, Munich, Germany, 8–14 September 2018; pp. 620–635.
49. Jo, I.; Kim, J.; Kang, H.; Kim, Y.-D.; Choi, S. Open Set Recognition by Regularising Classifier with Fake Data Generated by Generative Adversarial Networks. In Proceedings of the 2018 IEEE International Conference on Acoustics, Speech and Signal Processing (ICASSP), Calgary, AB, Canada, 15–20 April 2018; pp. 2686–2690.
50. Lee, K.; Lee, H.; Lee, K.; Shin, J. Training Confidence-calibrated Classifiers for Detecting Out-of-Distribution Samples. In Proceedings of the International Conference on Learning Representations, Vancouver, BC, Canada, 30 April–3 May 2018.
51. Denouden, T.; Salay, R.; Czarnecki, K.; Abdelzad, V.; Phan, B.; Vernekar, S. Improving Reconstruction Autoencoder Out-of-distribution Detection with Mahalanobis Distance. *arXiv* **2018**, arXiv:1812.02765.

52. Xia, Y.; Cao, X.; Wen, F.; Hua, G.; Sun, J. Learning Discriminative Reconstructions for Unsupervised Outlier Removal. In Proceedings of the 2015 IEEE International Conference on Computer Vision (ICCV), Santiago, Chile, 7–13 December 2015; pp. 1511–1519.
53. Oza, P.; Patel, V.M. C2AE: Class Conditioned Auto-Encoder for Open-Set Recognition. In Proceedings of the 2019 IEEE/CVF Conference on Computer Vision and Pattern Recognition (CVPR), Long Beach, CA, USA, 16–20 June 2019; pp. 2302–2311. [CrossRef]
54. Yoshihashi, R.; Shao, W.; Kawakami, R.; You, S.; Iida, M.; Naemura, T. Classification-Reconstruction Learning for Open-Set Recognition. In Proceedings of the 2019 IEEE/CVF Conference on Computer Vision and Pattern Recognition (CVPR), Long Beach, CA, USA, 16–20 June 2019; pp. 4011–4020.
55. Aytekin, C.; Ni, X.; Cricri, F.; Aksu, E. Clustering and Unsupervised Anomaly Detection with L2 Normalized Deep Auto-Encoder Representations. In Proceedings of the International Joint Conference on Neural Networks, Rio, Brazil, 8–13 July 2018.
56. Guo, C.; Pleiss, G.; Sun, Y.; Weinberger, K.Q. On Calibration of Modern Neural Networks. In Proceedings of the International Conference on Machine Learning, Sydney, Australia, 6–11 August 2017.
57. Omae, Y.; Mori, M.; Akiduki, T.; Takahashi, H. A Novel Deep Learning Optimization Algorithm for Human Motions Anomaly Detection. *Int. J. Innov. Comput. Inf. Control* **2019**, *15*, 199–208.
58. Goodfellow, I. NIPS 2016 Tutorial: Generative Adversarial Networks. In Proceedings of the Conference on Neural Information Processing Systems (NIPS), Barcelona, Spain, 5–10 December 2016.
59. Yang, Y.; Hou, C.; Lang, Y.; Guan, D.; Huang, D.; Xu, J. Open-set human activity recognition based on micro-Doppler signatures. *Pattern Recognit.* **2019**, *85*, 60–69. [CrossRef]
60. Vernekar, S.; Gaurav, A.; Denouden, T.; Phan, B.; Abdelzad, V.; Salay, R.; Czarnecki, K. Analysis of Confident-Classifiers for Out-of-distribution Detection. In Proceedings of the 7th International Conference on Learning Representations–ICLR, New Orleans, LA, USA, 6–9 May 2019.
61. Schlegl, T.; Seeböck, P.; Waldstein, S.M.; Langs, G.; Schmidt-Erfurth, U. f-AnoGAN: Fast unsupervised anomaly detection with generative adversarial networks. *Med. Image Anal.* **2019**, *54*, 30–44. [CrossRef]
62. Ren, J.; Liu, P.J.; Fertig, E.; Snoek, J.; Poplin, R.; DePristo, M.A.; Dillon, J.V.; Lakshminarayanan, B. Likelihood Ratios for Out-of-Distribution Detection. *arXiv* **2019**, arXiv:1906.02845.
63. Nguyen, A.; Yosinski, J.; Clune, J. Deep Neural Networks are Easily Fooled: High Confidence Predictions for Unrecognizable Images. In Proceedings of the 28th IEEE Conference on Computer Vision and Pattern Recognition, Boston, MA, USA, 7–12 June 2015.
64. Burns, D.M.; Whyne, C.M. Seglearn: A Python Package for Learning Sequences and Time Series. *J. Mach. Learn. Res.* **2018**, *19*, 1–7.
65. Li, D.; Chen, D.; Jin, B.; Shi, L.; Goh, J.; Ng, S.-K. MAD-GAN: Multivariate Anomaly Detection for Time Series Data with Generative Adversarial Networks. *arXiv* **2019**, arXiv:1901.04997.
66. Karpathy, A.; Johnson, J.; Fei-Fei, L. Visualizing and Understanding Recurrent Networks. *arXiv* **2015**, arXiv:1506.02078.
67. Wang, Z.; Yan, W.; Oates, T. Time series classification from scratch with deep neural networks: A strong baseline. In Proceedings of the 2017 International Joint Conference on Neural Networks (IJCNN), Anchorage, AK, USA, 14–19 May 2017; pp. 1578–1585.
68. Burns, D.M.; Whyne, C.M. Personalized Activity Recognition with Deep Triplet Embeddings. *arXiv* **2020**, arXiv:2001.05517.
69. Lima, W.S.; Bragança, H.L.; Souto, E.J. NOHAR—NOvelty discrete data stream for Human Activity Recognition based on smartphones with inertial sensors. *Expert Syst. Appl.* **2021**, *166*, 114093. [CrossRef]
70. Shoaib, M.; Bosch, S.; Incel, O.D.; Scholten, J.; Havinga, P.J.M. Complex Human Activity Recognition Using Smartphone and Wrist-Worn Motion Sensors. *Sensors* **2016**, *16*, 426. [CrossRef]
71. Shahmohammadi, F.; Hosseini, A.; King, C.E.; Sarrafzadeh, M. Smartwatch Based Activity Recognition Using Active Learning. In Proceedings of the 2017 IEEE/ACM International Conference on Connected Health: Applications, Systems and Engineering Technologies (CHASE), Philadelphia, PA, USA, 17–19 July 2017; pp. 321–329.
72. Shoeibi, A.; Ghassemi, N.; Alizadehsani, R.; Rouhani, M.; Hosseini-Nejad, H.; Khosravi, A.; Panahiazar, M.; Nahavandi, S. A comprehensive comparison of handcrafted features and convolutional autoencoders for epileptic seizures detection in EEG signals. *Expert Syst. Appl.* **2021**, *163*, 113788. [CrossRef]
73. Wang, H.; Zhang, Q.; Wu, J.; Pan, S.; Chen, Y. Time series feature learning with labeled and unlabeled data. *Pattern Recognit.* **2019**, *89*, 55–66. [CrossRef]
74. Li, F.; Shirahama, K.; Nisar, M.A.; Köping, L.; Grzegorzek, M. Comparison of Feature Learning Methods for Human Activity Recognition Using Wearable Sensors. *Sensors* **2018**, *18*, 679. [CrossRef]
75. Reiss, A.; Stricker, D. Introducing a New Benchmarked Dataset for Activity Monitoring. In Proceedings of the 16th International Symposium on Wearable Computers, Seattle, WA, USA, 7–10 October 2012; pp. 108–109. [CrossRef]
76. Weiss, G.M.; Yoneda, K.; Hayajneh, T. Smartphone and Smartwatch-Based Biometrics Using Activities of Daily Living. *IEEE Access* **2019**, *7*, 133190–133202. [CrossRef]
77. Fida, B.; Bernabucci, I.; Bibbo, D.; Conforto, S.; Schmid, M. Varying behavior of different window sizes on the classification of static and dynamic physical activities from a single accelerometer. *Med. Eng. Phys.* **2015**, *37*, 705–711. [CrossRef] [PubMed]
78. Noor, M.H.M.; Salcic, Z.; Wang, K.I.-K. Adaptive sliding window segmentation for physical activity recognition using a single tri-axial accelerometer. *Pervasive Mob. Comput.* **2017**, *38*, 41–59. [CrossRef]

9. Ma, C.; Li, W.; Cao, J.; Du, J.; Li, Q.; Gravina, R. Adaptive sliding window based activity recognition for assisted livings. *Inf. Fusion* **2020**, *53*, 55–65. [CrossRef]
0. Akbari, A.; Wu, J.; Grimsley, R.; Jafari, R. Hierarchical Signal Segmentation and Classification for Accurate Activity Recognition. In Proceedings of the 2018 ACM International Joint Conference and 2018 International Symposium on Pervasive and Ubiquitous Computing and Wearable Computers, Singapore, 8–12 October 2018; pp. 1596–1605.
1. Krizhevsky, A.; Sutskever, I.; Hinton, G.E. Imagenet classification with deep convolutional neural networks. In *Advances in Neural Information Processing Systems*; MIT Press: Cambridge, MA, USA, 2012; pp. 1097–1105.
2. Hassen, M.; Chan, P.K. Learning a Neural-network-based Representation for Open Set Recognition. *arXiv* **2018**, arXiv:1802.04365.
3. Pidhorskyi, S.; Almohsen, R.; Doretto, G. Generative Probabilistic Novelty Detection with Adversarial Autoencoders. *Adv. Neural Inf. Process. Syst.* **2018**, *31*, 6822–6833.

 *sensors* 

Article

Monitoring Physical Activity with a Wearable Sensor in Patients with COPD during In-Hospital Pulmonary Rehabilitation Program: A Pilot Study

Sebastian Rutkowski [1,*], Joren Buekers [2,3,4,5], Anna Rutkowska [1], Błażej Cieślik [6] and Jan Szczegielniak [1]

1 Faculty of Physical Education and Physiotherapy, Opole University of Technology, 45-758 Opole, Poland; a.rutkowska@po.edu.pl (A.R.); j.szczegielniak@po.edu.pl (J.S.)
2 Department of Biosystems, KU Leuven, 3001 Leuven, Belgium; joren.buekers@kuleuven.be
3 ISGlobal, 08002 Barcelona, Spain
4 Universitat Pompeu Fabra (UPF), 08002 Barcelona, Spain
5 CIBER Epidemiología y Salud Pública (CIBERESP), 08002 Barcelona, Spain
6 Faculty of Health Sciences, Jan Dlugosz University in Czestochowa, 42-200 Czestochowa, Poland; b.cieslik@ajd.czest.pl
* Correspondence: s.rutkowski@po.edu.pl; Tel.: +48-507-027-792

Abstract: Accelerometers have become a standard method of monitoring physical activity in everyday life by measuring acceleration in one, two, or three axes. These devices provide reliable and objective measurements of the duration and intensity of physical activity. We aimed to investigate whether patients undertake physical activity during non-supervised days during stationary rehabilitation and whether patients adhere to the rigor of 24 h monitoring. The second objective was to analyze the strengths and weaknesses of such kinds of sensors. The research enrolled 13 randomly selected patients, qualified for in-patient, 3 week, high-intensity, 5 times a week pulmonary rehabilitation. The SenseWear armband was used for the assessment of physical activity. Participants wore the device 24 h a day for the next 4 days (Friday–Monday). The analysis of the number of steps per day, the time spent lying as well as undertaking moderate or vigorous physical activity (>3 metabolic equivalents of task (METs)), and the energy expenditure expressed in kcal showed no statistically significant difference between the training days and the days off. It seems beneficial to use available physical activity sensors in patients with chronic obstructive pulmonary disease (COPD); measurable parameters provide feedback that may increase the patient's motivation to be active to achieve health benefits.

Keywords: COPD; wearable sensors; SenseWear Armband; physical activity; weekday-to-weekend; energy expenditure

Citation: Rutkowski, S.; Buekers, J.; Rutkowska, A.; Cieślik, B.; Szczegielniak, J. Monitoring Physical Activity with a Wearable Sensor in Patients with COPD during In-Hospital Pulmonary Rehabilitation Program: A Pilot Study. *Sensors* **2021**, *21*, 2742. https://doi.org/10.3390/s21082742

Academic Editors: Maria de Fátima Fonseca Domingues, Ayman Radwan and Andrea Sciarrone

Received: 2 March 2021
Accepted: 11 April 2021
Published: 13 April 2021

1. Introduction

Chronic obstructive pulmonary disease (COPD) is a progressive disease that limits airflow through the respiratory tract. It is estimated that the disease affects 210 million people worldwide [1]. COPD is a leading cause of morbidity and mortality worldwide and will become the fourth leading cause of death by 2030. The Global Initiative for Chronic Obstructive Lung Disease (GOLD) defines COPD as a disease state characterized by airflow limitation, causing shortness of breath and significant systemic effects involving the lung and likewise causing extrapulmonary adverse reactions, with a high disease rate, high disability rate, high mortality rate, and a long course of disease [2]. The occurrence of pain in the cervical and thoracic spine region is very common, this probably leads to changes in the muscle tone [3]. COPD has also been shown to impair coordination and reduce balance and agility. In comparison to healthy people, patients with COPD demonstrate significant deficiencies in performing motor tasks, as well as in postural balance [4]. COPD is characterized not only by shortness of breath, dyspnea, chronic cough, and sputum production but also by fatigue and reductions in both physical capacity and physical activity [5,6]. A study by Theander and Unosson reported that patients with COPD perceived significantly greater functional limitations in cognitive, physical and psychosocial functioning due to fatigue compared to those in a control group [7]. The prevalence of the symptom is high; in a study concerning the severity of fatigue in patients with stable, moderate-to-severe COPD, it was shown that almost half of all patients experienced abnormal fatigue: 23% mild fatigue and 24% severe fatigue [8]. Fatigue affected even greater proportions of patients than either depression or anxiety [9]. The negative effect of fatigue on the patient's daily life is manifested in many aspects. Individuals indicated that physical limitations were mainly focused on walking and moving and performing homework, and personal hygiene was sometimes too physically demanding. All these symptoms cause a limitation of the level of physical activity, which in turn causes deterioration of physical health.

The level of physical activity in patients with COPD is, therefore, lower than that in healthy individuals with respect to age [10,11] and lower than that in individuals with other chronic conditions, including cardiovascular disease, diabetes [12], and rheumatoid arthritis [13]. Low levels of physical activity can already be observed in the early stages of the disease [14]. Furthermore, patients with COPD generally walk slower than healthy age-matched controls and are more sedentary [15,16]. The amount and duration of physical activity bouts to perform daily activities decreases with increasing disease severity [17]. Nevertheless, the importance of adequate physical activity levels in patients with COPD cannot be overestimated. A low physical activity level is a strong predictor of poor quality of life and high mortality [14,15]. Consequently, regular physical activity has been shown to reduce the risk of hospital admissions and mortality in patients with COPD [18]. It has also been shown that patients who decreased their activity level had an increased risk of mortality and showed faster disease progression [19]. A recent meta-analysis revealed that any level of physical activity or a reduction of sedentary time is associated with a lower risk of premature mortality in middle-aged and older adults [20].

The characteristic airflow limitation and associated dyspnea of patients with COPD can limit their daily physical activities. This can subsequently lead to physical deconditioning and a further decline in lung function, which can be the start of a deleterious vicious circle of deconditioning [21]. However, the reduced physical activity levels in patients with COPD are not determined by impaired respiratory function alone; other factors such as age, peripheral muscle weakness, hyperinflation, and dyspnea also affect physical activity levels [22]. Alternatively, dog walking and grandparenting have been associated with higher amounts and intensities of physical activity in patients with COPD [23].

All these elements highlight the importance of increasing physical activity levels in patients with COPD. One way of accomplishing this is through comprehensive pulmonary rehabilitation. Rehabilitation belongs to the essential management components in COPD and applied at an early stage of the disease, plays a very important role. A comprehensive rehabilitation program, beyond the physical training components, also includes patient education components on self-management. Patient awareness of current symptom level (either the COPD Assessment Test (CAT) or Modified Medical Research Council (mMRC) scores) and exacerbation frequency assessment have also been found to be very important. Due to the chronic nature of the disease, systematic physical activity, i.e., fitness training on a cycle ergometer or treadmill at a specific intensity, is a key approach to slow down disease progression. Many studies and systematic literature reviews show the beneficial effect of pulmonary rehabilitation in patients with chronic respiratory diseases on exercise capacity [24], lung function [25], respiratory muscle strength [26], and quality of life [27]. The adopted models of pulmonary rehabilitation vary in terms of intensity, duration, and the form of physical activity taken by the patients. Many authors have decided to assess the effect of home rehabilitation, while others have analyzed the impact of early rehabilitation on the hospitalization rate in the next months [28–30].

In recent years, increasing attention has been given to evaluating physical activity level as an outcome in patients with COPD [21]. Mantoani et al. carried out a systematic review of 60 intervention studies that evaluated physical activity as an outcome in patients with COPD [31]. The authors concluded that programs combined with coaching interventions and pulmonary rehabilitation programs lasting >12 weeks have the greatest potential to modify physical activity behaviors. Furthermore, it was observed that pulmonary rehabilitation programs do not lead to improved physical activity levels after completion of the program. The majority of patients were unable to maintain an active lifestyle after a rapid increase in exposure to planned supervised physical activity during the rehabilitation program. Thus, it seems that during rehabilitation programs, the focus is mainly on increasing functional exercise capacity and improving symptoms rather than on improving physical activity [6,32].

Besides the increased recognition of the health effects associated with physical (in)activity and the high prevalence of physical inactivity in patients with COPD, the development of technologies and devices that enable objective physical activity assessment in a patient-friendly manner also contributed to the increased interest in physical-activity-related research. Although subjective methods (such as questionnaires) have practical value, wearable accelerometers are likely to provide more accurate information about daily physical activity levels [33]. These devices provide reliable and objective measurements of the duration and intensity of physical activity [34,35]. A combination of subjective and objective methods has also been proposed to obtain a broader assessment of physical activity levels [36]. Wearable accelerometers have, thus, become a standard method of monitoring physical activity in everyday life by measuring acceleration in one, two, or three axes. Triaxial accelerometers have been increasingly used over the years, as they are considered superior to uniaxial accelerometers [37]. Wearable sensors providing user feedback have also been used as a treatment component in numerous physical activity counseling interventions [6]. Additionally, they have been used to assess energy expenditure during walking tests of patients with COPD, where their accuracy of assessment has been positively evaluated [38].

Despite the increased interest in physical-activity-related research, we found a scarcity of literature evaluating physical activity during supervised (weekdays) and non-supervised (weekend) days of a pulmonary rehabilitation program. Therefore, this study used a wearable sensor (SenseWear Armband) to assess physical activity levels during four consecutive days (Friday–Monday) of a 3 week, in-hospital, pulmonary rehabilitation program. We aimed to investigate whether patients have similar physical activity levels during supervised and non-supervised days of a stationary rehabilitation program and whether patients adhere to the rigor of 24 h monitoring. The second objective was to analyze the strengths and weaknesses of such kinds of sensors. We hypothesize that patients present lower physical activity levels during non-supervised days compared to supervised training days.

2. Materials and Methods

2.1. Participants

The study was conducted among patients who participated in pulmonary rehabilitation at the Specialist Hospital in Glucholazy (Poland). The research enrolled 15 randomly selected patients aged 50–80 years old who met the inclusion criteria. The inclusion criteria were COPD as the main diagnosis and written consent to participate in the study. The exclusion criteria were a main diagnosis other than COPD; pneumonia, tuberculosis, or another respiratory inflammatory disease in all stages and forms; condition after a heart attack; diabetes; state after thoracic and cardiac surgery; heart failure (stage III, IV New York Heart Association (NYHA)); advanced hypertension; diseases and injuries that can impair the function of the musculoskeletal system of transportation; disturbances of consciousness; and psychotic symptoms or other serious psychiatric disorders. The main group characteristics are presented in Table 1. The study adhered to the Declaration of Helsinki [39], and ethical approval was obtained from the Bioethics Committee of the

Opole Chamber of Physicians based on Resolution No. 199 of 07 February 2013, and the study was registered in ClinicalTrials.gov (NCT04726384).

Table 1. Group characteristic.

Variables	Mean (SD)
Age (years), mean (SD)	63.8 (9.1)
Female, n (%)	7 (54%)
Height (cm), mean (SD)	168 (9.3)
Weight (kg), mean (SD)	79.7 (15.6)
BMI (kg/m^2), mean (SD)	28.1 (4.3)
Smokers, n (%)	2 (15%)
FEV1 (%), means (SD)	78.2 (14.9)

FEV1: forced expiratory volume for 1 second, SD: standard deviation, BMI: body mass index.

2.2. Pulmonary Rehabilitation Program

Patients included in the study were qualified for in-patient, 3 week, high-intensity pulmonary rehabilitation 5 times a week (Monday–Friday, supervised days). During the weekend, patients were encouraged to go for walks and engage in minor physical activity on their own, but during this time they did not take advantage of the organized rehabilitation (non-supervised days). This program has been found to exhibit clinically meaningful improvements in exercise capacity, dyspnea, quality of life, and lung function in patients with COPD [40,41] or lung cancer [42,43]. All procedures were performed under the supervision of a specialist with an M.Phty. degree. The pulmonary rehabilitation program consisted of the following components performed once a day, each for 20–30 min (depending on the task):

- Endurance exercise training on a cycle ergometer to obtain a training heart rate (HR) which was calculated as follows: HR ((max HR − resting HR) × 60%) + resting HR through the use of the results of the 6 min walk test [44], or Borg-rated dyspnea or a fatigue score 4 to 6 (moderate to severe).
- Fitness exercises, coordination, balance exercises, and stretching exercises. Exercises were performed in the following positions: standing; on the knees; and lying on the side, abdomen, and back.
- Specific respiratory exercises: relaxation exercises for breathing muscles, strengthening exercises of the diaphragm with resistance, exercises to increase costal or chest breathing, prolonged exhalation exercise, and chest percussion.
- Inhalation with a 3% NaCl isotonic solution administered with an ultrasonic device.

The rehabilitation program was provided from 8 a.m. to 3 p.m. with a one hour lunch break between 12:30 and 1:30 p.m. During leisure time (after 3 p.m.), patients were encouraged to undertake any physical activity, however, without access to the rehabilitation unit and equipment.

2.3. Measurement

The SenseWear armband (Body Media Inc., Pittsburgh, PA, USA) was used to assess physical activity. The device allows for measuring physiological parameters and motion status by using built-in sensors, including the three-axis accelerometer for measuring the number of steps. Using algorithms developed by the producer, the device computes the level of energy expenditure defined in metabolic equivalents of task (METs) and calories during physical activity and rest periods, as well as the total energy expenditure. Additionally, the device counts the total time (min) during lying and during being active (measured when energy expenditure > 3 METs). The device has been considered a reliable source for assessing the physical activity level [45].

The group was informed of the purpose of the study and asked to wear the device 24 h a day for the next 4 days (Friday–Monday) excluding bath time, no more than 30 min [46] (Figure 1). Patients received the device on Thursday afternoon and returned it on Tuesday. Patients were also asked to indicate their subjective observations when returning the device at the end of the experiment. For this purpose, we did not use any standardized satisfaction scale; we wanted to explore the strengths and weaknesses of the patients' feelings.

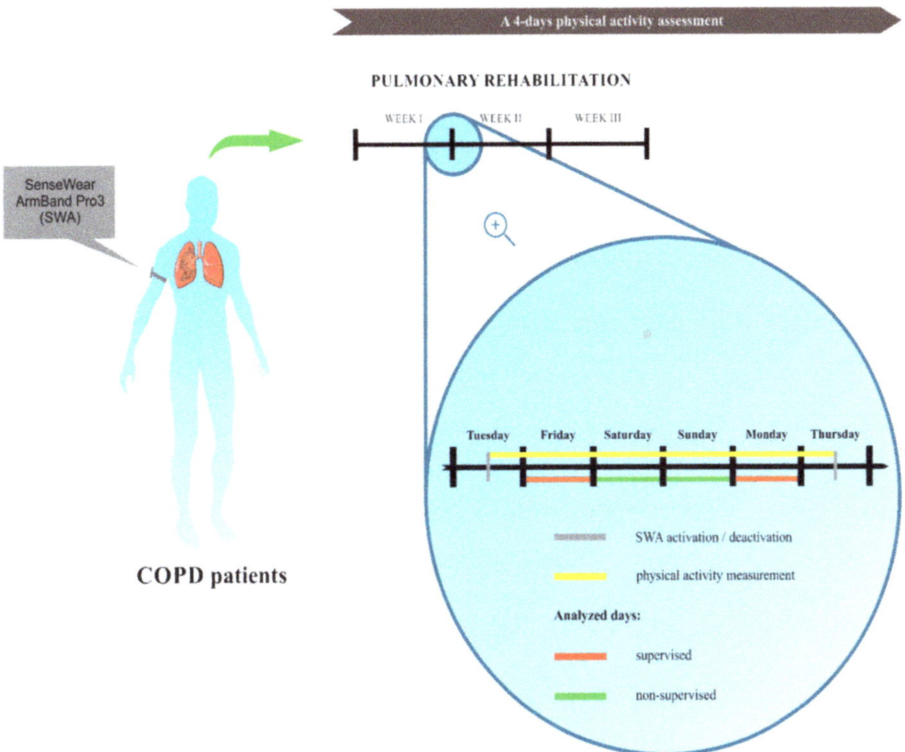

Figure 1. Study flow chart of physical activity assessment.

2.4. Statistical Analysis

The sample size was calculated based on the recommendation of the pilot study sample size in the medical field, according to Julious [47] and van Belle [48]; 12 participants were suggested. Considering a 20% drop-out rate, 15 patients were included in the study. Categorical variables were presented as numeric values and percentages, continuous variables as mean ± standard deviation (SD) or median and interquartile range [IQR], where appropriate, according to the Kolmogorov–Smirnov normality test. Differences in the energy expenditure between training days and off days were compared using Mann–Whitney U test or Student's paired *t*-test. Differences between consecutive days were assessed with Friedman's ANOVA. All statistical analyses were performed using Statistica 13 software (StatSoft, Cracow, Poland). The statistical significance level was set at $\alpha = 0.05$.

3. Results

The analyzed data were obtained from 13 patients; data from two patients were excluded due to failure to meet recommendations for wearing the armband for 95% of the day (armband off for no more than 30 min a day). We noted that both patients did not meet the requirements to wear the device during non-supervised days. In both cases, the armband was worn around 60% of the time. Results are presented as median [IQR] and mean (±SD).

The analysis of the number of steps per day, the time spent lying as well as undertaking moderate or vigorous physical activity (>3 METs), and the energy expenditure expressed in kcal showed no statistically significant difference between the supervised training days and the non-supervised days off (Table 2).

Table 2. Results of the study.

Variable		Training Days (n = 26)	Off Days (n = 26)	p
Steps (n)	Median [IQR] Mean (SD)	9153 [7744–12,524] 10,428 (3323)	8421 [5668–12,552] 9859 (5287)	0.57 *
Active time (min)	Median [IQR] Mean (SD)	110 [76–128] 112 (64)	119 [65–140] 117 (76)	0.75 *
Time lying (min)	Median [IQR] Mean (SD)	492 [444–526] 490 (63)	480 [449–547] 486 (76)	0.77 **
Energy expenditure (kcal)	Median [IQR] Mean (SD)	2616 [2089–2899] 2517 (455)	2491 [2011–2812] 2560 (646)	0.58 **

* According to Wilcoxon test, ** according to t-Student test. IQR: interquartile range.

The mean duration of physical activity > 3 METs was 112 min, which corresponds to the protocol of physical activity during supervised training days. Physical activity on non-supervised days must, therefore, have been generated by physical activities generating an energy expenditure greater than a leisurely walk.

Analysis of the results showed no statistically significant differences between the consecutive days of the study for all variables (Figure 2).

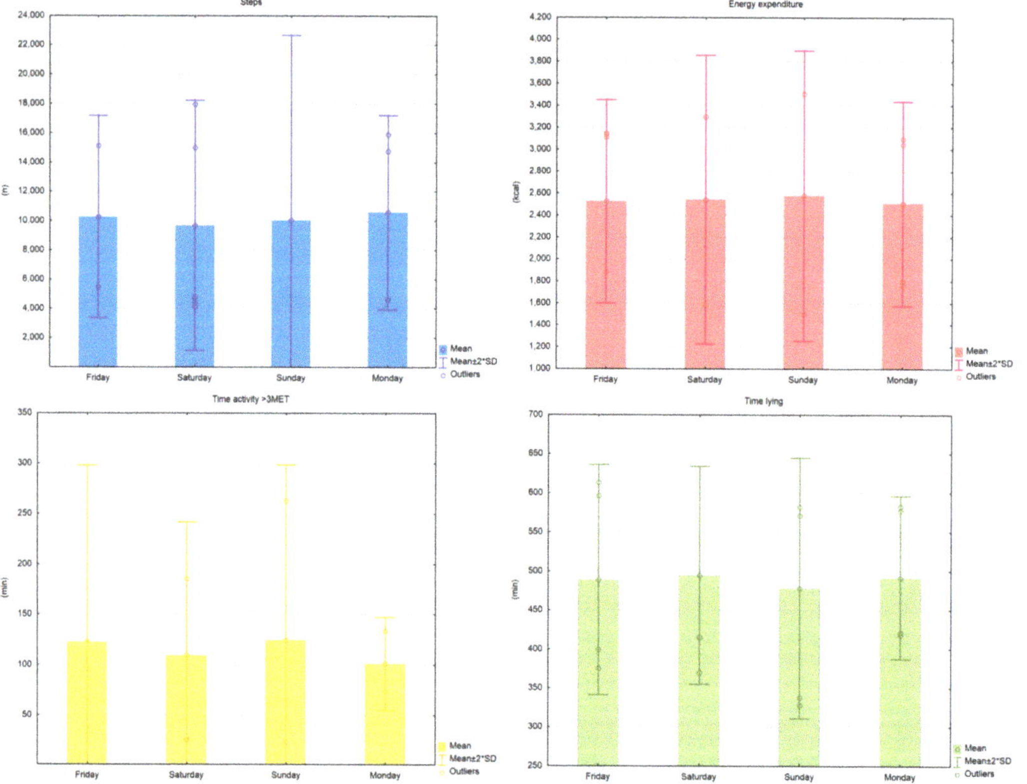

Figure 2. Examined parameters on consecutive days of the study.

4. Discussion

This study aimed to investigate the level of physical activity during four days of participation in the pulmonary rehabilitation program of patients with COPD and to compare non-supervised days (weekend) with supervised training days (weekdays). We hypothesized that the patients on non-supervised days will engage in less physical activity. The results showed no significant differences in physical activity levels between supervised and non-supervised days, expressed in energy expenditure (kcal), as well as time spent in moderate physical activity (>3 METs) or spent in a lying position. Thus, the results do not support the hypothesis. Moreover, there were no significant differences in the number of steps between supervised and non-supervised days. These results indicate similar levels of physical activity both on the weekdays and on the weekend. This type of control allows us to assess the involvement of people undergoing rehabilitation at a time when no one supervises them, which in turn is important in the context of the effectiveness of the entire treatment process. Thus, our results are of great clinical importance, it has been shown that modifications to patient behavior that enhance adherence to health-enhancing patient behavior and increase activity levels in everyday life [49] are key factors to maintaining the improved physical capacity achieved through participation in pulmonary rehabilitation. To our best knowledge, to date, this is the first study evaluating energy expenditure during two distinct activities: supervised activity during the pulmonary program and non-supervised days in patients with COPD during a 3 week, in-hospital, rehabilitation program.

Lahham et al. compared levels of physical activity during center- and home-based pulmonary rehabilitation in people with COPD using the SenseWear Armband device [50]. Differences in time spent in total physical activity (≥1.5 METs), time spent in moderate to vigorous–intensity physical activity (≥3 METs), and steps were compared. Home rehabilitation participants engaged in a mean of 310 (199–328) min per day of physical activity (29% moderate- to high-intensity physical activity) when compared to center-based rehabilitation participants who spent a mean of 300 (204–370) min per day (28% moderate- to high-intensity physical activity, p = 0.98). The daily number of steps did not differ between groups; home rehabilitation: 5232 [2067–7718], while for in-center rehabilitation, it was 4049 [1983–6040], p = 0.66). In our study, we noted a higher number of steps taken by patients. However, it is difficult to compare the time spent on physical activity because we assumed different levels of minimum energy expenditure, in our study ≥3 METs, while Lahham et al. [50] used ≥1.5 METs.

Ward et al. utilized a different type of activity monitor in their study, i.e., the Fitbit Zip. It was used in the study to measure the number of steps during a 6 week pulmonary rehabilitation intervention. The number of total steps taken per day between week 1 and week 6 of the intervention increased by 20% (week 1: 3565 [95% confidence interval (CI) 2779–4351] vs. week 6: 4447 [95% CI 3333–5561] steps/day, p = 0.036), whereas the number of steps taken during the recommended pulmonary rehabilitation exercise increased by 56% (week 1: 595 [95% CI 397–793] week 6: 927 [95% CI 599–1256] steps per day, p = 0.009) [51]. Geidl et al. analyzed a sample of 326 patients with COPD and their level of physical activity and time spent sitting during the 8 days before the pulmonary rehabilitation program using the ActiGraph wGT3X device [52]. The study group was divided into four subgroups based on time spent sitting and physical activity intensity. The daily step counts in that study ranged from 2749 (sedentary non-movers) to 5649 (sedentary occasional movers), to 7866 (sedentary movers), to 11,045 (sedentary exercisers). All four subgroups had a long sedentary daily routine (7.5–10.75 h). The mean age of the study group was 58 years, and most of the subjects were professionally active, most probably because of this, the daily step count results met the recommendations for patients with COPD who need to achieve >4580 steps per day [53] to avoid severe physical inactivity. The results show that patients with COPD have different levels of physical activity in free-living conditions. However, most patients with COPD spend a significant and unhealthy portion of their daily lives engaging in sedentary behavior.

A non-supervised method for stimulating patients with COPD to increase their physical activity levels in free-living conditions was presented by the Urban Training™ Study Group [23,54,55]. First, urban walking trails of different intensities and in different types of public spaces (e.g., beach or park) were designed and validated [54]. Afterward, a randomized controlled trial of 407 patients with COPD was performed, in which the intervention group was advised to walk on the developed urban trails but without any supervision. These patients furthermore received a pedometer and personalized calendar to monitor their physical activity, in combination with other behavioral strategies for increasing their physical activity levels (i.e., physical activity brochure, website, phone text messages, walking groups, and a phone number). This intervention was implemented for 12 months and proved to be efficacious in increasing physical activity levels, quantified by the amount of steps per day over the course of a week, in patients with COPD [55].

Based on the above-mentioned studies, the first attempts have already been made to assess the accurate estimation of physical activity levels of patients with COPD in either a supervised or non-supervised setting. However, another aspect seems to be the development of technology for this kind of study. Regarding the second objective of the study, i.e., the strengths and weaknesses of such kinds of sensors, the subjective acceptance by patients of such a monitoring system was noted. Patients indicated in their final reports that they were unaware of wearing the sensor, except when they over-tightened the device on the attachment strap after bathing. However, the authors noted a high frequency of returning dirty devices. In our opinion, this indicates that patients did not wash the devices although at the beginning of the study participants were informed about the possibility of washing with warm water the part of the sensors that are directly attached to the skin.

McNamara et al. evaluated the comfort of the SenseWear armband on a group of patients with COPD [35]. Results indicate that adverse effects may occur during the use of the device, most commonly in the form of skin itching, redness, and bruising. Moreover, 17% reported that the device was uncomfortable to wear at night, and 11% reported that it was uncomfortable to wear during the day. Despite this, compliance in wearing the SenseWear armband over 7 days was very high in this study (92%). Similarly, a one week observational study of patients with COPD reported no issues with using the SenseWear armband to provide contextual information about physical activity and sleep over the course of 7 days [56]. In a prospective study at three Northern European sites, the SenseWear armband was used to assess physical activity levels over 6 consecutive days in 134 patients with COPD and 46 controls. The authors defined a valid measurement period as a wearing time higher than 22 h per day, on at least 5 days. Excellent compliance with wearing the SenseWear armband was reported, with at least 94% of the patients in the three different sites having a valid measurement period [57].

An international team of investigators sought to validate six physical activity monitors in patients with COPD against a gold standard of indirect calorimetry in the form of oxygen uptake data from a portable metabolic system. The study used single-axis accelerometers Kenz Lifecorder Plus and Actiwatch, and triaxial accelerometers: RT3, ActiGraph GT3X, DynaPort® MiniMod, and SenseWear Armband. The study concluded that triaxial activity monitors were the best monitors to assess intensity physical activity for patients with COPD [58]. Patel et al. suggest that the SenseWear Pro armband may be a useful tool for assessing physical activity levels during therapeutic interventions [38]. Cavalheri et al. found it useful for assessing total energy expenditure during activities of daily living in patients with COPD [59]. Our observations support this conclusion. We noted 87% adherence to the study, where it was possible to obtain more than 95% of patient monitoring on 4 consecutive days. The individuals who were lost returned the device within the designated timeframe, but the device wear rate was below the accepted threshold. Visual assessment of the charts of these individuals indicated that the device was usually left in place for several hours, usually the evening hours (5–10 pm). In the authors' speculations, it seems possible that these actions were intentional, as there were pieces of information to hospital staff that patients attended "informal" evening meetings. As an alternative to the

SenseWear armband, the Polar A300™ can be worn as a wrist device similar to a watch. Boeselt et al. compared the two devices in regards to the number of steps, burned calories, daily activity time, and metabolic equivalents in patients with COPD over 3 days of daily life [60]. Data analysis over 3 days showed that 90% of the steps (95% CI over/under the means between Polar A300™ and SWA −4223–1887), 100% of the calories (95% CI −2798–1887), 90% of the daily activity data (95% CI −12.32–4065), and 95% of the MET (95% CI −3.11–2.75) were within the limits of agreement. The A300™ device is no worse at assessing physical activity time, step count, and calorie consumption in patients with COPD compared to SenseWear Armband.

Technological advances have, furthermore, allowed the combination of measurements of physical activity with other (physiological) measurements, such as heart rate. Joosen et al. implemented such a mobile health system, consisting of a smartphone and heart rate monitor, in a care home setting for 10 weeks [61]. Triaxial accelerometry data from the smartphone were converted into interpretable activity (e.g., steps per hour, time walking, walking distance) and stride (e.g., stride duration, stride speed, stride displacement) features, while heart rate measurements were converted into interpretable heart rate features (e.g., median heart rate, minimal heart rate, time constant of heart rate increase). Participants received weekly feedback about their activity and heart rate features. The implementation of this mobile health system was associated with increased physical activity levels during the first 5 weeks of the study, after which physical activity levels starting declining again. In addition, the calculated features were converted into a fitness score, which could predict the outcome of more labor-intensive exercise tests.

More recently, the combination of physical activity and heart rate measurements has been used to address the current COVID-19 pandemic. Quer et al. were able to discriminate between COVID-19 symptomatic positive and negative cases (area under the curve of 0.80) by combining self-reported symptoms with measurements of physical activity, sleep and heart rate [62]. Natarajan et al. obtained an area under the curve of 0.77 for the prediction of illness on a specific day, based on measurements with a Fitbit for that day and the preceding 4 days [63]. Mishra et al. observed that 26 out of 32 individuals who were infected with COVID-19 had alterations in their daily steps, time asleep, or heart rate [64]. These studies show that measurements with wearable sensors could be used for the early detection of COVID-19.

To the best of our knowledge, our study is the first to explore the weekday-to-weekend physical activity level among patients with COPD during in-hospital pulmonary rehabilitation. Although this study provides encouraging results, we recognize that some limitations should be considered. Firstly, the research included a small study group. Secondly, the number of observation days could be extended. Investigating only one weekend may introduce bias in the results since physical activity may have been influenced by, for example, good weather conditions. Finally, energy expenditure was assessed using a commercial activity monitor and stimulated estimation of energy expenditure using machine learning on multimodal data. To accurately measure the energy expenditure, there are methods such as doubly labeled water and direct and indirect calorimetry, but their cost and practical limitations make them suitable only for stationary research and professional sports.

5. Conclusions

Interest in objective measures of physical activity in patients with COPD due to the close relationship between physical activity levels and exercise tolerance, disease symptoms, disability incidence, and mortality continues to rise. Therefore, it seems beneficial to use available physical activity monitors in patients with COPD, as measurable parameters provide feedback that may increase the patient's motivation to be active to achieve health benefits. Portable, lightweight, skin sensors mounted on the arm or wrist appear to provide adequate comfort and meaningful measurements to monitor and modify patient behavior to enhance adherence to health-enhancing patient behavior and increase activity level in everyday life.

Author Contributions: Conceptualization, S.R.; methodology, S.R.; formal analysis, S.R., J.B., and B.C.; investigation, A.R.; resources, J.S.; writing—original draft preparation, S.R.; writing—review and editing, S.R., J.B., A.R., B.C., and J.S.; supervision, J.S.; project administration, S.R. All authors have read and agreed to the published version of the manuscript.

Funding: This research received no external funding.

Institutional Review Board Statement: The study was conducted according to the guidelines of the Declaration of Helsinki and approved by Bioethics Committee of the Opole Chamber of Physicians on the basis of Resolution No. 199 of 07 February 2013.

Informed Consent Statement: Informed consent was obtained from all subjects involved in the study.

Data Availability Statement: The data presented in this study are available on request from the corresponding author.

Acknowledgments: J.B. reports grants from the European Respiratory Society (ERS Long-Term Research Fellowship 2020) and other support from the Spanish Ministry of Science and Innovation through the "Centro de Excelencia Severo Ochoa 2019–2023" Program (CEX2018-000806-S) and the Generalitat de Catalunya through the CERCA program, during the conduct of the study.

Conflicts of Interest: The authors declare no conflict of interest.

References

1. Alfarroba, S.; Rodrigues, F.; Papoila, A.L.; Santos, A.F.; Morais, L. Pulmonary Rehabilitation in COPD According to Global Initiative for Chronic Obstructive Lung Disease Categories. *Respir. Care* **2016**, *61*, 1331–1340. [CrossRef]
2. Jobst, B.J.; Owsijewitsch, M.; Kauczor, H.U.; Biederer, J.; Ley, S.; Becker, N.; Kopp-Schneider, A.; Delorme, S.; Heussel, C.P.; Puderbach, M.; et al. GOLD stage predicts thoracic aortic calcifications in patients with COPD. *Exp. Ther. Med.* **2019**, *17*, 967–973. [CrossRef]
3. Bentsen, S.B.; Rustoen, T.; Miaskowski, C. Prevalence and characteristics of pain in patients with chronic obstructive pulmonary disease compared to the Norwegian general population. *J. Pain* **2011**, *12*, 539–545. [CrossRef] [PubMed]
4. Butcher, S.J.; Meshke, J.M.; Sheppard, M.S. Reductions in functional balance, coordination, and mobility measures among patients with stable chronic obstructive pulmonary disease. *J. Cardiopulm. Rehabil.* **2004**, *24*, 274–280. [CrossRef] [PubMed]
5. Al-Shair, K.; Kolsum, U.; Singh, D.; Vestbo, J. The Effect of Fatigue and Fatigue Intensity on Exercise Tolerance in Moderate COPD. *Lung* **2016**, *194*, 889–895. [CrossRef]
6. Langer, D.; Demeyer, H. Interventions to modify physical activity in patients with COPD: Where do we go from here? *Eur. Respir. J.* **2016**, *48*, 14–17. [CrossRef] [PubMed]
7. Theander, K.; Unosson, M. Fatigue in patients with chronic obstructive pulmonary disease. *J. Adv. Nurs.* **2004**, *45*, 172–177. [CrossRef]
8. Peters, J.B.; Heijdra, Y.F.; Daudey, L.; Boer, L.M.; Molema, J.; Dekhuijzen, P.N.; Schermer, T.R.; Vercoulen, J.H. Course of normal and abnormal fatigue in patients with chronic obstructive pulmonary disease, and its relationship with domains of health status. *Patient Educ. Couns.* **2011**, *85*, 281–285. [CrossRef]
9. Wong, C.J.; Goodridge, D.; Marciniuk, D.D.; Rennie, D. Fatigue in patients with COPD participating in a pulmonary rehabilitation program. *Int. J. Chron. Obstruct Pulmon. Dis.* **2010**, *5*, 319–326. [CrossRef]
10. Troosters, T.; Sciurba, F.; Battaglia, S.; Langer, D.; Valluri, S.R.; Martino, L.; Benzo, R.; Andre, D.; Weisman, I.; Decramer, M. Physical inactivity in patients with COPD, a controlled multi-center pilot-study. *Respir. Med.* **2010**, *104*, 1005–1011. [CrossRef]
11. Watz, H.; Waschki, B.; Meyer, T.; Magnussen, H. Physical activity in patients with COPD. *Eur. Respir. J.* **2009**, *33*, 262–272. [CrossRef]
12. Tudor-Locke, C.; Hart, T.L.; Washington, T.L. Expected values for pedometer-determined physical activity in older populations. *Int. J. Behav. Nutr. Phys. Act.* **2009**, *6*, 59. [CrossRef]
13. Arne, M.; Janson, C.; Janson, S.; Boman, G.; Lindqvist, U.; Berne, C.; Emtner, M. Physical activity and quality of life in subjects with chronic disease: Chronic obstructive pulmonary disease compared with rheumatoid arthritis and diabetes mellitus. *Scand. J. Prim. Health Care* **2009**, *27*, 141–147. [CrossRef] [PubMed]
14. van Helvoort, H.A.; Willems, L.M.; Dekhuijzen, P.R.; van Hees, H.W.; Heijdra, Y.F. Respiratory constraints during activities in daily life and the impact on health status in patients with early-stage COPD: A cross-sectional study. *NPJ Prim. Care Respir. Med.* **2016**, *26*, 16054. [CrossRef] [PubMed]
15. Pitta, F.; Troosters, T.; Spruit, M.A.; Probst, V.S.; Decramer, M.; Gosselink, R. Characteristics of physical activities in daily life in chronic obstructive pulmonary disease. *Am. J. Respir. Crit. Care Med.* **2005**, *171*, 972–977. [CrossRef] [PubMed]
16. Park, S.K.; Richardson, C.R.; Holleman, R.G.; Larson, J.L. Physical activity in people with COPD, using the National Health and Nutrition Evaluation Survey dataset (2003–2006). *Heart Lung* **2013**, *42*, 235–240. [CrossRef]

17. Donaire-Gonzalez, D.; Gimeno-Santos, E.; Balcells, E.; Rodriguez, D.A.; Farrero, E.; de Batlle, J.; Benet, M.; Ferrer, A.; Barbera, J.A.; Gea, J.; et al. Physical activity in COPD patients: Patterns and bouts. *Eur. Respir. J.* **2013**, *42*, 993–1002. [CrossRef] [PubMed]
18. Garcia-Aymerich, J.; Lange, P.; Benet, M.; Schnohr, P.; Anto, J.M. Regular physical activity reduces hospital admission and mortality in chronic obstructive pulmonary disease: A population based cohort study. *Thorax* **2006**, *61*, 772–778. [CrossRef] [PubMed]
19. Vaes, A.W.; Garcia-Aymerich, J.; Marott, J.L.; Benet, M.; Groenen, M.T.; Schnohr, P.; Franssen, F.M.; Vestbo, J.; Wouters, E.F.; Lange, P.; et al. Changes in physical activity and all-cause mortality in COPD. *Eur. Respir. J.* **2014**, *44*, 1199–1209. [CrossRef] [PubMed]
20. Ekelund, U.; Tarp, J.; Steene-Johannessen, J.; Hansen, B.H.; Jefferis, B.; Fagerland, M.W.; Whincup, P.; Diaz, K.M.; Hooker, S.P.; Chernofsky, A.; et al. Dose-response associations between accelerometry measured physical activity and sedentary time and all cause mortality: Systematic review and harmonised meta-analysis. *BMJ* **2019**, *366*, l4570. [CrossRef]
21. Watz, H.; Pitta, F.; Rochester, C.L.; Garcia-Aymerich, J.; ZuWallack, R.; Troosters, T.; Vaes, A.W.; Puhan, M.A.; Jehn, M.; Polkey, M.I.; et al. An official European Respiratory Society statement on physical activity in COPD. *Eur. Respir. J.* **2014**, *44*, 1521–1537. [CrossRef]
22. Gimeno-Santos, E.; Frei, A.; Steurer-Stey, C.; de Batlle, J.; Rabinovich, R.A.; Raste, Y.; Hopkinson, N.S.; Polkey, M.I.; van Remoortel, H.; Troosters, T.; et al. Determinants and outcomes of physical activity in patients with COPD: A systematic review. *Thorax* **2014**, *69*, 731–739. [CrossRef]
23. Arbillaga-Etxarri, A.; Gimeno-Santos, E.; Barberan-Garcia, A.; Benet, M.; Borrell, E.; Dadvand, P.; Foraster, M.; Marin, A.; Monteagudo, M.; Rodriguez-Roisin, R.; et al. Socio-environmental correlates of physical activity in patients with chronic obstructive pulmonary disease (COPD). *Thorax* **2017**, *72*, 796–802. [CrossRef] [PubMed]
24. Li, W.; Pu, Y.; Meng, A.; Zhi, X.; Xu, G. Effectiveness of pulmonary rehabilitation in elderly patients with COPD: A systematic review and meta-analysis of randomized controlled trials. *Int. J. Nurs. Pract.* **2019**, *25*, e12745. [CrossRef]
25. Salcedo, P.A.; Lindheimer, J.B.; Klein-Adams, J.C.; Sotolongo, A.M.; Falvo, M.J. Effects of Exercise Training on Pulmonary Function in Adults with Chronic Lung Disease: A Meta-Analysis of Randomized Controlled Trials. *Arch. Phys. Med. Rehabil.* **2018**, *99*, 2561–2569.e2567. [CrossRef]
26. Neves, L.F.; Reis, M.H.; Plentz, R.D.; Matte, D.L.; Coronel, C.C.; Sbruzzi, G. Expiratory and expiratory plus inspiratory muscle training improves respiratory muscle strength in subjects with COPD: Systematic review. *Respir. Care* **2014**, *59*, 1381–1388. [CrossRef]
27. Liao, W.H.; Chen, J.W.; Chen, X.; Lin, L.; Yan, H.Y.; Zhou, Y.Q.; Chen, R. Impact of Resistance Training in Subjects With COPD: A Systematic Review and Meta-Analysis. *Respir. Care* **2015**, *60*, 1130–1145. [CrossRef]
28. Ko, F.W.; Dai, D.L.; Ngai, J.; Tung, A.; Ng, S.; Lai, K.; Fong, R.; Lau, H.; Tam, W.; Hui, D.S. Effect of early pulmonary rehabilitation on health care utilization and health status in patients hospitalized with acute exacerbations of COPD. *Respirology* **2011**, *16*, 617–624. [CrossRef]
29. Revitt, O.; Sewell, L.; Morgan, M.D.; Steiner, M.; Singh, S. Short outpatient pulmonary rehabilitation programme reduces readmission following a hospitalization for an exacerbation of chronic obstructive pulmonary disease. *Respirology* **2013**, *18*, 1063–1068. [CrossRef]
30. Jacome, C.; Marques, A. Short- and Long-term Effects of Pulmonary Rehabilitation in Patients with Mild COPD: A Comparison with Patients With Moderate to Severe Copd. *J. Cardiopulm. Rehabil. Prev.* **2016**, *36*, 445–453. [CrossRef]
31. Mantoani, L.C.; Rubio, N.; McKinstry, B.; MacNee, W.; Rabinovich, R.A. Interventions to modify physical activity in patients with COPD: A systematic review. *Eur. Respir. J.* **2016**, *48*, 69–81. [CrossRef]
32. Troosters, T.; van der Molen, T.; Polkey, M.; Rabinovich, R.A.; Vogiatzis, I.; Weisman, I.; Kulich, K. Improving physical activity in COPD: Towards a new paradigm. *Respir. Res.* **2013**, *14*, 115. [CrossRef]
33. Pitta, F.; Troosters, T.; Probst, V.S.; Spruit, M.A.; Decramer, M.; Gosselink, R. Quantifying physical activity in daily life with questionnaires and motion sensors in COPD. *Eur. Respir. J.* **2006**, *27*, 1040–1055. [CrossRef]
34. Fruin, M.L.; Rankin, J.W. Validity of a multi-sensor armband in estimating rest and exercise energy expenditure. *Med. Sci. Sports Exerc.* **2004**, *36*, 1063–1069. [CrossRef]
35. McNamara, R.J.; Tsai, L.L.; Wootton, S.L.; Ng, L.W.; Dale, M.T.; McKeough, Z.J.; Alison, J.A. Measurement of daily physical activity using the SenseWear Armband: Compliance, comfort, adverse side effects and usability. *Chron. Respir. Dis.* **2016**, *13*, 144–154. [CrossRef]
36. Laeremans, M.; Dons, E.; Avila-Palencia, I.; Carrasco-Turigas, G.; Orjuela, J.P.; Anaya, E.; Brand, C.; Cole-Hunter, T.; de Nazelle, A.; Gotschi, T.; et al. Physical activity and sedentary behaviour in daily life: A comparative analysis of the Global Physical Activity Questionnaire (GPAQ) and the SenseWear armband. *PLoS ONE* **2017**, *12*, e0177765. [CrossRef]
37. Plasqui, G.; Joosen, A.M.; Kester, A.D.; Goris, A.H.; Westerterp, K.R. Measuring free-living energy expenditure and physical activity with triaxial accelerometry. *Obes. Res.* **2005**, *13*, 1363–1369. [CrossRef]
38. Patel, S.A.; Benzo, R.P.; Slivka, W.A.; Sciurba, F.C. Activity monitoring and energy expenditure in COPD patients: A validation study. *COPD* **2007**, *4*, 107–112. [CrossRef]
39. World Medical, A. World Medical Association Declaration of Helsinki: Ethical principles for medical research involving human subjects. *JAMA* **2013**, *310*, 2191–2194. [CrossRef]

40. Greulich, T.; Koczulla, A.R.; Nell, C.; Kehr, K.; Vogelmeier, C.F.; Stojanovic, D.; Wittmann, M.; Schultz, K. Effect of a Three-Week Inpatient Rehabilitation Program on 544 Consecutive Patients with Very Severe COPD: A Retrospective Analysis. *Respiration* **2015**, *90*, 287–292. [CrossRef]
41. Rutkowski, S.; Rutkowska, A.; Jastrzebski, D.; Racheniuk, H.; Pawelczyk, W.; Szczegielniak, J. Effect of Virtual Reality-Based Rehabilitation on Physical Fitness in Patients with Chronic Obstructive Pulmonary Disease. *J. Hum. Kinet* **2019**, *69*, 149–157. [CrossRef]
42. Rutkowska, A.; Jastrzebski, D.; Rutkowski, S.; Zebrowska, A.; Stanula, A.; Szczegielniak, J.; Ziora, D.; Casaburi, R. Exercise Training in Patients With Non-Small Cell Lung Cancer During In-Hospital Chemotherapy Treatment: A RANDOMIZED CONTROLLED TRIAL. *J. Cardiopulm. Rehabil. Prev.* **2019**, *39*, 127–133. [CrossRef] [PubMed]
43. Jastrzebski, D.; Zebrowska, A.; Rutkowski, S.; Rutkowska, A.; Warzecha, J.; Ziaja, B.; Palka, A.; Czyzewska, B.; Czyzewski, D.; Ziora, D. Pulmonary Rehabilitation with a Stabilometric Platform after Thoracic Surgery: A Preliminary Report. *J. Hum. Kinet* **2018**, *65*, 79–87. [CrossRef] [PubMed]
44. Szczegielniak, J.; Latawiec, K.J.; Luniewski, J.; Stanislawski, R.; Bogacz, K.; Krajczy, M.; Rydel, M. A study on nonlinear estimation of submaximal effort tolerance based on the generalized MET concept and the 6MWT in pulmonary rehabilitation. *PLoS ONE* **2018**, *13*, e0191875. [CrossRef] [PubMed]
45. Johannsen, D.L.; Calabro, M.A.; Stewart, J.; Franke, W.; Rood, J.C.; Welk, G.J. Accuracy of armband monitors for measuring daily energy expenditure in healthy adults. *Med. Sci. Sports Exerc.* **2010**, *42*, 2134–2140. [CrossRef] [PubMed]
46. Demeyer, H.; Burtin, C.; Van Remoortel, H.; Hornikx, M.; Langer, D.; Decramer, M.; Gosselink, R.; Janssens, W.; Troosters, T. Standardizing the analysis of physical activity in patients with COPD following a pulmonary rehabilitation program. *Chest* **2014**, *146*, 318–327. [CrossRef] [PubMed]
47. Julious, S.A. Sample size of 12 per group rule of thumb for a pilot study. *Pharm. Stat.* **2005**, *4*, 287–291. [CrossRef]
48. Van Belle, G. Statistical Rules of Thumb. 2008. Willey. Available online: https://www.wiley.com/en-us/Statistical+Rules+of-Thumb%2C+2nd+Edition-p-9780470144480 (accessed on 26 February 2021).
49. Cruz, J.; Brooks, D.; Marques, A. Walk2Bactive: A randomised controlled trial of a physical activity-focused behavioural intervention beyond pulmonary rehabilitation in chronic obstructive pulmonary disease. *Chron. Respir. Dis.* **2016**, *13*, 57–66. [CrossRef] [PubMed]
50. Lahham, A.; McDonald, C.F.; Mahal, A.; Lee, A.L.; Hill, C.J.; Burge, A.T.; Cox, N.S.; Moore, R.; Nicolson, C.; O'Halloran, P.; et al. Participation in Physical Activity During Center and Home-Based Pulmonary Rehabilitation for People With COPD: A Secondary Analysis of a Randomized Controlled Trial. *J. Cardiopulm. Rehabil. Prev.* **2019**, *39*, E1–E4. [CrossRef]
51. Ward, S.; Orme, M.; Zatloukal, J.; Singh, S. Adherence to walking exercise prescription during pulmonary rehabilitation in COPD with a commercial activity monitor: A feasibility trial. *BMC Pulm. Med.* **2021**, *21*, 30. [CrossRef]
52. Geidl, W.; Carl, J.; Cassar, S.; Lehbert, N.; Mino, E.; Wittmann, M.; Wagner, R.; Schultz, K.; Pfeifer, K. Physical Activity and Sedentary Behaviour Patterns in 326 Persons with COPD before Starting a Pulmonary Rehabilitation: A Cluster Analysis. *J. Clin. Med.* **2019**, *8*, 1346. [CrossRef]
53. Depew, Z.S.; Novotny, P.J.; Benzo, R.P. How many steps are enough to avoid severe physical inactivity in patients with chronic obstructive pulmonary disease? *Respirology* **2012**, *17*, 1026–1027. [CrossRef]
54. Arbillaga-Etxarri, A.; Torrent-Pallicer, J.; Gimeno-Santos, E.; Barberan-Garcia, A.; Delgado, A.; Balcells, E.; Rodriguez, D.A.; Vilaro, J.; Vall-Casas, P.; Irurtia, A.; et al. Validation of Walking Trails for the Urban Training of Chronic Obstructive Pulmonary Disease Patients. *PLoS ONE* **2016**, *11*, e0146705. [CrossRef]
55. Arbillaga-Etxarri, A.; Gimeno-Santos, E.; Barberan-Garcia, A.; Balcells, E.; Benet, M.; Borrell, E.; Celorrio, N.; Delgado, A.; Jane, C.; Marin, A.; et al. Long-term efficacy and effectiveness of a behavioural and community-based exercise intervention (Urban Training) to increase physical activity in patients with COPD: A randomised controlled trial. *Eur. Respir. J.* **2018**, *52*, 1800063. [CrossRef]
56. Buekers, J.; Theunis, J.; De Boever, P.; Vaes, A.W.; Koopman, M.; Janssen, E.V.; Wouters, E.F.; Spruit, M.A.; Aerts, J.M. Wearable Finger Pulse Oximetry for Continuous Oxygen Saturation Measurements During Daily Home Routines of Patients With Chronic Obstructive Pulmonary Disease (COPD) Over One Week: Observational Study. *JMIR Mhealth Uhealth* **2019**, *7*, e12866. [CrossRef] [PubMed]
57. Waschki, B.; Spruit, M.A.; Watz, H.; Albert, P.S.; Shrikrishna, D.; Groenen, M.; Smith, C.; Man, W.D.; Tal-Singer, R.; Edwards, L.D.; et al. Physical activity monitoring in COPD: Compliance and associations with clinical characteristics in a multicenter study. *Respir. Med.* **2012**, *106*, 522–530. [CrossRef] [PubMed]
58. Van Remoortel, H.; Raste, Y.; Louvaris, Z.; Giavedoni, S.; Burtin, C.; Langer, D.; Wilson, F.; Rabinovich, R.; Vogiatzis, I.; Hopkinson, N.S.; et al. Validity of six activity monitors in chronic obstructive pulmonary disease: A comparison with indirect calorimetry. *PLoS ONE* **2012**, *7*, e39198. [CrossRef]
59. Cavalheri, V.; Donaria, L.; Ferreira, T.; Finatti, M.; Camillo, C.A.; Cipulo Ramos, E.M.; Pitta, F. Energy expenditure during daily activities as measured by two motion sensors in patients with COPD. *Respir. Med.* **2011**, *105*, 922–929. [CrossRef] [PubMed]
60. Boeselt, T.; Spielmanns, M.; Nell, C.; Storre, J.H.; Windisch, W.; Magerhans, L.; Beutel, B.; Kenn, K.; Greulich, T.; Alter, P.; et al. Validity and Usability of Physical Activity Monitoring in Patients with Chronic Obstructive Pulmonary Disease (COPD). *PLoS ONE* **2016**, *11*, e0157229. [CrossRef]

1. Joosen, P.; Piette, D.; Buekers, J.; Taelman, J.; Berckmans, D.; De Boever, P. A smartphone-based solution to monitor daily physical activity in a care home. *J. Telemed. Telecare* **2019**, *25*, 611–622. [CrossRef] [PubMed]
2. Quer, G.; Radin, J.M.; Gadaleta, M.; Baca-Motes, K.; Ariniello, L.; Ramos, E.; Kheterpal, V.; Topol, E.J.; Steinhubl, S.R. Wearable sensor data and self-reported symptoms for COVID-19 detection. *Nat. Med.* **2021**, *27*, 73–77. [CrossRef]
3. Natarajan, A.; Su, H.W.; Heneghan, C. Assessment of physiological signs associated with COVID-19 measured using wearable devices. *NPJ Digit. Med.* **2020**, *3*, 156. [CrossRef]
4. Mishra, T.; Wang, M.; Metwally, A.A.; Bogu, G.K.; Brooks, A.W.; Bahmani, A.; Alavi, A.; Celli, A.; Higgs, E.; Dagan-Rosenfeld, O.; et al. Pre-symptomatic detection of COVID-19 from smartwatch data. *Nat. Biomed. Eng.* **2020**, *4*, 1208–1220. [CrossRef]

 sensors

Article

HRV Features as Viable Physiological Markers for Stress Detection Using Wearable Devices

Kayisan M. Dalmeida and Giovanni L. Masala *

Department of Computing and Mathematics, Manchester Metropolitan University, Manchester M15 6BH, UK; kayisan.m.dalmeida@stu.mmu.ac.uk
* Correspondence: g.masala@mmu.ac.uk; Tel.: +44-(0)161-247-1407

Abstract: Stress has been identified as one of the major causes of automobile crashes which then lead to high rates of fatalities and injuries each year. Stress can be measured via physiological measurements and in this study the focus will be based on the features that can be extracted by common wearable devices. Hence, the study will be mainly focusing on heart rate variability (HRV). This study is aimed at investigating the role of HRV-derived features as stress markers. This is achieved by developing a good predictive model that can accurately classify stress levels from ECG-derived HRV features, obtained from automobile drivers, by testing different machine learning methodologies such as K-Nearest Neighbor (KNN), Support Vector Machines (SVM), Multilayer Perceptron (MLP), Random Forest (RF) and Gradient Boosting (GB). Moreover, the models obtained with highest predictive power will be used as reference for the development of a machine learning model that would be used to classify stress from HRV features derived from heart rate measurements obtained from wearable devices. We demonstrate that HRV features constitute good markers for stress detection as the best machine learning model developed achieved a Recall of 80%. Furthermore, this study indicates that HRV metrics such as the Average of normal-to-normal (NN) intervals (AVNN), Standard deviation of the average NN intervals (SDNN) and the Root mean square differences of successive NN intervals (RMSSD) were important features for stress detection. The proposed method can be also used on all applications in which is important to monitor the stress levels in a non-invasive manner, e.g., in physical rehabilitation, anxiety relief or mental wellbeing.

Keywords: stress; wearable device; machine learning; smart watch; heart rate variability; electrocardiogram

Citation: Dalmeida, K.M.; Masala, G.L. HRV Features as Viable Physiological Markers for Stress Detection Using Wearable Devices. *Sensors* **2021**, *21*, 2873. https://doi.org/10.3390/s21082873

Academic Editor: Maria de Fátima Domingues

Received: 21 March 2021
Accepted: 14 April 2021
Published: 19 April 2021

Publisher's Note: MDPI stays neutral with regard to jurisdictional claims in published maps and institutional affiliations.

1. Introduction

Stress can be defined as a biological and psychological response to a combination of external or internal stressors [1,2], which could be a chemical or biological agent or an environmental stimulus that causes stress to an organism [3]. Stress is, in essential, the body's coping mechanism to any kind of foreign demand or threat. At the molecular level, in a stressful situation the Sympathetic Nervous System (SNS) produces stress hormones, such as cortisol, which then, via a cascade of events, lead to the increase of available sources of energy [4]. This large amount of energy is used to fuel a series of physiological mechanisms such as: increasing the metabolic rate, increasing heart rate and causing the dilation of blood vessels in the heart and other muscles [5], while decreasing non-essential tasks such as immune system and digestion. Once stressors no longer impose a threat to the body, the brain fires up the Parasympathetic Nervous System (PSN) which is in charge of restoring the body to homeostasis. However, if the PSN fails to achieve homeostasis, this could lead to chronic stress; thus, causing a continual and prolonged activation of the stress response [6]. Conversely, during acute stress, the stress response develops immediately, and it is short-lived.

Studies carried out in this field suggest that stress can lead to abnormalities in the cardiac rhythm, and this could lead to arrythmia [7]. Additionally, stress does not only have physical implications, but it can also be detrimental to one's mental health; in fact chronic stress can enhance the chances of developing depression. For these reasons, it is important to develop a system that can detect and measure stress in an individual in a non-invasive manner in such way that stress can be regulated or relieved via personalised medical interventions or even by just alerting the user of their stressful state.

Furthermore, stress has been identified as one of the major causes of automobile crashes which then lead to high rates of fatalities and injuries each year [8]. As reported by Virginia Tech Transportation Institute (VTTI) and the National Highway Traffic Safety Administration (NHTSA), lack of attention and stress were the leading cause of traffic accidents in the US, with a rate of ~80%. Therefore, being able to accurately monitor stress in drivers could significantly reduce the amount of road traffic accidents and consequently increase public road safety.

Given that stress is regulated by the Autonomous Nervous System, it can be measured via physiological measurements such as Electrocardiogram (ECG), Galvanic Skin Response (GSR), electromyogram (EMG), heart rate variability (HRV), heart rate (HR), blood pressure breathe frequency, Respiration Rate and Temperature [9]. These are considered to be an accurate methodology for bio signal recording as they cannot be masked or conditioned by human voluntary actions. However, this study will be mainly focusing on HRV, which is controlled by PSN and SNS; therefore, an imbalance in any functions regulated by these two nervous system branches will affect HRV [10]. HRV is the variation in interval between successive normal RR (or NN) intervals [11]; it is derived from an ECG reading and it is measured by calculating the time interval between two consecutive peaks of the heartbeats [12]. As explained in [11] the RR intervals are obtained by calculating the difference between two R waves in the QRS complex.

HRV can be subdivided into time domain and frequency domain metrics as described in Table 1.

Table 1. Time and Frequency Metrics derived from Heart Rate Variability.

	Time Domain Metrics
SDNN	Standard deviation of all NN intervals
SDANN	Standard deviation of the average NN intervals
AVNN	Average of NN intervals
RMSSD	Square root of the mean squared differences of successive RR intervals
pNN50	Percentage differences of successive RR intervals larger than 50 ms
	Frequency Domain Metrics
TP	Total Power—total spectral power of all NN intervals up to 0.004 Hz
LF	Low Frequency—total spectral power of all NN intervals with frequency ranging from 0.04 Hz to 0.15 Hz
HF	High Frequency—total spectral power of all NN intervals with frequency ranging from 0.15 Hz to 0.4 Hz
VLF	Very Low Frequency—total spectral power of all NN intervals with frequencies >0.004 Hz
ULF	Ultra-Low Frequency—total spectral power of all NN intervals with frequencies <0.003 Hz
LF/HF	Ratio of low to high frequency

HRV is traditionally obtained from ECG and requires the use of computational software for calculation; this is a process is limited to laboratory or clinical settings and requires a certain degree of technical knowledge for interpretation and calculation. Thanks to the advancement of technology, however, commercially available portable devices and wearables have the capacity to monitor and record HRV measurements. Dobs et al. (2019) performed a systemic review and meta-analysis on the numerous studies that compared the quality of HRV measurements acquired from ECG and obtained from portable devices, such as Elite HRV, Polar H7 and Motorola Droid [13]. Twenty-three studies revealed that HRV measurements obtained from portable devices resulted in a small amount of absolute error when compared to ECG; however, this error is acceptable, as this method of acquiring HRV is more practical and cost-effective, as no laboratory or clinical apparatus are required [13].

Furthermore, the Apple Watch is one of the most best-selling and popular smartwatches in the market. Studies, carried out by Shcherbina and colleagues [14], demonstrated that the Apple Watch was the best HR estimating smartwatch with one-minute granularity and with the lowest overall median error (below 3%) while Samsung Gear S2 reported the highest error. In addition, it is also important to validate the HRV estimation of the Apple Watch. Currently, the best way to obtain RR raw values from the Apple Watch is via the Breathe app developed by Apple. Authors in [15] conducted an investigation that validated the Apple Watch in relation to HRV measurements derived during mental stress in 20 healthy subjects. In this study, the RR interval series provided by the Apple watch was validated using the RR interval obtained from Polar H7 [15]. Successively, the HRV parameters were compared and their ability to identify the Autonomous Nervous System (ANS) response to mild mental stress was analysed [15]. The results revealed that the Apple Watch HRV measurements had good reliability and the HRV parameters were able to indicate changes caused by mild mental stress as it presented a significant decrease in HF power and RMSSD in stress condition compared to the relax state [15]. Therefore, this study suggests that the Apple Watch presents a potential non-invasive and reliable tool for stress monitoring and detection. In this study, raw RR intervals, from beat-to-beat measurements obtained from the Breathe app, are considered for stress classification.

This study is aimed at developing a good predictive model that can accurately classify stress levels from ECG-derived HRV features, obtained from automobile drivers, testing different machine learning methodologies such as K-Nearest Neighbour (KNN), Support Vector Machines (SVM), Multilayer Perceptron (MLP), Random Forest (RF) and Gradient Boosting (GB). Moreover, the models obtained with highest predictive power will be used as a reference for the development of a machine learning model that would be used to classify stress from HRV features derived from heart rate measurements obtained from wearable devices in a unsupervised system-based web application.

The paper is organised as follows. Section 2 provides a discussion of related work conducted in the literature. Section 3 describes the experimental methodology of the study, including a description of the dataset, pre-processing, hyperparameter tuning and the design protocol used for the development of a simple stress detection web application based on Apple Watch derived data. Section 4 presents the experimental results and Section 5 an intensive discussion of the results obtained. Lastly, Section 6 provides the concluding remarks of the study, as well as proposed future work.

2. Related Work

As stress level changes so does the HRV and it has been proven that HRV decreases as stress increases [11]. This is possible because HRV provides a measure to monitor the activity of the ANS and, therefore, can provide a measure of stress [16]. Authors in [16] explored the interaction between HRV and mental stress. Here they took ECG recordings during rest and mental task conditions, which was meant to reflect a stressful state. Linear HRV measures were then analysed in order to provide information on how the heart responds to a stressful task. The results demonstrated that the mean RR interval was significantly lower during a mental task than in the rest condition [16]. This difference was significant only when time domain parameters (pNN50) and the mean RR interval were analysed; while the frequency domain measure did not show a significant difference, although there was an elevated LF/HF in the stressed condition [16]. As LF is associated with the SNS and HF with PNS, the increased LF/HF ratio does suggest that there is a higher sympathetic activity in the stress condition compared to the resting state [16].

Furthermore, investigations have been carried out in order to accurately classify stress in drivers via HRV measurements. For example, authors in [17] aimed to classify ECG data using extracted parameters into highly stressed and normal physiological states of drivers. In this study, they extracted time domain, frequency domain and nonlinear domain parameters from HRV obtained by extracting RR intervals from QRS complexes. These extracted features were fed into the following machine learning classifiers: K Nearest Neighbor (KNN), radial basis function (RBF) and Support Vector Machine (SVM. The results showed that SVM with RBF kernel gave the highest results, with 83.33% accuracy when applied to time and non-linear parameters, while giving an accuracy of 66.66% with frequency parameter [17]. This was in concordance with the result obtained by [16] as the frequency domain parameters did not give a significant difference between rest and mental tasks.

In this study, instead of analysing how each HRV measure is affected by the onset of stress, we took into consideration the combination of both time and frequency domain HRV features and how these aid stress classification with the use of machine learning models. The performance of the machine learning models was evaluated, taking into consideration the following metrics: Area Under Receiver Operator Characteristic Curve (AUROC), Recall/Sensitivity and F1 score, without relying only on accuracy. Furthermore, we detected stress in a non-invasive manner using the Apple Watch, from which we extracted heart rate data, obtained from volunteers subjected to different mental state conditions.

3. Materials and Methods

3.1. Datasets

The first part of this study consists in the development of a good stress predictive model from ECG-derived HRV measurements. The dataset used was collected at Massachusetts Institute of Technology (MIT) by Healey and Picard [18], which is freely available from PhysioNet [19]. The dataset consists of a collection of multi-parameter recordings obtained from 27 young and healthy individuals while they were driving on a designated route in the city and highways around Boston, Massachusetts. The driving protocol involved a route that was planned to put the driver though different levels of stress; specifically, the drive consisted of periods of rest, highway driving and city driving which were presumed to induce low, medium and high stress, respectively [18]. This investigation measured four types of physiological signals: ECG, EMG, GSR and respiration. The dataset is available in the PhysioNet waveform format containing 18 *.dat* and 18 *.hea* files with a *.txt* metadata file. Each bio-signal *.dat* file contains the original recording for ECG, EMG, GSR, HR and Respiration. As the aim of this study is to classify stress based on HRV metrics, a beat annotation file was created from *.dat* files by using the WQRS tool that works by locating QRS complexes in the ECG signal using and gives an annotation file as the output [20]. The annotation file serves the purpose of extracting RR intervals together with its corresponding timestamp using the PhysioNet HRV toolkit. HRV features were extracted from the RR intervals by splitting the dataset in windows of 30 s. Time domain features were calculated using a C implementation that connects Python to the PhysioNet HRV toolkit and by calling the *get_hrv* method which returns the HRV metrics. While frequency domain metrics were obtained by applying the Lomb Periodogram which determines the power spectrum at any given frequency [21]. GSR signals were used to determine and label the stress states in drivers, as the marker in the dataset was mainly made of missing values. The median GSR values were used as the cut-off point, thus, values above the median were labelled as stress while the values below the median were labelled as no stress. For clarity reasons, this dataset will be referred to as 'original-dataset'.

The second portion of this investigation aimed to develop machine learning models that would classify stress from HRV features derived from HRV measurements obtained from the Apple Watch. For this purpose, data was collected from 4 Apple Watch users, who were asked not to exercise or intake caffeine before and on the day of the experiment. The volunteers were subjected to 2 different conditions. The first condition was a 15-min relaxation period where they listened to relaxing lo-fi music. The second was a stressful condition experienced after an 8-h shift of work. Immediately after each task, the volunteers were asked to record their beat-to-beat measurements 5 times, using the Apple Breathe App available on their Apple Watch. The subjects were subjected to these two conditions on separate days. Thanks to the Breathe App, it was possible to obtain raw RR intervals from beat-to-beat measurements and all the data was accessible from the user's Personal Health Record, which can be exported in XML format via Apple's Health App. The beat-to-beat measurements of interests, mapped into the <InstantaneousBeatsPerMinute> tag, were extracted from the XML file in Python using *xml* and *pandas* modules. Successively, the raw RR intervals (in seconds) were derived from the beat per minute (bpm) readings using the following equation:

$$RR = \frac{60}{bpm} \tag{1}$$

Moreover, HRV features were extracted from the calculated *RR* intervals using the *NumPy* library, for time domain, and the *pyhrv* library, specifically the *frequency_domain* module and the Welch's Method for frequency domain features [22]. This dataset will be used as a blind test for the obtained classifier, in order to measure its predictive power on unseen data; hence, this dataset will be referred to as a 'blind-dataset' throughout this paper. The stress prediction of the blind-dataset was performed by a simple web application, developed using *Streamlit*. This experimental procedure is illustrated in Figure 1.

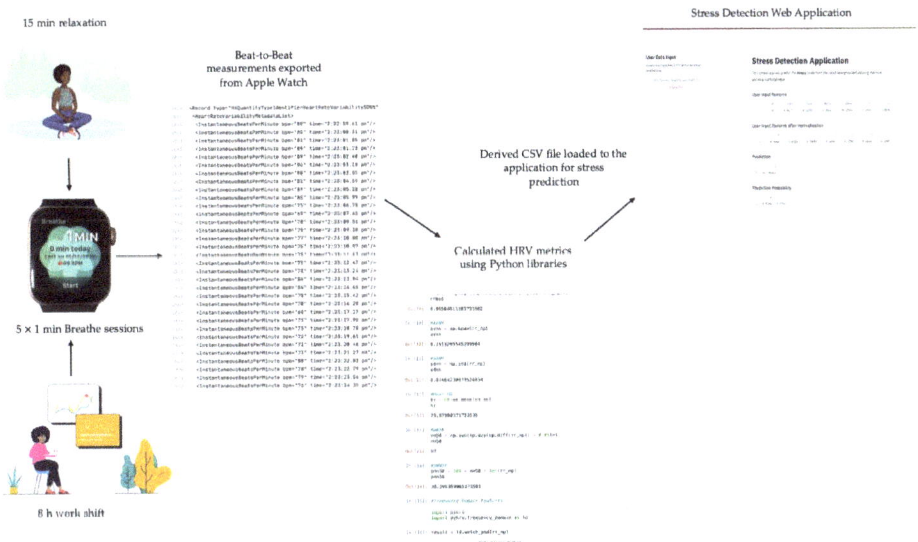

Figure 1. Illustration of the experimental procedure followed for stress detection on data obtained from Apple Watch users.

3.1.1. Data Pre-Processing

Firstly, missing values in original-dataset were replaced with the mean value of each column. Then the data was further split into training and testing datasets with an 80:20 (training:testing) split. From this point onwards, the testing and training data were treated separately as different entities in order to prevent data overfitting and data leakage. Data normalisation was done separately on the training and testing set instead of the whole dataset that could leak information about the test into the train set. Normalisation was performed using the *scikit-learn* library, where continuous values are rescaled in a range between 0 and 1 with the aim of having all numeric columns in the same range, as there are features that are in different ranges such as ECG, HR, EMG, seconds and HF.

3.1.2. Feature Selection

Features were selected based on their relevance to the classification task that this study proposed. This was accomplished using three techniques: Pearson's Correlation, Recursive Feature Elimination (RFE) [23] and Extra Tree Classifier [24], used to estimate feature importance. The common least important features from each method were dropped from both training and testing datasets; Figure 2 illustrates this process.

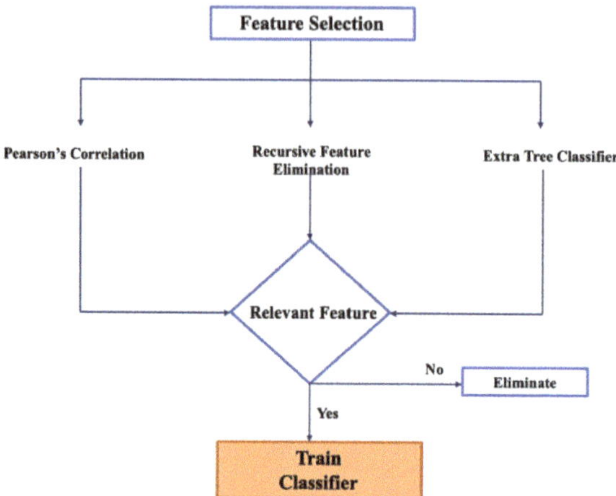

Figure 2. Flow chart illustrating the Feature Selection Process implemented in this study.

Pearson's Correlation calculates the correlation coefficient between each feature and the target class (stress) and this value ranges between −1 and 1. Low correlation is represented by values close to 0, with 0 being no correlation, and high positive and negative correlations are achieved with values closer to 1 and −1, respectively. In this study, relevant features were chosen based on their highly positive and highly negative correlations with the target. Feature Importance using Extra Trees Classifier, is an ensemble-based learning algorithm that aggregates the results of multiple decision trees to output a classification result [24]. In each decision, a Gini Importance of the feature is calculated which determines the best feature to split the data on based on the Gini Index mathematical criteria. RFE functions by recursively eliminating attributes and building the Linear Regression machine learning model on the basis of the selected attributes. It then uses the accuracy of the model that contributes the most to the predictive output of the algorithm. RFE will then rank each feature based on importance with 1 being the most important.

As the second goal of this study was to develop a classification model that would classify stress from data obtained from wearable devices, a 'modified-dataset' was created from tailoring original-dataset to present features that were purely relevant to the attributes calculated from the RR intervals recorded from the device. This also aimed to further test the classifiers' performance on a dataset resembling that generated from the wearable device. Therefore, the relevant features for the modified-dataset were: HR, AVNN, SDNN, RMSSD, pNN50, TP and VLF. The modified-dataset was also the reference dataset for the stress detection application which was used to validate the predictive power of the trained algorithms in a unsupervised system.

3.2. Parameter Tuning

In order to achieve the most efficient classification model, hyper-parameter tuning was performed on each algorithm used in this study to determine the best choice of parameters that would yield the highest performance. After generating the baseline for each classifier, where the parameters were set to their default values, a *scikit-learn* library [25] function that loops through a set of predefined hyperparameters and fit the model on the training set was used to perform parameter tuning. Different ranges of each parameter were used in each grid. The outputs from the grid search are the best parameter combinations that give the highest predictive performance which were then compared to their corresponding baseline models. All algorithms in this study were created with the *scikit-learn* library.

3.2.1. K-Nearest Neighbour

K-Nearest Neighbour (KNN) performs classification based on the closest neighbouring training points in a given region [26]; thus, the classification of new test data is dependent on the number of neighbouring labelled examples present at that given location. In order to obtain the best KNN classification model, different values for k (number of nearest neighbours) and the p value (the power parameter equivalent to the Euclidean distance or Manhattan distance) were investigated. The k values investigated ranged from 1 to 30 inclusive, while p values could either be 1 (Manhattan distance) or 2 (Euclidean distance). The best parameter values resulted from the grid search are as follows: $k = 25$ and $p = 1$, uniform weights was also selected meaning that all points in each neighbourhood are weighted equally.

3.2.2. Support Vector Machine

The function of the Support Vector Machine (SVM) algorithm is to locate the hyper-plane in N-dimensional space (where N represents the number of features) that classifies the data instances into their corresponding class [27] The performance of this algorithm is affected by hyperparameters such as the soft margin regularization parameter (C) and kernel, a function that transforms low dimensional inputs space into a higher dimensional space making the data linearly separable.

For the SVM classification model, different C values (0.001, 0.01, 0.1, 1, 10, 100 and 1000) and kernels, such as Linear kernel, Polynomial (poly) kernel and Gaussian Radial Basis Function (RBF) kernel were tested. As RBF and poly kernel depends on the gamma (γ, that determines the distance of influence of a single training point) and degree (the degree used to find the hyperplane) parameters respectively, 3 grid searches were carried out for each kernel with γ values of 0.001, 0.01, 0.1, 1, 10, 100 and 1000 and degree values ranged from 1 to 6 inclusive. The best parameter settings resulted to be RBF kernel with $\gamma = 10$ and $C = 100$.

3.2.3. Multilayer Perceptron

Multilayer Perceptron (MLP) is a feedforward artificial neural network that was developed to circumvent the drawbacks and limitations imposed by the single-layer perceptron [28]. MLPs are made of at least 3 layers of nodes (input layer, hidden layer and output layer), where each node is connected to every node in the subsequent layer with a certain weight. MLP's performance, like other machine learning algorithms, is highly dependent on hyperparameter tuning of the following parameters: learning rate coefficient (h), momentum (μ) and the size of the hidden layer. h determines the size of the weight's adjustments made at each iteration; h values of 0.3, 0.25, 0.2, 0.15, 0.1, 0.1, 0.005, 0.01 and 0.001 were investigated in the grid search. μ controls the speed of training and learning rate; this parameter was set to a range between 0 and 1 with intervals of 0.1. Finally, the size of the hidden layer corresponds to the number of layers and neurones in the hidden layer; the following hidden layer sizes were analysed (10, 30, 10), (4, 6, 3, 2), (20), (4, 6, 3), (10, 20) and (100, 100, 400), where each value represent the number of neurons at its corresponding layer position. A configuration of $h = 0.001$, $\mu = 0.1$ and three hidden layers of 100, 10 and 400 nodes, respectively, proved to be the optimal settings for the model following the grid search.

3.2.4. Random Forest

Random Forest (RF) is an ensemble-based learning algorithm consisting of a combination of randomly generated decision tree classifiers, the results of which are aggregated to obtain a better predictive performance [26]. Based on the parameter tuning grid search performed, the optimal configuration for this algorithm was when the number of trees in the forest (estimators) was set to 300, out of the values 1, 2, 3, 4, 8, 16, 32, 64 and 100 that were tested, with the maximum number of features set to the square root of the total number of features, while the log base 2 of the number of features gave a lower prediction performance.

3.2.5. Gradient Boosting

Gradient Boosting (GB) is also an ensemble-based algorithm composed of multiple decision trees trained to predict new data and where each tree is dependent on one another This model, which is trained in a gradual, sequential and additive manner, is highly dependent on the learning rate parameter that regulates the shrinkage of the contribution of each tree to the model. The optimal value for this parameter was found to be 0.14 as other learning rate values of 1, 0.5, 0.25, 0.1, 0.05 and 0.01 were also tested in the grid search

A Naïve Bayes probabilistic algorithm [26] was used as the baseline model for performance comparison between the other more complex algorithms. The configuration for this model was kept as simple as possible by utilising the parameters in their default values as presented by the *GuassianNB* python model.

Furthermore, in order to determine whether there were statistical differences between the investigated models and the baseline model, a One-Way ANOVA statistical test with Tukey's *post Hoc* comparison was performed on the mean AUROC scores. The null and alternate hypothesis formulated were:

Hypothesis 1 (H1). *Null Hypothesis: The mean AUROC score for the compared 2 models are equal.*

Hypothesis 2 (H2). *Alternative Hypothesis: The mean AUROC score for the 2 compared models are not equal, at least AUROC value of one model is different from the other.*

4. Results

All results, related to original-dataset and modified-dataset, are described in terms of machine learning metrics such as Area Under Receiver Operator Characteristic Curve (AUROC), Recall/Sensitivity and F1 score [26], including their standard deviation. Every machine learning algorithm was run with a five-fold cross validation. Meanwhile, results from stress classification from data obtained from the Apple Watch are expressed in terms of prediction probability.

4.1. Feature Selection on Original-Dataset

Feature Selection was performed in order to determine the attributes in the dataset that most contribute to the classification task. Figure 3 represents the heat map plot obtained from Pearson's Correlation. Feature selection scores from RFE, shown in Table 2, indicate that the most relevant features are those with the lowest score. This also shows that the best features (score of 1) were time domain HRV metrics such as RMSSD and AVNN, and frequency domain metrics like TP and ULF, followed by SDNN with a score of 4 (Table 2). Furthermore, Figure 4 illustrates a histogram of the feature importance scores based on the Extra Trees Classifier. Figure 4 shows the Gini Importance of each feature, where the greater the value, the greater the importance of the feature in stress classification.

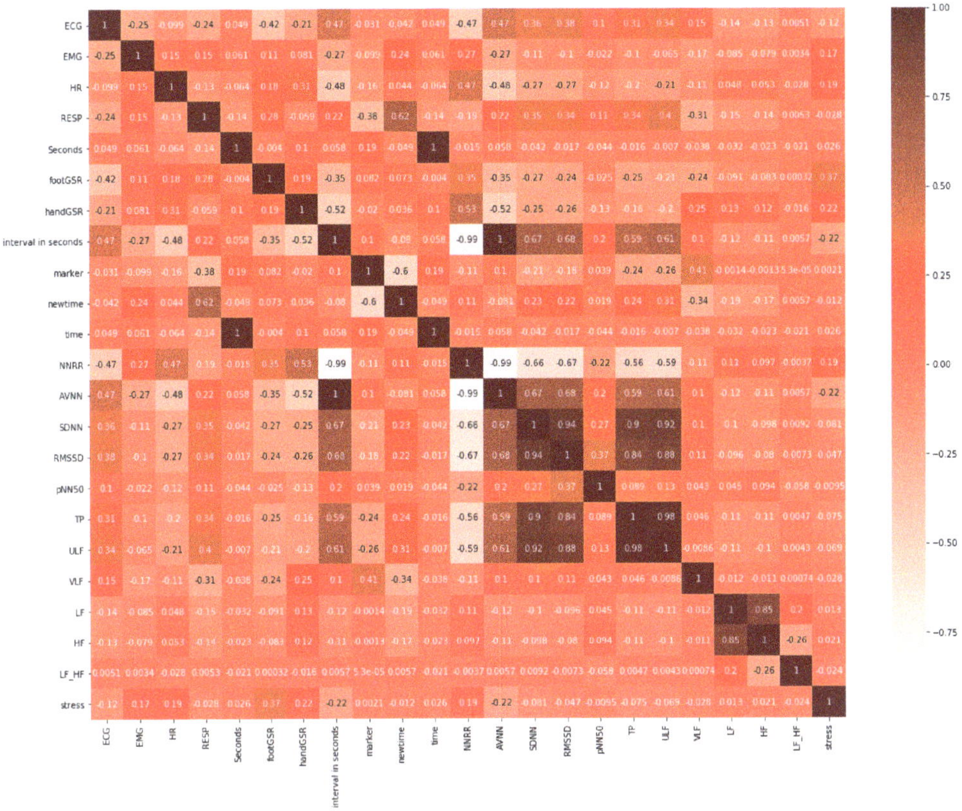

Figure 3. Heat map plot of Pearson's Correlation Feature Selection performed on original-dataset.

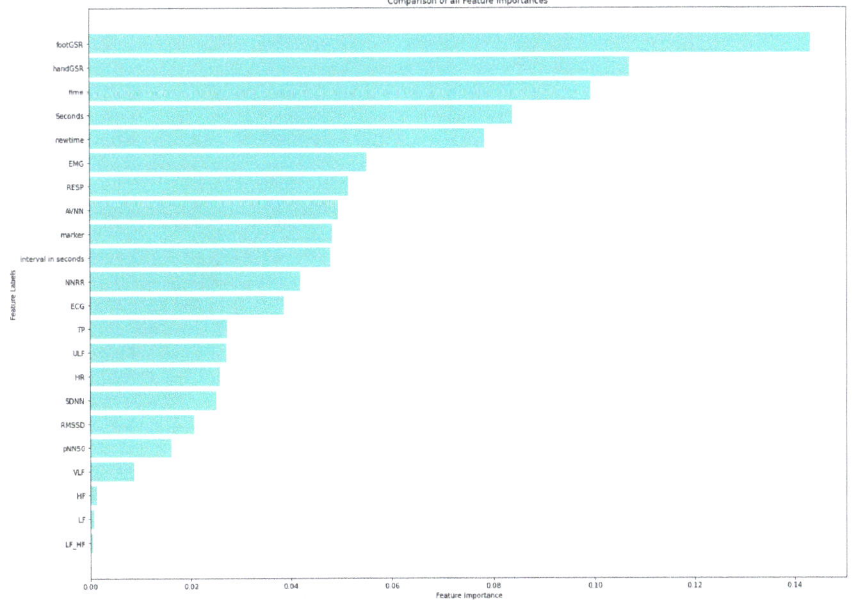

Figure 4. Feature Importance of features from original-dataset using Extra Trees Classifier.

Table 2. RFE feature importance score on original-dataset, the most relevant features have the lowes RFE score.

Feature	RFE Score
EMG	1
HR	1
footGSR	1
handGSR	1
Interval in seconds	1
NNRR	1
AVNN	1
RMSSD	1
TP	1
ULF	1
RESP	2
marker	3
SDNN	4
LF_HF	5
LF	6
ECG	7
pNN50	8
HF	9
newtime	10
time	11
VLF	12
Seconds	13

The common features, from each method, that least contributed to classification or that had the lowest score were dropped from the dataset; these were LF_HF, LF and HF Additionally, GSR attributes were also dropped because they presented a very strong correlation with stress classification as these were used for stress labelling. Thus, in order to avoid data leakage and overfitting, they were eliminated. Moreover, intuitively redundant features were also dropped like the time related features marker, due to its high number of missing values and EMG, given that it is irrelevant in the context of the smart watch.

4.2. Stress Classification on Original-Dataset

In this experiment, stress was classified from bio-signals obtained from subjects who drove under different stress conditions. The results obtained from hyperparameter tuning illustrated in Table 3, showed that the three best models for the classification task imposed by this dataset were MLP, RF and GB which yielded an AUROC of 83%, 85% and 85% respectively. Thus, the models have more than 83% probability of correctly classifying data instances.

Table 3. Comparison of the predictive performance of the best classifiers obtained from the grid search (trained on original-dataset).

Algorithm	AUROC	Recall	F1 Score
NB	0.60 ± 0.02	0.63 ± 0.04	0.61 ± 0.02
KNN	0.80 ± 0.01	0.76 ± 0.02	0.74 ± 0.01
SVM	0.81 ± 0.01	0.79 ± 0.03	0.77 ± 0.01
MLP	0.83 ± 0.01	0.81 ± 0.07	0.77 ± 0.02
RF	0.85 ± 0.01	0.81 ± 0.03	0.78 ± 0.02
GB	0.85 ± 0.01	0.80 ± 0.02	0.79 ± 0.01

NB, Naïve Bayes; KNN, K Nearest Neighbour; SVM, Support Vector Machine; MLP, Multilayer Perceptron; RF, Random Forest; GB, Gradient Boosting. NB represent the baseline model used as means of comparison for the other complex machine learning algorithm.

Moreover, MLP and RF presented a Recall of 81% while GB 80% (Table 3); this indicates that at least 80% of the predicted Tue Positive instances are actual positives. Therefore, at least 80% of the instances predicted to be in the stress class have been correctly classified as such. Finally, the F1 scores for MLP, RF and GB are 77%, 78% and 79%, respectively; thus, the model has at least 77% accuracy on the dataset. Figure 3 illustrates the Receiver Operating Characteristics (ROC) curve for all the classifiers investigated in this study.

Figure 5 consolidates the findings shown in Table 3, illustrating that the models with the greatest ROC area are GB, RF and MLP. It is also visible that this NB model serves as a good baseline model as its ROC curve suggest that its classification is nearly due to chance. A statistical analysis was performed to measure the significance of these results (Table 4).

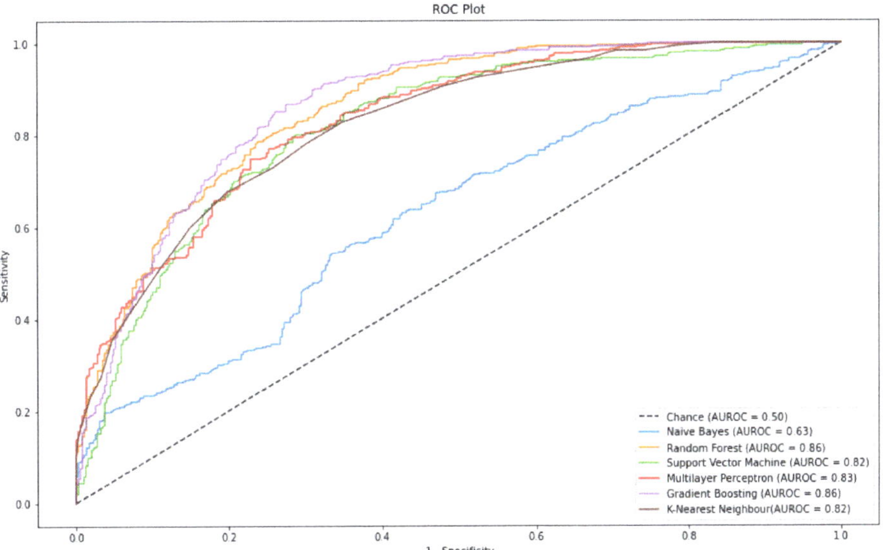

Figure 5. ROC curve plot of each classification model trained on original-dataset. The AUROC scores were achieved by the models during stress prediction of the test dataset from the original-dataset.

Table 4. Statistical Evaluation of the machine learning models.

Model A	Model B	mean (A)	mean (B)	diff	se	p-Tukey [1]
GB	KNN	0.852	0.800	0.052	0.009	**0.001**
GB	MLP	0.852	0.825	0.027	0.009	**0.039**
GB	NB	0.852	0.603	0.249	0.009	**0.001**
GB	RF	0.852	0.853	−0.001	0.009	0.9
GB	SVM	0.852	0.813	0.039	0.009	**0.001**
KNN	MLP	0.800	0.825	−0.025	0.009	0.077
KNN	NB	0.800	0.603	0.197	0.009	**0.001**
KNN	RF	0.800	0.853	−0.053	0.009	**0.001**
KNN	SVM	0.800	0.813	−0.013	0.009	0.671
MLP	NB	0.825	0.603	0.222	0.009	**0.001**
MLP	RF	0.825	0.853	−0.028	0.009	**0.036**
MLP	SVM	0.825	0.813	0.012	0.009	0.732
NB	RF	0.603	0.853	−0.250	0.009	**0.001**
NB	SVM	0.603	0.813	−0.210	0.009	**0.001**
RF	SVM	0.853	0.813	0.040	0.009	**0.001**

[1] p values in bold represent statistical significance, where $p < 0.05$.

Table 4 shows that there was a statistical difference between the AUROC means of all hyperparameter-tuned models and the baseline (NB–AUROC = 60%) as the $p < 0.05$. This confirms that the parameter tuning did improve the model's performance significantly, and thus, H1 is accepted. Moreover, the Tukey's comparison test showed that there is a statistically significant difference between the AUROC values of GB and MLP and between MLP and RF ($p < 0.05$). However, the differences between GB and RF are not statistically significant ($p = 0.9$). Figure 6 summarises the results obtained during this experimental series, by illustrating the performance comparison between the hyperparameter-tuned models and the baseline NB model.

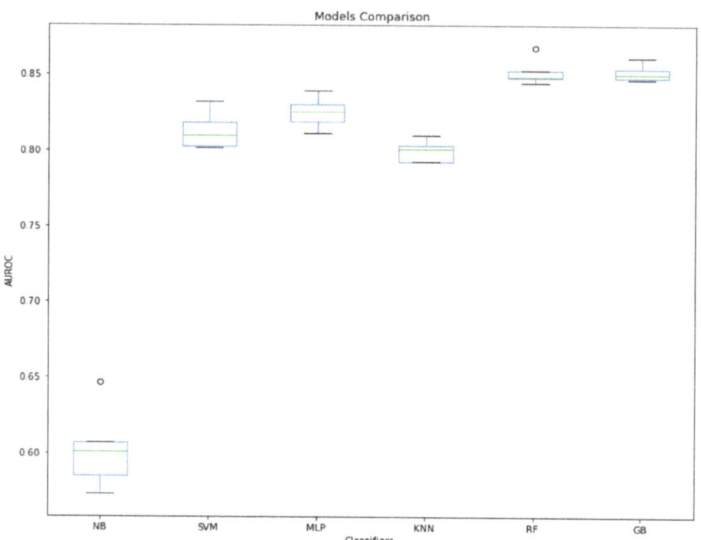

Figure 6. Model performance comparison of machine learning algorithms trained on original-dataset.

4.3. Stress Classification on Modified-Dataset

The other objective of this study was to develop a classification model that would classify stress from HRV data obtained from wearable devices. To achieve this, classifiers from Table 3 were used for stress classification of a modified-dataset, which is a modification of the original-dataset but with features that mimic those obtained from the wearable device. Table 5 shows the results obtained during the classification task.

Table 5. Predictive performance of machine learning classifiers on modified-dataset.

Algorithm	AUROC	Recall	F1 Score
NB	0.60 ± 0.02	0.69 ± 0.04	0.63 ± 0.02
KNN	0.74 ± 0.02	0.76 ± 0.02	0.71 ± 0.02
SVM	0.74 ± 0.01	0.79 ± 0.02	0.74 ± 0.01
MLP	0.75 ± 0.01	0.80 ± 0.06	0.72 ± 0.02
RF	0.77 ± 0.01	0.74 ± 0.01	0.72 ± 0.01
GB	0.73± 0.01	0.70 ± 0.02	0.70 ± 0.01

As shown in Table 5, MLP seems to be the overall best performing classifier with 75% AUROC, 80% Recall and 72% F1 score.

In addition, Figure 7 illustrates the Receiver Operating Characteristics (ROC) curve for all the classifiers used for the classification of modified-dataset.

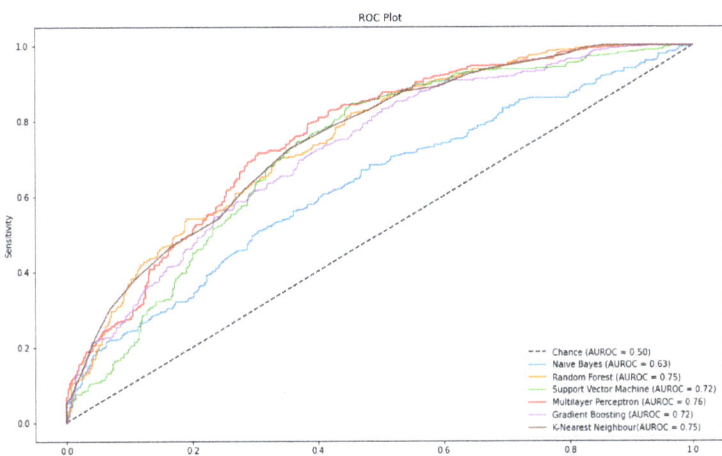

Figure 7. ROC curve plot of each classification model tested on modified-dataset. The AUROC scores were achieved by the models during stress prediction of the test dataset from the modified-dataset.

Figure 7 shows that the ROC curve from the MLP classifiers seems to be the furthest away from the chance curve and to have the largest area under the curve.

Additionally, a statistical analysis (One-Way ANOVA statistical test with Tukey's *post Hoc*) of the top three best performing algorithms, obtained from the original-dataset, and their corresponding algorithms, from the modified-dataset, was performed in order to determine their statistical difference. Moreover, these results provided additional insight into which model would be best suited to be implemented in the stress detection application. As shown in Table 6, it is evident that the machine learning algorithms trained on original-dataset are statistically the better performing models ($p < 0.05$), which is expected due to the fact that more information on the dataset is being fed to the model during training.

Table 6. Statistical Evaluation of the machine learning models. Numbers 1 and 2 correspond to original-dataset and modified-dataset respectively.

Model A	Model B	mean (A)	mean (B)	diff	se	*p*-Tukey [1]
GB 1	GB 2	0.852	0.731	0.121	0.008	**0.001**
GB 1	MLP 1	0.852	0.825	0.027	0.008	**0.011**
GB 1	MLP 2	0.852	0.752	0.100	0.008	**0.001**
GB 1	RF 1	0.852	0.853	−0.001	0.008	0.9
GB 1	RF 2	0.852	0.768	0.084	0.008	**0.001**
GB 2	MLP 1	0.731	0.825	−0.094	0.008	**0.001**
GB 2	MLP 2	0.731	0.752	−0.021	0.008	0.088
GB 2	RF 1	0.731	0.853	−0.122	0.008	**0.001**
GB 2	RF 2	0.731	0.768	−0.037	0.008	**0.001**
MLP 1	MLP 2	0.825	0.752	0.073	0.008	**0.001**
MLP 1	RF 1	0.825	0.853	−0.028	0.008	**0.01**
MLP 1	RF 2	0.825	0.768	0.057	0.008	**0.001**
MLP 2	RF 1	0.752	0.853	−0.101	0.008	**0.001**
MLP 2	RF 2	0.752	0.768	−0.016	0.008	0.313
RF 1	RF 2	0.853	0.768	0.085	0.008	**0.001**

[1] *p* values in bold represent statistical significance, where $p < 0.05$.

Furthermore, Table 6 indicates that there is no significant difference in the AUROC values between RF2 and MLP2 ($p = 0.31$). MLP2 was then chosen as the model that will be implemented in the stress detection web application due to its 80% recall score and overall performance. Additionally, another One-Way ANOVA statistical test with Tukey's *post*

Hoc comparison was performed to determine whether there were statistical differences between the models and the Naïve Bayes baseline model. The results determined that there was a statistical difference between the AUROC means of the models and the baseline as the $p = 0.001$ (results not shown).

4.4. Stress Classification from HRV Measurements Obtained from Apple Watch

A simple web application that would perform stress classification on HRV data uploaded by the user (blind-dataset) was developed with the aim to analyse data extracted using wearable devices. The aim of this process was to test the predictive power of the chosen model on data obtained from real participants. The application was developed in Python using the *Streamlit* framework and it is programmed in such way that the user can upload a csv format data, which will be first normalised and then classified as "stress" or "no stress" using the saved MLP model with Recall 80% and AUROC of 75%. Firstly, the application will prompt the user to insert the csv file in the side menu bar. Secondly, the backend code will normalise the input data, so all data instances are within the same range and display the inserted and normalised data in a tabular format. Thirdly, the normalised data undergoes classification, and the results are displayed as Prediction Probability, shown in Figure 8.

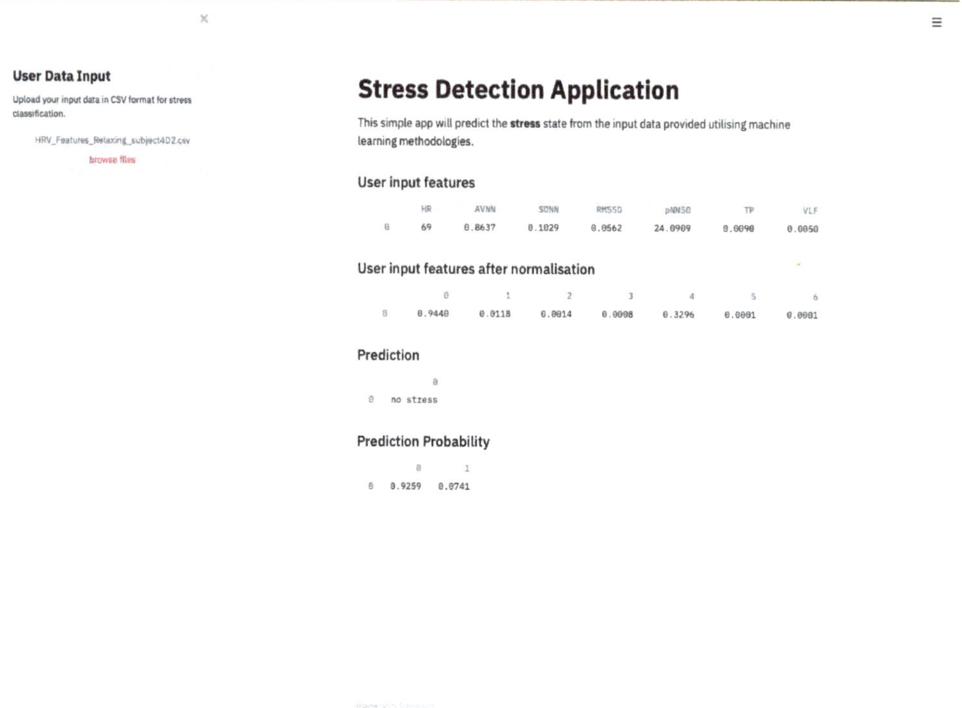

Figure 8. User Interface of the Stress Detection Web Application developed using *Streamlit*.

After running the program with the input data derived from the volunteers, the prediction probabilities for the model to predict an instance as stress or no stress were recorded for the different stress scenarios. Figure 9 summarises the results of this investigation in a bar chart presenting the mean prediction probabilities.

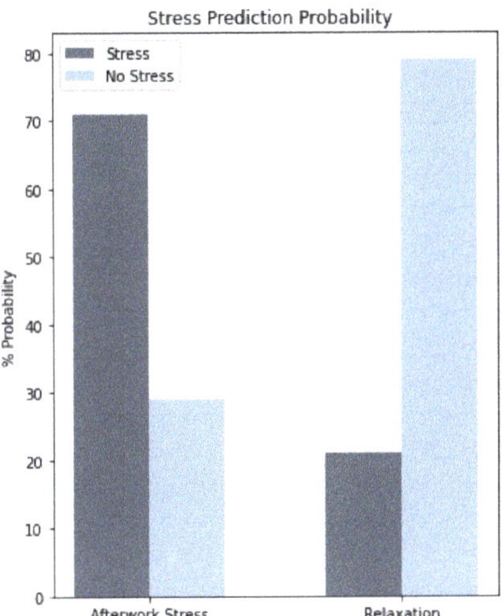

Figure 9. Mean Prediction Probability obtain from the stress detection app with volunteers input data, who were subjected to different stress conditions (after work stress and relaxation).

As displayed in Figure 9, the model was able to correctly classify a stress state with a prediction probability of 71 ± 0.1%. Additionally, it was able to achieve a prediction probability of 79 ± 0.3% when the model was presented with a relaxing situation.

5. Discussion

Stress has been identified as one of the major causes of automobile crashes [8] and an important player in the development of cardiac arrythmia [7]; therefore, it is important to be able to detect and measure stress in a non-invasive and efficient manner. In this study, to accomplish this, we address the stress detection problem by using traditional machine learning algorithms which were trained on ECG-derived HRV metrics obtained from automobile drivers [18,19].

In this paper, stress classification was performed mainly using HRV-derived features as studies have shown that HRV is impacted during changes in stress levels, given that it is highly controlled by the ANS [10]. Moreover, other investigations proved that RMSSD, AVNN and SDNN were evaluated as being the most reliable HRV metrics in distinguishing between stressful and non-stressful situations [28]. Those findings were also confirmed in this study as shown in Table 2, where AVNN, RMSSD and SDNN were classified as the HRV features with the highest RFE feature importance scores. Therefore, they were considered to be the features that contribute the most in the stress classification performance of the model. This further confirms that HRV features are viable markers for stress detection.

Following hyperparameter tuning, we were able to produce stress classification models with high predictive power. As shown in Table 3, the best 3 models for the classification task imposed by original-dataset were MLP, RF and GB with AUROC of 83%, 85% and 85%, respectively; thus, these classifiers have ~84% probability of successfully distinguishing between the stress and no stress class. In addition, MLP and RF gave Recall scores of 81% while GB of 80%; indicating that ~80% of the predicted positive instances are actual positives. Furthermore, these scores were statistically greater than the Naïve Bayes baseline model ($p < 0.05$) as illustrated in Table 4.

There are very few studies performed on stress classification in drivers using HRV derived features [17,18], although each study took a different approach to the classification problem, the classification yielded similar results. For instance, [17] investigated KNN, SVM-RBF and Linear SVM as their potential classifiers for stress detection. Their results suggested that SVM with RBF kernel was the best performing model by giving an accuracy of 83% [17]. However, more extensive investigation is necessary to corroborate this finding by also considering other classification metrics.

It is also imperative to discuss the fact that stress is a result of a combination of external (environment) and internal factors (e.g., mental health). Thus, stress could be perceived as a subjective mental state; for example, certain situations like a drive in the city or in the highway might not induce the same level of stress in every individual. For instance individuals suffering from anxiety could feel stressed in such conditions. Additionally stress could be induced from the invasive apparatus used such as the electrodes placed in different parts of their body and the sensor placed around their diaphragm in [18]; the fact that the subject is aware that they are being monitored for changes in their mental state could also impact their stress levels. For this reason, is important to use less intrusive and everyday devices such as smart watches or mobile phones that are already an essential part of life in this modern society.

In this paper we also aimed to develop a classification model that would detect stress from data obtained from the Apple Watch. For this purpose, the best classifiers trained on original-dataset were tested for the classification of the modified-dataset which presented features that mimic those derived from the wearable device. Table 5 demonstrates that the overall ideal model for the stress classification of HRV features derived from wearable-obtained RR intervals, is MLP with a AUROC of 75% and a Recall of 80%. This was determined based on the Recall score, as in this stress classification task there is a high cost associated with False Negatives. For instance, if an individual's condition, which is actually stressed, is predicted as not stressed, the cost associated with this False Negative can be high, especially in a medical or driving context which could then lead to a misdiagnosis or a car accident respectively. Therefore, it is imperative to select the model with the highest sensitivity.

Figure 8 shows the user interface (UI) of the simple stress detection web application The purpose of this was simply to provide a visual UI to demonstrate the software functionality. This could then be implemented into a mobile or car application where the user would be alerted when stress is detected and would prompt them to relax or take breaks.

The blind-dataset, obtained from the volunteers, served as a blind test for the MLP classifier in order to measure its predictive power on unseen data in an unsupervised application system.

When classifying a stressful task, the web application was able to correctly predict stress conditions with a 71% prediction probability. Additionally, it was able to achieve a prediction probability of 79% when the model was presented with a relaxing state. However, it is important to further improve the model's performance by investigating multiple stress levels in order to obtain more accurate stress detection.

6. Conclusions

In this paper, we developed a comparative study to determine the viability of HRV features as physiological markers for stress detection. This was achieved by computing different supervised machine learning models to determine which model can be used to analyse data extracted using wearable devices. The MLP model was considered to be an ideal algorithm for stress classification due to its 80% sensitivity score. The predictive power of this classifier was found to be statistically greater compared to the baseline model created with the Naïve Bayes algorithm with a p value of 0.001. This model was then implemented in the unsupervised stress detection application where stress can be detected from blind dataset of HRV features, and extracted from real users using wearable devices under different stress conditions.

A benefit of this study is that there is a need for technologies that would monitor stress in drivers in order to reduce car crashes, as nearly 80% of road incidents are due to drivers being under stress. This project could be the initial steps for tackling this problem. In fact, the algorithm produced in this model could be implemented in smart cars. So, when drivers are experiencing episodes of stress, the automobile could switch to autopilot as well as alert the driver of their state. This implementation could massively reduce traffic accidents as well as reduce the number of fatalities and injuries caused by car crashes.

However, the benefit of this study can also be extended to all applications in which it is important to monitor stress levels e.g., in physical rehabilitation post incident, in temporary or chronic anxiety, in mental health disease, as well as in many ageing conditions. The distribution of smart watches is growing in the population and people appreciate their functionalities. Therefore, wearable devices offer a big opportunity to extract health parameters without an uncomfortable and invasive approach.

We plan that future work should involve the improvement of the classification models by exploring a wider range of parameter values during the hyperparameter tuning process. Additionally, the Deep Learning approach could also be implemented in order to compare its performance in comparison to the supervised models used in this study.

Moreover, another future work we propose is the development of a classifier that would be able to distinguish between different levels stress: high, medium and low. In addition to this, we suggest collecting new real-world ECG data, from which HRV features could be extracted, in order to gain a better insight on the predictive power of the models obtained in this study. This would also provide a more updated dataset compared to that used in this study, dated 2005 [18]. As technologies have advanced, a more accurate ECG recording could be acquired; thus, this would make the classification more accurate and relevant to real world implementations.

Therefore, a natural evolution of this work will require the acquisition of a large dataset through smart watches and in an extensive number of tests involving human subjects e.g., through a driving simulator. Furthermore, it will be important to test the model considering other domains focused to the elderly and health care.

Author Contributions: Conceptualization, K.M.D. and G.L.M.; methodology, K.M.D. and G.L.M.; software, K.M.D.; validation, K.M.D. and G.L.M.; formal analysis, K.M.D. and G.L.M.; investigation, K.M.D. and G.L.M.; resources, K.M.D. and G.L.M.; data curation, K.M.D. and G.L.M.; writing—original draft preparation, K.M.D.; writing—review and editing, K.M.D.; visualization, K.M.D.; supervision, G.L.M.; project administration, G.L.M.; funding acquisition, G.L.M. All authors have read and agreed to the published version of the manuscript.

Funding: Authors of this paper acknowledge the funding provided by the EU Interreg 2 Seas Mers Zeeën AGE'In project (2S05-014) to support the work described in this publication.

Institutional Review Board Statement: The study was conducted in accordance with the Declaration of Helsinki, and the protocol was approved by the Manchester Metropolitan University, UK, EthOS Application 24084, for the MSc thesis "Machine Learning for Stress Levels Classification from ECG Signals in Automobile Drivers".

Informed Consent Statement: All subjects gave their informed consent for inclusion before they participated in the study and data have been anonymized.

Data Availability Statement: The main dataset used was public "PhysioNet [18]", as described in the Materials and Methods section; only a few people were involved in the test to validate the feature extraction from apple watch in real condition and the modality of the experiments is reported in this work, therefore the tests can be replicated on further subjects.

Conflicts of Interest: The authors declare no conflict of interest.

References

1. McLeod, S.A. What is the stress Response. Simply Psychology. 2010. Available online: https://www.simplypsychology.org/stress-biology.html (accessed on 9 October 2020).
2. Moldagulova, A.; Sulaiman, R.B. Using KNN algorithm for classification of textual documents. In Proceedings of the 8th International Conference on Information Technology (ICIT), Amman, Jordan, 17–18 May 2017; pp. 665–671.
3. Sato, T.; Yamamoto, H.; Sawada, N.; Nashiki, K.; Tsuji, M.; Muto, K.; Kume, H.; Sasaki, H.; Arai, H.; Nikawa, T.; et al. Restraint stress alters the duodenal expression of genes important for lipid metabolism in rat. *Toxicology* **2016**, *227*, 248–261. [CrossRef] [PubMed]

4. Salai, M.; Vassányi, I.; Kósa, I. Stress detection using low-cost heart rate sensors. *J. Healthc. Eng.* **2016**. [CrossRef] [PubMed]
5. Ahmed, M.U.; Begum, S.; Islam, M.S. Heart rate and inter-beat interval computation to diagnose stress using ECG sensor signa *Mrtc Rep.* **2010**, *4*. Available online: http://www.es.mdh.se/pdf_publications/1929.pdf (accessed on 16 April 2021).
6. Herman, J. Neural control of chronic stress adaptation. *Front. Behav. Neurosci.* **2013**, *7*, 61. [CrossRef] [PubMed]
7. Pramod, B.; Vani, M. Stress Detection with Machine Learning and Deep Learning using Multimodal Physiological Data. In Proceedings of the Second International Conference on Inventive Research in Computing Applications (ICIRCA), Coimbatore India, 15–17 July 2020; pp. 51–57.
8. Rastgoo, M.N.; Nakisa, B.; Maire, F.; Rakotonirainy, A.; Chandran, V. Automatic driver stress level classification using multimoda deep learning. *Expert Syst. Appl.* **2019**, *138*, 112793. [CrossRef]
9. Goel, S.; Tomar, P.; Kaur, G. ECG Feature Extraction for Stress Recognition in Automobile Drivers. *Electron. J. Biol.* **2016**, *12*, 156–165.
10. Londhe, A.N.; Atulkar, M. Heart rate variability analysis: Application overview. In Proceedings of the 2018 Second Internationa Conference on Inventive Communication and Computational Technologies (ICICCT), Coimbatore, India, 20–21 April 2018 pp. 1518–1523.
11. Reijmerink, I.; van der Laan, M.; Cnossen, F. Heart rate variability as a measure of mental stress in surgery: A systematic review *Int. Arch. Occup. Environ. Health* **2020**, *25*, 1–17.
12. Khairuddin, M.K.N.B.; Nakamoto, K.; Nakamura, H.; Tanaka, K.; Nakashima, S. Heart rate and heart rate variability measuring system by using Smartphone. In Proceedings of the 5th Intl Conf on Applied Computing and Information Technology/4th Intl Conf on Computational Science/Intelligence and Applied Informatics/2nd Intl Conf on Big Data, Cloud Computing, Data Science (ACIT-CSII-BCD), Hamamatsu, Japan, 9–13 July 2017; pp. 47–52.
13. Dobbs, W.C.; Fedewa, M.V.; MacDonald, H.V.; Holmes, C.J.; Cicone, Z.S.; Plews, D.J.; Esco, M.R. The accuracy of acquiring hear rate variability from portable devices: A systematic review and meta-analysis. *Sports Med.* **2019**, *49*, 417–435. [CrossRef] [PubMed
14. Shcherbina, A.; Mattsson, C.M.; Waggott, D.; Salisbury, H.; Christle, J.W.; Hastie, T.; Wheeler, M.T.; Ashley, E.A. Accuracy in wrist-worn, sensor-based measurements of heart rate and energy expenditure in a diverse cohort. *J. Pers. Med.* **2017**, *7*, 3 [CrossRef] [PubMed]
15. Hernando, D.; Roca, S.; Sancho, J.; Alesanco, Á.; Bailón, R. Validation of the apple watch for heart rate variability measurements during relax and mental stress in healthy subjects. *Sensors* **2018**, *18*, 2619. [CrossRef] [PubMed]
16. Taelman, J.; Vandeput, S.; Spaepen, A.; Van Huffel, S. Influence of mental stress on heart rate and heart rate variability. In *4th European Conference of the International Federation for Medical and Biological Engineering*; Springer: Berlin/Heidelberg, Germany 2009; pp. 1366–1369.
17. Munla, N.; Khalil, M.; Shahin, A.; Mourad, A. Driver stress level detection using HRV analysis. In Proceedings of the 2015 international conference on advances in biomedical engineering (ICABME), Beirut, Lebanon, 16–18 September 2015; pp. 61–64.
18. Healey, J.A.; Picard, R.W. Detecting stress during real-world driving tasks using physiological sensors. *IEEE Trans. Intell. Transp Syst.* **2005**, *6*, 156–166. [CrossRef]
19. Goldberger, A.; Amaral, L.; Glass, L.; Hausdorff, J.; Ivanov, P.C.; Mark, R.; Mietus, J.E.; Moody, G.B.; Peng, C.K.; Stanley, H.E PhysioBank, PhysioToolkit, and PhysioNet: Components of a new research resource for complex physiologic signals. *Circulation* **2000**, *101*, e215–e220. [CrossRef] [PubMed]
20. Zong, W.; Moody, G.B.; Jiang, D. A robust open-source algorithm to detect onset and duration of QRS complexes. *Comput. Cardiol* **2003**, *30*, 737–740.
21. Li, K.; Rüdiger, H.; Ziemssen, T. Spectral analysis of heart rate variability: Time window matters. *Front. Neurol.* **2019**, *10*, 545 [CrossRef] [PubMed]
22. Gomes, P.; Silva, H.; Margaritoff, P. pyHRV-Open-Source Python toolbox for heart rate variability. 2018. Available online https://pypi.org/project/pyhrv/ (accessed on 16 April 2021).
23. Brownlee, J. Feature Selection in Python with Scikit-Lear. Machine Learning Mastery. 2020. Available online: https:// machinelearningmastery.com/feature-selection-in-python-with-scikit-learn/ (accessed on 21 March 2021).
24. Sharaff, A.; Gupta, H. Extra-tree classifier with metaheuristics approach for email classification. In *Advances in Computer Communication and Computational Sciences*; Springer: Singapore, 2019; pp. 189–197.
25. Pedregosa, F.; Varoquaux, G.; Gramfort, A.; Michel, V.; Thirion, B.; Grisel, O.; Blondel, M.; Prettenhofer, P.; Weiss, R.; Dubourg, V. et al. Scikit-learn: Machine Learning in Python. *JMLR* **2011**, *12*, 2825–2830.
26. Duda, R.O.; Hart, P.E.; Stork, D.G. *Pattern Classification and Scene Analysis*; Wiley: New York, NY, USA, 1973; Volume 3.
27. Gandhi, R. Support vector machine, Introduction to machine learning algorithms. Towards Data Science. 2018. Available online: https://towardsdatascience.com/support-vector-machine-introduction-to-machine-learning-algorithms-934a444fca47 (accessed on 16 April 2021).
28. Amato, F.; Mazzocca, N.; Moscato, F.; Vivenzio, E. Multilayer perceptron: An intelligent model for classification and intrusion detection. In Proceedings of the 2017 31st International Conference on Advanced Information Networking and Applications Workshops (WAINA), Taipei, Taiwan, 27–29 March 2017; pp. 686–691.

Article

Short- and Long-Term Effects of a Scapular-Focused Exercise Protocol for Patients with Shoulder Dysfunctions—A Prospective Cohort

Cristina dos Santos [1,2], Mark A. Jones [3,4] and Ricardo Matias [1,5,*,†]

1 Escola Superior Saúde—Instituto Politécnico de Setúbal, 2910-761 Setúbal, Portugal; cristina.santos@ess.ips.pt
2 Escola Superior de Saúde do Alcoitão, 2649-506 Alcabideche, Portugal
3 Allied Health and Human Performance, University of South Australia, Adelaide 5001, Australia; mark.jones@unisa.edu.au
4 International Centre for Allied Health Evidence, University of South Australia, Adelaide 5001, Australia
5 Champalimaud Research and Clinical Centre, Champalimaud Centre for the Unknown, 1400-038 Lisbon, Portugal
* Correspondence: ricardo.matias@neuro.fchampalimaud.org; Tel.: +351-210-480-200 (ext. 4479)
† Current Address: Champalimaud Research, Champalimaud Center for the Unknown, Av. De Brasília, 1400-038 Lisbon, Portugal.

Abstract: Current clinical practice lacks consistent evidence in the management of scapular dyskinesis. This study aims to determine the short- and long-term effects of a scapular-focused exercise protocol facilitated by real-time electromyographic biofeedback (EMGBF) on pain and function, in individuals with rotator cuff related pain syndrome (RCS) and anterior shoulder instability (ASI). One-hundred and eighty-three patients were divided into two groups (n = 117 RCS and n = 66 ASI) and guided through a structured exercise protocol, focusing on scapular dynamic control. Values of pain and function (shoulder pain and disability index (SPADI) questionnaire, complemented by the numeric pain rating scale (NPRS) and disabilities of the arm, shoulder, and hand (DASH) questionnaire) were assessed at the initial, 4-week, and 2-year follow-up and compared within and between. There were significant differences in pain and function improvement between the initial and 4-week assessments. There were no differences in the values of DASH 1st part and SPADI between the 4-week and 2-year follow-up. There were no differences between groups at the baseline and long-term, except for DASH 1st part and SPADI ($p < 0.05$). Only 29 patients (15.8%) had a recurrence episode at follow-up. These results provide valuable information on the positive results of the protocol in the short- and long-term.

Keywords: scapula neuromuscular activity and control; rotator cuff related pain syndrome; anterior shoulder instability; scapular dyskinesis; electromyographic biofeedback

Citation: dos Santos, C.; Jones, M.A.; Matias, R. Short- and Long-Term Effects of a Scapular-Focused Exercise Protocol for Patients with Shoulder Dysfunctions—A Prospective Cohort. *Sensors* **2021**, *21*, 2888. https://doi.org/10.3390/s21082888

Academic Editor: Maria de Fátima Domingues

Received: 17 February 2021
Accepted: 12 April 2021
Published: 20 April 2021

1. Introduction

The rotator cuff related pain syndrome (RCS) [1,2] and anterior shoulder instability (ASI) are the two most prevalent shoulder dysfunctions [3,4]. They are characterized by the presence of pain [5–9], decreased function [5,7,9], muscle weakness [5,6,10–13], altered range of motion (ROM) [5,6,9], altered scapula neuromuscular control [12–14], and scapular dyskinesis [12,15,16].

Research investigating the scapular orientation and kinematics in RCS compared to asymptomatic controls concluded that no irrefutable relationship could be found between the scapula orientation and RCS [17]. However, scapular-focused stabilization and motor control exercise is promoted to address scapular dyskinesis, reduce pain [18], and restore function [11] and have been included in most studies demonstrating the benefit of exercise for RCS [19,20]. Reijneveld et al. [21] found no evidence effectiveness on a scapular-focused treatment approach in patients with RCS.

When it comes to shoulder instability, there is limited research in the management to guide therapists [7]. For traumatic instability, the current literature recommends surgical treatment [8], but for atraumatic instability, physiotherapy remains the recommended course of treatment [22] in the form of exercise to improve muscle strength and proprioception [7]. Yet, the lack of specific detail about the exercises used and the low-quality studies available is a concern [3,7,13]. Both cohort [22] and randomized controlled trials [9,23] studied the effect of specific exercise programs in patients with shoulder instability. They mostly found a significant benefit in reducing pain [9,22], increasing stability [22,23], muscle strength [9,23], ROM [23], and function [22,23]. Eshoj et al. [23] reported that a neuromuscular shoulder exercise program incorporating strength, coordination, balance, proprioception, and functional kinetic chain work was superior to the standard care exercise program emphasizing strength training to increase muscle mass in patients with traumatic shoulder instability.

Several randomized studies [5,18,24] have investigated the effects of motor control and muscle strengthening exercises in patients with RCS. Above all, they observed that scapular-focused exercise leads to higher patient-rated outcomes [18], including reduction in pain level [5,18,24] and improvement in function [5,18,24], ROM [18], and strength [18,25]. Other studies investigated the effect of scapular-focused exercises on electromyographic measures of muscle activity [18,24,26–29] and the timing of onset [30,31] with no uniformity in results [31]. Studies incorporating electromyographic biofeedback (EMGBF) to guide exercise performance also reported inconsistent findings regarding its effect. Huang et al. [27] found that the use of EMGBF improves motor control in both symptomatic and asymptomatic subjects where Juul-Kristensen et al. [24] found EMGBF made no difference to pain and function outcomes. Larsen et al. [28] proposed that individuals with subacromial impingement syndrome may benefit from incorporating EMGBF to improve the neuromuscular function.

However, to date, the scapular-focused exercise incorporated in research interventions has been quite varied with a lack of clarification about the intensity, frequency, and progression of exercises and a lack of explicit objective scapular related criteria for the success and progression of exercise [16,25]. Moreover, despite being referred to as an aid in shoulder intervention [3], the exercise programs mostly have not emphasized biofeedback as a learning strategy or an objective measure of motor control. Given the conflicting results of the value and need for scapular-focused exercise, further research is needed incorporating more explicit criteria for the administration of scapular-focused exercise before the call to abandon this intervention can be heeded.

The main objective of this study was to describe the short- and long-term effects of a scapular-focused exercise protocol supported by real-time EMGBF on the level of pain and function in individuals with shoulder dysfunctions. Additionally, scapular neuromuscular activity and control, ROM, and glenohumeral flexor and abductor isometric muscle strength (GMS) were assessed to explore the mechanisms of recovery.

It was hypothesized that:

i. After 4-weeks of treatment, the protocol would lead to an amelioration in both groups in pain and function (decrease in the shoulder pain and disability index (SPADI) [32] levels with a minimal clinically important difference (MCID) ranging from 8 to 13 points) [33]; decrease in the numeric pain rating scale (NPRS) level of at least a MCID of 2.17 points [34]; and decrease in the disabilities of the arm, shoulder, and hand (DASH) levels with a MCID of 10.2 points [35].
ii. Primary outcome ameliorations (pain reduction and function improvement) made at the 4-week assessment would be retained at the 2-year assessment in both groups.

2. Materials and Methods

2.1. Study Design

A prospective cohort was developed to implement the scapular-focused exercise protocol, with initial, 4-week, and 2-year follow-up assessments.

2.2. Sample

From 213 patients recruited consecutively from an outpatient orthopaedic clinic, 183 were included and 30 unable to commit to the schedule of treatments were excluded before commencing. These 183 patients were divided into two groups according to the diagnostic categorization: RCS group (n = 117) and ASI group (n = 66). All patients had a prior consultation with an orthopaedic physician who made the diagnosis and recommended physiotherapy. The mean (±standard deviation) age for the RCS group was 41.1 (±12.2) and for the ASI group 26.7 (±10.3) years. Patient symptoms originated mostly from overuse in the RCS group (59.0%) and trauma in the ASI group (48.5%). Most patients in both groups were in the chronic stage of the condition (length of symptoms for more than 6 weeks) (81.2% for the RCS group and 71.2% for the ASI group). Sample demographics and clinical information are presented in Table 1. All patients were included based on the following criteria: 1. Age between 18 and 60 years; 2. read, write, and speak Portuguese; 3. primary complaint of shoulder pain; 4. RCS or ASI clinical diagnosis. Patients were excluded if they had: 1. Neurological symptoms [36]; 2. positive thoracic outlet syndrome (screened with Allen's and Adson's tests) [36,37]; 3. history of shoulder surgery or fracture [38]; 4. structural injuries confirmed by imaging (e.g., ligaments and labrum); 5. symptoms reproduced by cervical examination [37,38]; 6. unable to commit to the scheduled treatments; 7. anti-inflammatory drug use.

Table 1. Patient characteristics.

		RCS Group (n = 117)	**ASI Group (n = 66)**
Age (mean (SD))		41.1 (12.2)	26.7 (10.3) **
Sex (%)	Female	48 (41.0)	37 (56.1)
	Male	69 (59.0)	29 (43.9)
Origin of symptoms (%)	Trauma	30 (25.6)	32 (48.5) **
	Non-traumatic	18 (15.4)	0 (0.0) **
	Overuse	69 (59.0)	27 (40.9) **
	Sub or Dislocation	0 (0.0)	7 (10.6) **
Length of symptoms (%)	Acute (0–2 weeks)	3 (2.6)	6 (9.1)
	Sub-acute (2–6 weeks)	19 (16.2)	13 (19.7)
	Chronic (+6 weeks)	95 (81.2)	47 (71.2)
Symptomatic side (%)	Dominant	80 (68.4)	53 (80.3) *
	Non-Dominant	34 (29.0)	9 (13.6) *
	Bilateral	3 (2.6)	4 (6.1) *

Abbreviations: RCS: Rotator cuff related pain syndrome; AS: Anterior shoulder instability; SD: Standard deviation; * $p < 0.05$; ** $p < 0.001$ between-groups.

2.3. Diagnostic Criteria

For RCS classification, patients were required to have current anterolateral acromial area pain [39], pain with active shoulder elevation [38], pain with passive or isometric resisted shoulder external rotation [40,41], and at least two positive results from the Neer test [42], Hawkins test [43], and Jobe/Empty can test [44]. Despite the poor diagnostic accuracy of these tests [45], they were included as assessments of impairment clinically associated with this syndrome [2]. Patients were classified ASI if they presented with current anterior or anterosuperior shoulder pain [46], pain with passive, active or resisted shoulder movement at 90° abduction combined with external rotation, and a positive apprehension-relocation-surprise test as this continuum has demonstrated the best overall diagnostic discriminative performance [47]. All patients gave written informed consent before data collection. This research had the approval of the Ethics Committee for Research of the School of Healthcare—Setúbal Polytechnic Institute.

2.4. Testing Procedure

The primary outcome measure of pain and function was the SPADI [33], complemented by the NRPS [48] and DASH [35]. The secondary outcome measures of scapular neuromuscular activity and control were a combination of surface electromyography and clinical observation. The surface electromyography (Physioplux system version 1.06 comprised of four pairs of 24-mm-diameter silver chloride gel surface electrodes, a ground electrode of the same type, four electrode pair cables connected to miniaturized differential amplifiers, and a main HUB unit that communicates via Bluetooth™ to a computer) enabled both patients and the physiotherapist to assess, monitor, and correct in real-time the muscular activation and behavior during the exercises. Clinical observation of the scapula's medial and inferior borders was used to detect scapular dyskinesis [classified as present if one or both scapular prominences (medial and inferior border) were observed during the glenohumeral movement or classified as absent if no prominence was observed [14]], using these specifications to increase the validity of the observation. Range of motion (ROM) was measured using a standard plastic goniometer (following the procedures for the glenohumeral joint motion measurements [49] recognizing the limitation of measurement without stabilization [50]). Graded glenohumeral flexors and abductors isometric muscle strength (GMS) was measured through isometric manual muscle testing (acknowledging the reduced sensitivity compared to dynamometry [51]). Outcome measures are presented in Table 2. Assessments and interventions were performed by the same examiner. All outcome assessments were carried out prior to the start of the weekly scheduled treatment (Figure 1), at 4-weeks and 2-years after the patient was discharged, hereinafter referred to as initial (baseline), 4-week (short-term), and follow-up (long-term) assessments, respectively (Figure 2).

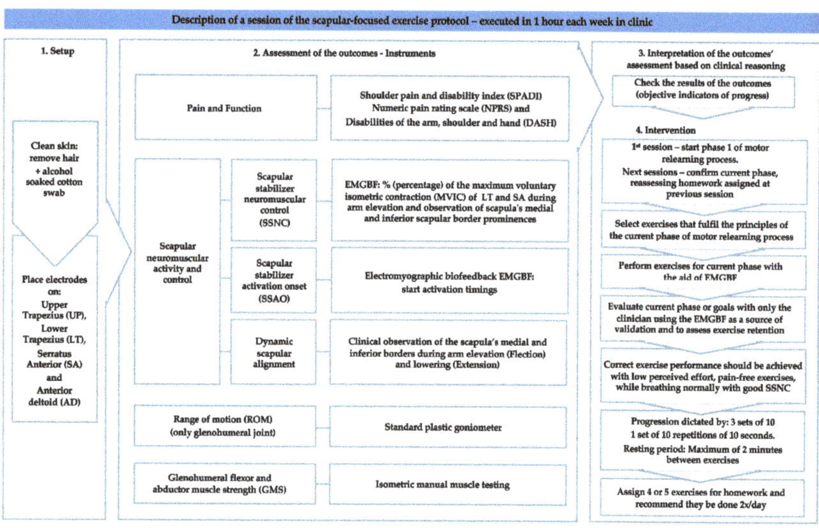

Figure 1. Resume of a session of the scapular-focused exercise protocol.

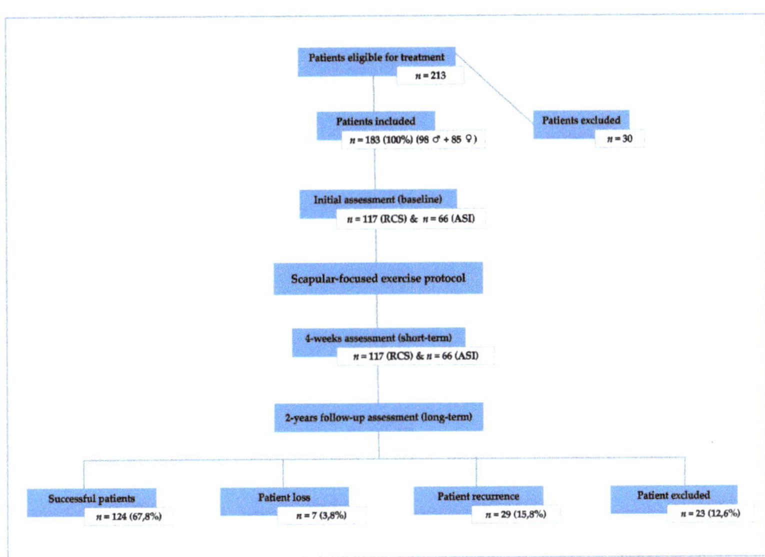

Figure 2. Scapular-focused exercise protocol flow diagram.

Table 2. Resume of testing procedure.

Outcome		Goal	Instrument	MCID	Assessment Procedures
Pain and Function		Determine pain intensity between assessment moments and measure and monitor function and symptoms over time	SPADI [32]	ranging from 8 to 13 points [33]	Filling in the SPADI questionnaire
			NPRS [48]	2.17 [34]	Patient asked to report the worst pain felt in the last week
			DASH [35]	10.2 [33]	Filling in the DASH questionnaire
Scapular neuromuscular activity and control	SSNC	Assess the muscular percentage of MVIC activity of LT, SA, and UT during arm elevation and lowering	EMGBF, Physioplux™ system version 1.06	N/A	Actively raise (Flexion) then lower (Extension) the arm at a controlled self-paced velocity through maximum painless ROM in the sagittal, scapular, and frontal planes from a natural standing position for one set of three repetitions with a 20-s pause between repetitions
	SSAO	Assess muscular activation onset during rapid active shoulder elevation	EMGBF, Physioplux™ system version 1.06	N/A	Actively raise (Flexion) the arm as rapidly as possible, without exacerbating pain or discomfort, to a maximum arm elevation angle of 45° in the sagittal, scapular, and frontal planes from a natural standing position for one set of three repetitions with a 20-s pause between repetitions
	Dynamic Scapular Alignment	Detect scapular dyskinesis	Clinical observation of the scapular medial and inferior border [14]	N/A	Clinical observation of the scapular medial and inferior border behavior during the arm elevation (Flexion) and lowering (Extension)
ROM		Assess glenohumeral ROM	Standard goniometer [49]	N/A	Normative ROM assessment with a standard goniometer
GMS		Assess glenohumeral flexor and abductor muscle strength	Isometric manual muscle testing [52]	N/A	Measured in a sitting position with the arm at 90° in the sagittal and frontal planes, respectively. Manual resistance was applied against the forearm with the elbow extended.

Abbreviations: MCID: Minimal Clinically Important Difference; SPAD: Shoulder pain and disability index; NPRS: Numeric pain rating scale; DASH: Disabilities of the arm, shoulder, and hand; SSNC: Scapular stabilizer neuromuscular control; MVIC: Maximum voluntary isometric contraction; LT: Lower trapezius; SA: Serratus anterior; UT: Upper trapezius; EMGBF: Electromyographic biofeedback; ROM: Range of motion; SSAO: Scapular stabilizer activation onset; GMS: Glenohumeral flexor and abductor muscle strength; N/A: Non applicable.

All outcomes were assessed in the initial, weekly, 4-week, and follow-up moments as summarized in Table 2 and described in detail in the Appendix A.

2.5. Treatment Protocol

The treatment protocol was developed using the sequential stages of motor relearning, cognitive, associative, and autonomous [53], as a framework, while promoting the integration of local and global muscle function [54]. The treatment was divided into three phases (Appendix A) and conducted in weekly sessions to both: 1. Objectively assess the progress towards the outcomes and 2. Treat patients using exercises for the main purpose of increasing scapular neuromuscular activity and control.

2.6. Statistics

Descriptive statistics (means and frequency) were used to characterize the groups and variables' distribution. The Mann-Whitney U test and Wilcoxon signed-rank test were used to compare the quantitative outcomes. Fisher's exact test and McNemar exact test were used to compare the qualitative outcomes. Regarding the missing values (present only at follow-up) a complete-case analysis approach was adopted, assuming the missing data is completely random and unrelated to any of the variables involved in the study. The significance level was set at $p < 0.05$ and all statistical analysis was performed using the Python Software Foundation, Python Language Reference, version 3.7, available at http://www.python.org (accessed on 3 May 2020).

3. Results

At baseline, in the initial assessment, both RCS and ASI groups had high levels of pain and poor levels of function (SPADI, NPRS, and DASH), decreased scapular neuromuscular activity and control (SSNC, SSAO, and scapular alignment), decreased ROM and GMS. There was a difference in the scores of SPADI and DASH 1st and 3rd parts ($p < 0.05$) but none in any of the secondary outcome measures (Table 3).

After completion of the 4-weeks intervention, all outcomes improved compared with the baseline ($p < 0.05$) in both groups and the pain and function MCID values were met. Differences were found between the groups in the outcome SPADI, NPRS, and DASH 1st part at this short-term assessment ($p < 0.05$) (Table 3).

At the 2-year follow-up assessment, for the RCS group, there were no differences with the 4-week assessment in the level of SPADI, NPRS, DASH 1st and 3rd parts, SSAO, ROM, and GMS, reflecting the maintenance of the results in the long-term. However, differences were found in the DASH 2nd part, SSNC, and dynamic scapular alignment, which indicate a loss of the gains in these outcomes in the long-term ($p < 0.05$) (Table 3).

For the ASI group, there were no differences with the 4-week assessment in the level of SPADI, NPRS, DASH 1st part, SSAO, ROM, and GMS but differences were found in the DASH 2nd and 3rd parts, SSNC, and dynamic scapular alignment at the long-term ($p < 0.05$). At the 2-year follow-up, the two groups were only different in the levels of SPADI and DASH 1st part (Table 3).

At the 2-year follow-up, five (2.7%) patients were unable to return for an objective re-assessment and instead were contacted by either email or phone to answer the outcomes not requiring their presence, seven (3.8%) patients were unreachable, 29 (15.8%) patients reported returning to physiotherapy between the treatment protocol and follow-up to seek new treatment due to the same shoulder problem (recurrence), and 23 (12.6%) were not included in the 2-year follow-up as they reported having had new traumatic incidents unrelated to their treatment in the study that resulted in them seeking further health care services (e.g., shoulder surgery, fractures, muscle or tendons ruptures, etc.).

Table 3. Comparison of outcomes within groups and between-groups.

		RCS Group			ASI Group		
		Initial (n = 117)	**4-Weeks** (n = 117)	**2-Year Follow-Up** (n = 93)	**Initial** (n = 66)	**4-Weeks** (n = 66)	**2-Year Follow-Up** (n = 54)
	SPADI (0–100)	42.07 ± 18.64	9.03 ± 8.21 **	8.62 ± 15.12	32.74 ± 19.50 ‡‡	4.80 ± 5.66 **‡‡	7.24 ± 15.78 ‡
	NPRS (0–10) (Worst Pain felt)	5.85 ± 1.97	1.58± 1.29 **	1.46± 2.05	5.27 ± 2.34	0.91 ± 1.16 **‡‡	1.21 ± 1.96
	DASH 1st part (0–100 point)	33.55 ± 16.53	7.63 ± 6.85 **	7.51 ± 12.92	28.47 ± 15.48 ‡	4.93 ± 5.78 **‡‡	4.37 ± 9.02 ‡
	DASH 2nd part (0–100 point)	10.69 ± 19.25	2.83 ± 6.84 **	1.58 ± 7.54 *	8.60 ± 18.16	1.80 ± 4.63 **	0.22 ± 1.17 *
	DASH 3rd part (0–100 point)	45.88 ± 29.01	12.50± 14.27 **	10.00± 17.59	53.80 ± 31.01 ‡	9.66 ± 12.61 **	8.15 ± 16.47 *
SSNC	Diminished (poor or moderate)	117 (100.00)	78 (66.67) **	61 (65.59) *	66 (100.00)	39 (59.09) **	22 (47.74) *
	Good	0 (0.00)	39 (33.33) **	32 (34.41) *	0(0.00)	27 (40.91) **	32 (59.26) *
SSAO (ms)	Feedback	59 (50.43)	22 (18.80) **	18 (19.35)	32 (48.48)	8 (12.12) *	7 (12.96)
	Feedforward	58 (49.57)	95 (81.20) **	75 (80.65)	34 (51.52)	58 (87.88) *	47 (87.04)
Dynamic Scapular Alignment	"YES" scapula dyskinesis (IB, MB or both prominences)	117 (100.00)	85 (72.65) **	49 (52.69) *	100 (100.00)	43 (65.15) **	32 (59.26) *
	"NO" scapula dyskinesis (no prominences)	0 (0.00)	32 (27.35) **	44 (47.31) *	0 (0.00)	23 (34.85) **	22 (40.74) *
ROM	Decreased	102 (87.18)	13 (11.11) **	9 (9.68)	51 (77.27)	1 (1.52) **	1 (1.85)
	Normal	15 (12.82)	104 (88.89) **	84 (90.32)	15(22.73)	65 (98.48) **	53 (98.15)
GMS	Decreased	114 (97.44)	30 (25.64) **	19 (20.43)	65 (98.48)	13 (19.70) **	8 (14.81)
	Normal	3 (2.56)	87 (74.36) **	74 (75.57)	1 (1.52)	53 (80.30) **	46 (85.19)

Abbreviations: RC: Rotator cuff related pain syndrome; ASI: Anterior shoulder instability; SPADI: Shoulder pain and disability index; NPRS: Numeric pain rating scale; DASH: Disabilities of the arm shoulder, and hand; DASH 1st part: Daily life activities questions; DASH 2nd part: Work optional module; DASH 3rd part: Sport/performing arts optional module; SSNC: Scapular stabilizer. Neuromuscular control; SSAO: Scapular stabilizer activation onset; IB: Inferior border of the scapula; MB: Medial border of the scapula; ROM: Range of motion; GMS: Glenohumeral flexors and abductors isometric muscle strength; * $p < 0.05$; ** $p < 0.001$ within groups; ‡ $p < 0.05$; ‡‡ $p < 0.001$ between-groups.

Repeated measures for time were unfeasible due to the non-normal distribution of outcomes. The power analysis by t-tests was computed for pain and function outcomes considering the difference between two dependent means. The results obtained showed an excellent power for all variables (0.99 < d < 1.00), given the sample size of 183 participants. These results boost confidence in the outcomes reported, reinforcing the relevance of the intervention and assessment methods on the recovery efficacy of these patients, between their initial and short-time assessments and between their short-time and long-term assessments.

4. Discussion

In this study of the measures taken at the initial assessment, the RCS and ASI groups were different for age, and the outcome of SPADI, DASH 1st, and 3rd part, with higher mean age and higher SPADI and DASH disability scores for the RCS group. Older patients usually present with worse function levels than younger patients [25,29]. The between group analysis, comparing the short-term results for pain and function demonstrated differences between the two groups (Table 3).

The within-group analysis, comparing results between the initial and 4-weeks (short term) assessments, showed that clinically meaningful changes were achieved for pain and function over time in both groups. Both outcomes reached their predefined MCID and the other outcomes presented meaningful improvements. While both groups improved significantly, it was not the diagnostic category that determined the specific exercises, rather it was an assessment of the patients' movement/control impairments. This is consistent with the view that, even when following a protocol or recommended guidelines, the management should be tailored to the patients' pain and disability presentations rather than the hypothesized clinical diagnostic categorization [55,56].

Two years after discharge, despite a slight loss in the outcomes, the scores of SPADI NPRS, and DASH 1st part for the ASI group and the scores of SPADI, DASH 1st, and 3rd part for the RCS group as well as the results of ROM and GMS for both groups were not different, demonstrating that the protocol of good results was not temporary. For the outcome scapular neuromuscular activity and control, only the SSAO component maintained the 4-week results through to the long-term. The SSNC and dynamic scapular alignment components presented differences.

For pain and function, the results of SPADI, NPRS, and DASH at 4-weeks were very good for both groups. NRPS at the short-term had a mean of 1.58 ($\pm$1.29) for the RCS group and 0.91 ($\pm$1.16) for the ASI group, which is better than most studies incorporating scapular exercise to treat RCS or ASI associated shoulder pain and dysfunction [9,18,23–25,28,29,34]. Disability improvement presented similar gains with this study compared to others (SPADI [5,26]; DASH [29,41]). These results corroborated studies that suggested a rehabilitation program incorporating motor control exercises is effective for reducing pain and disability for patients with RCS [5,18] and ASI [9,24].

The initial results of SSNC of decreased activity in LT and SA muscles corroborated the presupposition that shoulder dysfunctions comprise an alteration in the scapulothoracic stabilizer function [37], consistent with the findings of DeMey et al. [26]. Contrarily, Larsen et al. [28] reported a non-significant tendency to a higher level of mean UT, LT, and SA muscle activity in RCS patients compared to those without RCS. Collectively, these findings support the view that diagnostic categorization does not predict the muscle function, rather it is the presence of muscle dysfunction that represents either a risk variable that may contribute to pain and disability or a central nervous system response to pain and threat [57]. The initial SSNC findings in this study may reflect dysfunction in the feedforward processing present even before the onset of movement [58]. This general initial motor plan is expected to be fine-tuned using real-time internal feedback mechanisms. With a planning-control model underpinning the assessment and management of motor control/function, two principles guided the management of abnormal neuromuscular activity and motion in this study: (1) Treatment strategies to re-educate neuromuscular activity and control incorporating criteria for a preferred pattern of muscle activation prior to and during the execution of a motor command; (2) optimization of internal feedback mechanisms, so a deviation or perturbation of predicted movement can be effectively detected and corrected in real-time. Roy et al. [59,60] showed that conscious movement training with feedback causes immediate effects on motor strategies and can restore the force-couple activation in the scapular muscles, especially the stabilizers, consistent with the improvement in LT and SA activity in both groups of this study.

Concerning SSAO, half of the sample in this study already presented a feedforward mechanism [61] rather than a feedback mechanism found in other studies [36,62,63]. This highlights that the pattern of activation alone is not responsible for the patients' symptoms and disability. This is not surprising as physical impairments, whether they are of posture, mobility, motor control or others, do not predict pain and disability [55] and motor responses to pain are variable [64]. Rather, physical impairments, in this case in SSAO, can only be judged as potential predisposing or contributing factors that may contribute to some patients' disabilities depending on their lifestyle behaviors and requirements. Through the exercise protocol, patients who initially presented with a feedback mechanism changed to a feedforward one, as in other studies [65,66] where it is defended that the muscle pattern of onset can be improved by therapeutic exercises [65], and that the mechanisms can be trained, shifting from feedback to feedforward, while the movement is trained and repeated [66]. Contrary to these findings, DeMey et al. [26] observed no change in the recruitment timing after the treatment and Larsen et al. [28] saw no significant differences in muscle activation onset between patients with and without RCS, however neither of those studies incorporated biofeedback or motor performance criteria for facilitating learning and guiding the progression of exercise.

Contemporary neuroscience and motor control theory hold that pain alters motor patterning/control variably in response to the individual's conscious and unconscious perception of threat, leading to changes in movement and motor function to provide protection from further pain, injury or threat [57,64,67]. Strategies that reduce pain, dysfunction and threat generally will, in turn, alter central processing, motor control, and disability [55,62,65]. As such, the reduction in the level of pain and the improvement in the level of function found in this study cannot be attributed to a single variable such as motor control. However, the approach to the scapular-focused exercise, emphasizing non-aggravating controlled progression of exercise with feedback, encouragement, and guidance in load management, likely contributed to reduced threat alongside improved control/strength leading to improvement.

The dynamic scapular alignment showed significant differences between the initial and 4-week assessments with very good results, but around 40% of the patients lost their gains at the follow-up, despite the great results of the pain and function, SSAO, ROM, and GMS outcomes. This supports the previous literature challenging the relationship between scapular alignment and RCS [1,17,68]. While a scapular-focused exercise protocol has been demonstrated in this study to be effective at reducing pain and disability; improving dynamic scapular alignment alone is not predictive of disability; and strategies to evaluate the contribution of scapular and other malalignments, such as the shoulder symptom modification procedure, described by Lewis [1], may prove helpful in predicting the potential contribution of dynamic scapular alignment to the individual patients' pain and dysfunction. Moreover, the kinematic analysis would provide a more objective analysis of scapular alignment in any future study.

High recurrence rates are common in shoulder dysfunctions, particularly in sport activities [68]. At a 3-month follow-up, Struyf et al. [18] found maintenance of the effects of a scapular-focused treatment in patients with RCS. Given the increasing body of evidence from studies demonstrating no increased clinical benefit from surgery compared with exercise [69], it seems reasonable that patients with RCS or ASI associated shoulder pain and dysfunction should undergo a conservative trial of rehabilitation before considering surgical options. In the current study, only 29 patients (15.8%) had a recurrence episode (new symptoms due to the same problem that brought them to physiotherapy in the first place).

The results of this study support other research [13,27,40,69–73] that a progressive scapular-focused approach incorporating feedback and home management can significantly reduce pain and increase function in RCS and ASI associated shoulder pain. Whether the specific attention is to motor control, in particular, SSNC requires further research.

Both 1st and 2nd hypotheses were confirmed successfully with a reduction of pain and an increase of function with differences at the short-term assessment, and no differences between the short- and the long-term. Some limitations should be considered that restrict the generalizability of results: (1) No direct cause-and-effect relationship can be drawn from this protocol and these results as it did not include a control group. Further studies are needed to assess the effectiveness of this protocol against other rehabilitation approaches and clarify the contribution of EMGBF and possibly the kinematic feedback [74,75]; (2) although the diagnostic criteria reflect commonly used clinical features, the lack of gold standard diagnostic criteria compromises the RCS and ASI cohort distinctions of this study; (3) all procedures were conducted by the same researcher, although bias was minimized by the principal outcomes of pain and function being patient-rated. For the scapular neuromuscular activity and control outcome, bias was minimized by assessing SSNC and SSAO with the real-time EMGBF automatically recorded by the system. Additionally, data collection by the same researcher with extensive experience with shoulder patients and a standardized exercise approach using the EMGBF software provides consistency in procedures and measures. Both the usability and learnability of the EMGBF software and the protocol's procedures should be assessed in the future, using a range of both novice and expert physiotherapists.

5. Conclusions

The presented findings suggest that a well-described scapular-focused exercise protocol, with the aid of real-time EMGBF feedback and home management, can reduce pain and increase function, as well as scapular neuromuscular activity and control, ROM, and GMS in patients with shoulder dysfunctions in the short-term. At the long-term, it appears to maintain the gains of pain and function, and the gains of SSAO, ROM, and GMS, but not for SSNC and dynamic scapular alignment. The inclusion of both ASI and RCS impairment associated groups adds evidence to the limited body of knowledge on the effect of physiotherapy on these types of shoulder dysfunctions.

Author Contributions: Conceptualization, C.d.S. and R.M.; methodology, C.d.S. and R.M.; formal analysis, C.d.S.; writing—original draft preparation, C.d.S., M.A.J., and R.M.; writing—C.d.S., M.A.J. and R.M.; supervision, R.M. All authors have read and agreed to the published version of the manuscript.

Funding: This research received no external funding.

Institutional Review Board Statement: The study was conducted according to the guidelines of the Declaration of Helsinki, and approved by the Ethics Committee for Research of the School of Healthcare—Setúbal Polytechnic Institute. Reference: 0107-2014 / ECR-SH-SPI.

Informed Consent Statement: Informed consent was obtained from all subjects involved in the study. Written informed consent has been obtained from the patient(s) to publish the data.

Data Availability Statement: Individual de-identified participant data that underlie the results reported in this article will be made available to investigators whose proposed use of the data has been approved by an independent and identified review committee. Proposals should be sent to the corresponding author and requesters will need to sign a data access agreement.

Acknowledgments: We thank Lucian Radu from the Department of Sciences and Technologies of the Autonomous University of Lisbon for his help and discussions on the statistical analysis.

Conflicts of Interest: The authors declare no conflict of interest.

Appendix A. Detailed Description of the Scapular-Focused Exercise Protocol

In the initial, weekly, 4-week, and follow-up moments, the following assessment procedures were performed:

To assess the primary outcome of pain and function, the self-administered questionnaire shoulder pain and disability index (SPADI) [32] was used. The reported minimal clinically important difference (MCID) ranging from 8 to13 points [33] was used to determine the clinical significance of the results. To complement the assessment of this primary outcome, the numeric pain rating scale (NPRS) [48] was used to determine the pain intensity between assessment moments with a MCID of 2.17 points [34], and the self-administered questionnaire disabilities of the arm, shoulder, and hand (DASH) was used to monitor the function with a MCID of 10.2 points [33]. DASH is divided into three parts (1st: Daily life activities, 2nd: work activity, and 3rd: Sports/arts activity, respectively) [35].

To assess the scapular stabilizer activation onset (SSAO) and scapular stabilizer neuromuscular control (SSNC), the electromyographic biofeedback (EMGBF), Physioplux system version 1.06 was used. The EMGBF system was comprised of four pairs of 24-mm-diameter silver chloride gel surface electrodes, a ground electrode of the same type, four electrode pair cables connected to miniaturized differential amplifiers, and a main HUB unit that communicates via Bluetooth™ to a computer. Each amplifier had a voltage gain of 1000, input impedance higher than 100 MΩ, a common mode rejection ratio of 110 dB, and a bandwidth (−3 dB) of 25 to 500 Hz. The four amplified electromyographic (EMG) signals were then collected by the main HUB unit and converted to a digital format with a 12-bit resolution at a sampling rate of 1000 Hz. An envelope function was applied to each EMG channel using the root mean square of the mean of the absolute signal value over the last 100 milliseconds (ms). The muscle onset was determined when the EMG signal amplitude was 3 standard deviation points above the baseline signal for a 25 ms window. The baseline signal was determined by the resting EMG signal during 500 ms, collected before each activity. Prior to the surface electrode application, the patients' skin was shaved (if necessary) and cleaned with alcohol to reduce skin impedance. The placement of the surface electrodes and the normalization of EMG data and muscle testing positions were based on the work of Ekstrom et al. [76] and Hermens et al. [77] (Table A1).

Table A1. Placement of the electrodes and normalization of EMG data.

Muscle	Placement of the Electrodes	Position	Normalization: Muscular Action to Measure the Maximum Voluntary Isometric Contraction
Upper Trapezius [76,77]	Between C7 spinous process and the lateral tip of the acromion	Sitting position with no back support. Shoulder abducted to 90° (no abduction in the case of pain) with the neck side-bent to the same side, rotated to the opposite side	Pressure applied to extend the head above the elbow (or to shoulder elevation in the case of pain)
Lower Trapezius [76]	At 2/3 on the line from the root of the spine of the scapula to the 8th thoracic vertebra	Sitting position with no back support Arm raised above the head in line with the lower trapezius muscle	Pressure applied against the arm elevation
Serratus Anterior [76,77]	Vertically along the mid-axillary line at the 6th rib through the 8th rib	Sitting position with no back support. Shoulder abducted to 125° in the scapular plane	Pressure applied above the elbow and at the inferior angle of the scapula attempting to de-rotate the scapula
Anterior Deltoid [76]	At one finger width distal and anterior to the acromion	Sitting position with no back support. Place the humerus in a slight external rotation to increase the effect of gravity on the anterior fibers	Pressure applied on the antero-medial surface of the arm, against abduction and flexion

All electrodes were placed over the belly, in line with the fiber directions, with an inter-electrode distance of 2.5 cm, and a ground electrode was placed over the contra-lateral clavicle. After measuring three 5-s maximum voluntary isometric contractions (MVIC) for each muscle [65] with a 20-s pause between MVIC, patients raised their arm in the sagittal plane from their standing postural natural position for two sets of three repetitions, again with a 20-s pause between repetitions. In the first set, patients were asked to perform the movement as rapidly as possible, without exacerbating pain or discomfort, to a maximum arm elevation angle of 45°, which was intended to record SSAO. The second set was performed in a controlled self-paced velocity through the patients' maximum painless ROM both concentrically and eccentrically to assess SSNC.

SSAO can be classified as being a feedforward or feedback onset. The feedforward activation onset represents the anticipatory muscle activation that occurs prior to the mobilizer muscles and the feedback activation onset is a muscle activation that occurs after the designated feedforward period [61]. By definition, and used for this study, a feedforward activation pattern (considered as normal) was the activation of lower trapezius (LT) and serratus anterior (SA) 100 ms before to 50 ms after the anterior deltoid (AD) activation onset [61]. This outcome was computed using an accurate statistical-based method for the muscle onset detection [78]. A feedback pattern was an activation of LT and SA greater than 50 ms after AD activation [61].

SSNC levels were classified as follows: (i) Reduced when observing LT and SA activity between 0–10% of MVIC; (ii) moderate when observing LT and SA activity between 10–30% of MVIC and less than 20% of the upper trapezius (UT) MVIC activity; (iii) good when observing LT and SA activity greater than 30% of MVIC and less than 20% of UT MVIC. These levels were determined while patients concentrically flexed their arm to 90° of elevation or within their non-painful available ROM, and eccentrically returned to the initial position. The muscle MVIC percentages considered for the "good" classification were extracted from Ludewig and Cook's [36] published results.

To assess the dynamic scapular alignment during active arm elevation and lowering, a clinical observation of the scapular medial border and the inferior angle was used to detect scapular dyskinesis. The dynamic scapular alignment was defined as normal when no prominence of the scapula medial and inferior borders was observed. The adopted dichotomous classification of scapula dyskinesis ("yes" when observing scapula medial and inferior scapular border or scapula medial border prominence or "no" when none is observed) was based on McClure et al. [14].

To assess the range of motion (ROM) a standard plastic goniometer was used and graded normal when the values corresponded with the normative ROM values expected for each movement and age group [49].

The glenohumeral flexor and abductor isometric muscle strength (GMS) was assessed by the isometric manual muscle testing [52], in a sitting position with the arm at 90° in the sagittal and frontal planes, respectively. Manual resistance was applied against the forearm with the elbow extended, graded normal (level 5 on a scale of 1 to 5) when the patient withstood the test position against a strong pressure [52], for 3 s, without losing the testing position.

Both evaluations of the outcomes and exercises intervention were recorded in the assessment, reassessment, and treatment form (Figures A1 and A2):

Assessment and treatment form

Date | Name

Personal data:
Birthday: / / Occupation:
Phone number: Sports/Hobbies:
Sex: M F Frequency: /week
Dominant side: Right Left Duration: h min

Shoulder Dysfunction
Shoulder Impingement Syndrome Anterior Shoulder Instability

Pain assessment: Localization and behavior of the symptoms
Morning:
During the day:
End of the day:
During night:
Moment Pain /10 NPRS Worst Pain felt since last week /10 NPRS
Objective *

Clinical History
Present: Past:
Traumatic Non-traumatic
Acute Sub-acute Chronic
Subluxation Dislocation Social-economics:
Repetitive movements
Other

Complementary information
Medication: Exams: X-ray Eco MRI CTscan
Painkiller NSAIDs Results:
Other General Health

Range of motion (ROM)

Passive movements		Active movements	
Flexion	º Painful	Flexion	º Painful
Abduction	º Painful	Abduction	º Painful
Ext Rot	º Painful	Ext Rot	º Painful
Int Rot	º Painful	Int Rot	º Painful

Glenohumeral flexor and abductor isometric muscle strength (GMS)
Flexors /5 Painful
Abductors /5 Painful

Special tests

RCS:	Positive	Negative	ASI:	Positive	Negative
Neer test			Apprehension		
Hawkins test			Relocation		
Jobe test			Surprise		
Others:			Others:		

Questionnaires
DASH 1st part /100 SPADI 1st part /100
DASH 2nd part /100 SPADI 2nd part /100
DASH 3rd part /100 SPADI total /100

Dynamic Scapular Alignment
Clinical observation of the medial border and inferior angle during arm elevation and lowering::
Medial border prominence
Inferior border prominence Yes No Scapular dyskinesis
No prominence

Scapular stabilizer activation onset
Lower trapezius (LT): Feedforward Feedback
Serratus Anterior (SA): Feedforward Feedback

Scapular stabilizer neuromuscular control
Reduced [0-10%] of MVIC of LT + SA
Moderated [10-30%] of MVIC of LT + SA
Good [>30%] of MVIC of LT + SA

Phase of motor relearning process And treatment
Cognitive Phase 1 (i) (ii)
Associative Phase 2 (i) (ii)
Autonomous Phase 3 (i) (ii)

Exercises: sets, duration or repetition

Exercise 1	Set	Reps	Duration	sec/min
Exercise 2	Set	Reps	Duration	sec/min
Exercise 3	Set	Reps	Duration	sec/min
Exercise 4	Set	Reps	Duration	sec/min
Exercise 5	Set	Reps	Duration	sec/min

Home exercises

Exercise 1	Set	Reps	Duration	sec/min	Daily
Exercise 2	Set	Reps	Duration	sec/min	Daily
Exercise 3	Set	Reps	Duration	sec/min	Daily
Exercise 4	Set	Reps	Duration	sec/min	Daily
Exercise 5	Set	Reps	Duration	sec/min	Daily

Figure A1. Assessment and treatment form.

Reassessment and next treatment sessions

Date	Name
Pain assessment: Localization and behavior of the symptoms	Morning: ___ During the day: ___ End of the day: ___ During night: ___ Moment Pain ___/10 NPRS Worst Pain felt since last week ___/10 NPRS Objective * ___
Range of motion (ROM)	Passive movements: Flexion ___° Painful ☐; Abduction ___° Painful ☐; Ext Rot ___° Painful ☐; Int Rot ___° Painful ☐ Active movements: Flexion ___° Painful ☐; Abduction ___° Painful ☐; Ext Rot ___° Painful ☐; Int Rot ___° Painful ☐
Glenohumeral flexor and abductor isometric muscle strength (GMS)	Flexors ___ /5 Painful ☐ Abductors ___ /5 Painful ☐
Special tests	RCS: Positive / Negative — Neer test ☐ ☐; Hawkins test ☐ ☐; Jobe test ☐ ☐; Others: ☐ ___ ASI: Positive / Negative — Apprehension ☐ ☐; Relocation ☐ ☐; Surprise ☐ ☐; Others ☐ ___
Questionnaires	DASH 1st part ___/100; DASH 2nd part ___/100; DASH 3rd part ___/100 SPADI 1st part ___/100; SPADI 2nd part ___/100; SPADI total ___/100
Dynamic Scapular Alignment	Clinical observation of the medial border and inferior angle during arm elevation and lowering: Medial border prominence ☐; Inferior border prominence ☐; No prominence ☐ > Yes ☐ No ☐ Scapular dyskinesis
Scapular stabilizer activation onset	Lower trapezius (LT): Feedforward ☐ Feedback ☐ Serratus Anterior (SA): Feedforward ☐ Feedback ☐
Scapular stabilizer neuromuscular control	Reduced [0–10%] of MVIC of LT + SA ☐ Moderated [10–30%] of MVIC of LT + SA ☐ Good [>30%] of MVIC of LT + SA ☐
Phase of motor relearning process And treatment	Cognitive Phase 1 ☐ (i) ☐ (ii) ☐ Associative Phase 2 ☐ (i) ☐ (ii) ☐ Autonomous Phase 3 ☐ (i) ☐ (ii) ☐
Exercises sets, duration or repetition	Exercise 1 ___ Set ___ Reps ___ Duration ___ sec/min Exercise 2 ___ Set ___ Reps ___ Duration ___ sec/min Exercise 3 ___ Set ___ Reps ___ Duration ___ sec/min Exercise 4 ___ Set ___ Reps ___ Duration ___ sec/min Exercise 5 ___ Set ___ Reps ___ Duration ___ sec/min Exercise 6 ___ Set ___ Reps ___ Duration ___ sec/min Exercise 7 ___ Set ___ Reps ___ Duration ___ sec/min Exercise 8 ___ Set ___ Reps ___ Duration ___ sec/min
Home exercises	Exercise 1 ☐ Set ___ Reps ___ Duration ___ sec/min Daily ___ Exercise 2 ☐ Set ___ Reps ___ Duration ___ sec/min Daily ___ Exercise 3 ☐ Set ___ Reps ___ Duration ___ sec/min Daily ___ Exercise 4 ☐ Set ___ Reps ___ Duration ___ sec/min Daily ___ Exercise 5 ☐ Set ___ Reps ___ Duration ___ sec/min Daily ___ Exercise 6 ☐ Set ___ Reps ___ Duration ___ sec/min Daily ___ Exercise 7 ☐ Set ___ Reps ___ Duration ___ sec/min Daily ___ Exercise 8 ☐ Set ___ Reps ___ Duration ___ sec/min Daily ___

Figure A2. Reassessment and next treatment form.

The treatment protocol was developed using the sequential stages of motor relearning, cognitive, associative, and autonomous [53], into three phases (Table A2), as a framework, while promoting the integration of local and global muscle function [54].

Table A2. Motor relearning phases of the treatment protocol.

Motor Relearning Phase	Phases Description and Purpose	Progression
Phase 1	Facilitate patient pain-free awareness and dynamic control of scapulothoracic neutral zone through its stabilizers' co-activation, namely, LT and SA, with a minimum participation of UT (or other scapulothoracic, glenohumeral, and spinal muscles)	(i) Patient should be able to activate scapular stabilizer muscles and dissociate their activation from other scapulothoracic, glenohumeral, and spinal muscles without pain provocation; (ii) Be capable of moving the scapula from different postural orientations and positions to its neutral zone (maintaining this position) through the co-activation of its scapular stabilizers in low-load exercises without pain provocation.
Phase 2	Progressively integrate scapular neuromuscular activity and control skills gained in Phase 1 during pain-free directional shoulder movements. It is currently accepted that the scapula axis of rotation changes with the increasing arm elevation and plane of movement [79]. This implies the integration of LT and SA co-activation with other scapulothoracic force generators, such as UT, and simultaneous coordination with glenohumeral muscles.	(i) Maintain scapulothoracic neutral zone by activating its stabilizers while raising (fexion) the arm (<30°) in different elevation planes, the primary aim of this stage being the focus on the scapular neuromuscular activity and control setting phase [80,81]; (ii) Arm elevation movements (>30°) should be chosen so that their primary elevation plane or direction matches that of symptom producing movements and progressively explored through the available pain-free glenohumeral ROM (concentrically and eccentrically).
Phase 3	Expected learning transfer of motor skills acquired in Phases 1 and 2 to functional activities.	(i) Fragmenting daily living activities into less complex achievable movements that can be progressively trained; (ii) and During normal function, occupational, recreational, and sports activities.

Abbreviations: LT: Lower trapezius; SA: Serratus anterior; UT: Upper trapezius; ROM: Range of motion.

The general principles for exercise prescription [82] recommend the use of variables such as the number of exercises, series, repetitions, recovery time, and the use of a periodization model [83] to support the exercise program prescription and progression. In this study, the magnitude of stimulus and progression (either in the same exercise or to progress to the next exercise or phase) were tailored to each patient's performance and re-assessment, while operating within the protocol's structure.

The progression guidelines were the following, as described in Table A3:

Table A3. Progression guidelines.

Progression Guidelines:	
Exercise complexity	Two possible sources: (i) Mechanical load, which included exercise variations that required greater arm elevation angles or the use of weights;(ii) Task or motor planning-control difficulty, which involved tasks and exercises in which it is necessary to incorporate both feedforward and feedback mechanisms of motor performance [58,84].
Feedback from the EMGBF	Provided during all sessions to facilitate the best performance at each step. However, to progress to the next exercise or phase, the patient had to demonstrate their capability to reproduce the same performance without visual feedback. At this stage, EMGBF was used by the clinician to confirm the correct exercise performance.
Perceived effort	Although a high-perceived effort is acceptable at the beginning of each phase or while increasing exercise complexity, correct exercise performance should be achieved with low perceived effort, pain-free exercise performance, and with normal breathing.
Sets, repetitions and endurance	In the absence of normative data for endurance, exercises for this population were progressed when the patient could perform three sets of 10 repetitions or hold the specified position for one set of 10 repetitions of 10 s with no pain, low perceived effort (although a high-perceived effort is acceptable at the beginning of each phase or while increasing exercise complexity), normal breathing, and good SSNC. Note, while this arbitrary performance criteria was effective for this population, the number of sets, repetitions or holding time goal for progression will vary with different patient groups according to sport, work, and lifestyle requirements.
Resting time between exercises	Although patients were encouraged to rest the least time possible between exercises, they could rest for a maximum of 2 min between exercises (especially high-loaded) but not between sets or repetitions [65].

Abbreviations: EMGBF: Electromyographic biofeedback; SSNC: Scapular stabilizer neuromuscular control.

The treatment protocol was conducted in weekly sessions to both: 1. Objectively assess the progress towards the outcomes and 2. treat patients using exercises for the main purpose of increasing scapular neuromuscular activity and control. EMGBF was used to provide patients with a real-time quality indicator for their exercise performance. SSNC was defined as a threshold (% of MVIC) of muscle activation. A minimum level of activity of LT and SA and a maximum level of activity of UT were initially set so that patients were able to achieve the objectives easily, and then thresholds were progressively increased towards their target cut-off points with a maximum step increase of 5% of MVIC. The EMGBF software was used as a form of augmented feedback to continually provide exercise performance feedback and software parameters modeled to display a green or red bar when muscle activity levels were, respectively, correctly and incorrectly attained.

At the end of each session, five homework exercises with print outs regarding sets, repetitions, and recovery time, to be completed twice daily, were assigned to the patient based upon the exercises correctly performed during the session. A schema of the scapular-focused treatment protocol, with some examples of the typical exercises executed can be seen, as follows, in Figure A3.

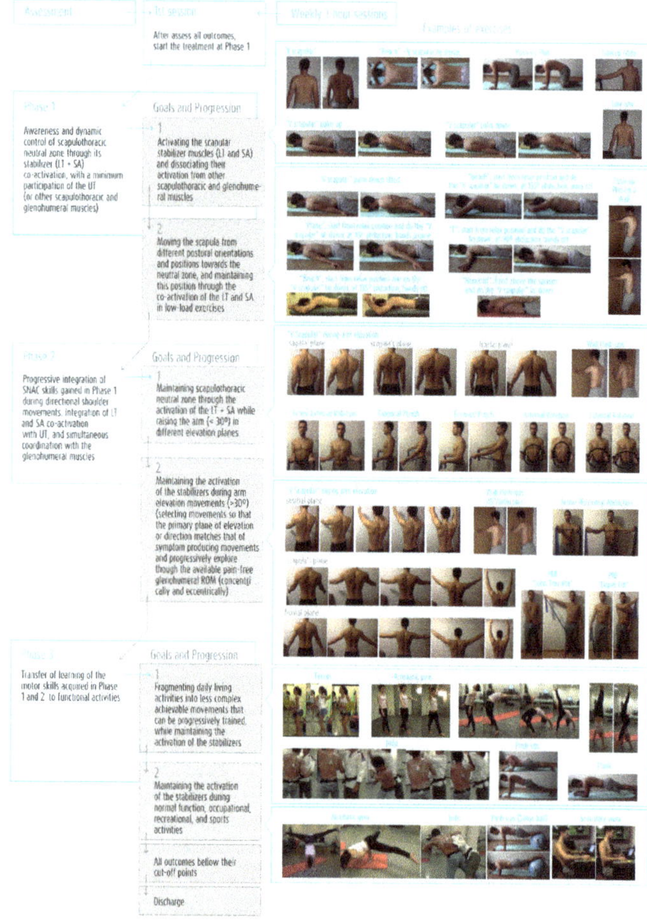

Figure A3. Scapular-focused treatment protocol.

References

1. Lewis, J. Rotator cuff related shoulder pain: Assessment, management and uncertainties. *Man. Ther.* **2016**, *23*, 57–68. [CrossRef] [PubMed]
2. Diercks, R.; Bron, C.; Dorrestijn, O.; Meskers, C.; Naber, R.; de Ruiter, T.; Willems, J.; Winters, J.; van der Woude, H.J. Guideline for diagnosis and treatment of subacromial pain syndrome. *Acta Orthop.* **2014**, *85*, 314–322. [CrossRef] [PubMed]
3. Gibson, K.; Growse, A.; Korda, L.; Wray, E.; MacDermid, J.C. The effectiveness of rehabilitation for nonoperative management of shoulder instability: A systematic review. *J. Hand Ther.* **2004**, *17*, 229–242. [CrossRef]
4. Michener, L.; McClure, P.; Karduna, A. Anatomical and biomechanical mechanisms of subacromial impingement syndrome. *Clin. Biomech.* **2003**, *18*, 369–379. [CrossRef]
5. Bae, Y.; Lee, G.; Shin, W. Effect of motor control and strengthening exercises on pain, function, strength and the range of motion of patients with shoulder impingement syndrome. *J. Phys. Ther. Sci.* **2011**, *17*, 687–692. [CrossRef]
6. Lee, S.; Savin, D.; Shah, N.; Bronsnick, D.; Goldberg, B. Scapular Winging: Evaluation and Trearment. *J. Bone Joint Surg. Am.* **2015**, *97*, 1708–1716. [CrossRef]
7. Bateman, M.; Smith, B.; Osborne, S.; Wilkes, S. Physiotherapy treatment for atraumatic recurrent shoulder instability: Early results of a specific exercise protocol using pathology-specific outcome measures. *Shoulder Elbow* **2015**, *7*, 282–288. [CrossRef] [PubMed]
8. Friedman, L.; Lafosse, L.; Garrigues, G. Global Perspectives on Management of Shoulder Instability: Decision Making and Treatment. *Orthop. Clin. N. Am.* **2019**, *51*, 241–258. [CrossRef] [PubMed]
9. Warby, S.; Ford, J.; Hahne, A.; Watson, L.; Balster, S.; Lenssen, R.; Pizzari, T. Comparison of 2 Exercise Rehabilitation Programs for Multidirectional Instability of the Glenohumeral Joint: A Randomized Controlled Trial. *Am. J. Sports Med.* **2018**, *46*, 87–97. [CrossRef] [PubMed]
10. Galace De Freitas, D.; Marcondes, F.; Monteiro, R.; Gonçalves Rosa, S.; de Moraes Barros Fucs, P.M.; Yukio Fukuda, T. Pulsed electromagnetic field and exercises in patients with shoulder impingement syndrome: A randomized, double-blind, placebo-controlled clinical trial. *Arch. Phys. Med. Rehabil.* **2014**, *95*, 345–352. [CrossRef]
11. Beaudreuil, J.; Lasbleiz, S.; Richette, P.; Seguin, G.; Rastel, C.; Aout, M.; Vicaut, E.; Cohen-Solal, M.; Lioté, F.; de Vernejoul, M.-C.; et al. Assessment of dynamic humeral centering in shoulder pain with impingement syndrome: A randomised clinical trial. *Ann. Rheum. Dis.* **2011**, *70*, 173–179. [CrossRef]
12. Cools, A.; Borms, D.; Castelein, B.; Vanderstukken, F.; Johansson, F.R. Evidence-based rehabilitation of athletes with glenohumeral instability. *Knee Surg. Sports Traumatol. Arthrosc.* **2016**, *24*, 382–389. [CrossRef] [PubMed]
13. Warby, S.; Pizzari, T.; Ford, J.; Hahne, A.J.; Watson, L. The effect of exercise-based management for multidirectional instability of the glenohumeral joint: A systematic review. *J. Shoulder Elbow Surg.* **2014**, *23*, 128–142. [CrossRef] [PubMed]
14. McClure, P.; Greenberg, E.; Kareha, S. Evaluation and management of scapular dysfunction. *Sports Med. Arthrosc.* **2012**, *20*, 39–48. [CrossRef]
15. Kibler, W.; Ludewig, P.; McClure, P.; Uhl, T.L.; Sciascia, A. Scapular Summit 2009: Introduction, 16 July 16, 2009, Lexington, Kentucky. *J. Orthop. Sports Phys. Ther.* **2009**, *39*, A1–A13. [CrossRef] [PubMed]
16. Kibler, W.; Ludewig, P.; McClure, P.; Michener, L.A.; Bak, K.; Sciascia, A.D. Clinical implications of scapular dyskinesis in shoulder injury: The 2013 consensus statement from the 'scapular summit'. *Br. J. Sports Med.* **2013**, *47*, 877–885. [CrossRef] [PubMed]
17. Ratcliffe, E.; Pickering, S.; McLean, S.; Lewis, J. Is there a relationship between subacromial impingement syndrome and scapular orientation? A systematic review. *Br. J. Sports Med.* **2014**, *48*, 1251–1256. [CrossRef] [PubMed]
18. Struyf, F.; Nijs, J.; Mollekens, S.; Jeurissen, I.; Truijen, S.; Mottram, S.; Meeusen, R. Scapular-focused treatment in patients with shoulder impingement syndrome: A randomized clinical trial. *Clin. Rheumatol.* **2013**, *32*, 73–85. [CrossRef] [PubMed]
19. Haahr, J.P.; Østergaard, S.; Dalsgaard, J.; Norup, K.; Frost, P.; Lausen, S.; Holm, E.A.; Andersen, J.H. Exercises versus arthroscopic decompression in patients with subacromial impingement: A randomised, controlled study in 90 cases with a one year follow up. *Ann. Rheum. Dis.* **2005**, *64*, 760–764. [CrossRef]
20. Holmgren, T.; Oberg, B.; Sjöberg, I.; Johansson, K. Supervised strengthening exercises versus home-based movement exercises after arthroscopic acromioplasty: A randomized clinical trial. *J. Rehabil. Med.* **2012**, *44*, 12–18. [CrossRef]
21. Reijneveld, E.; Noten, S.; Michener, L.; Cools, A.; Struyf, F. Clinical outcomes of scapular-focused treatment in patients with subacromial pain syndrome: Systematic review. *Br. J. Sports Med.* **2017**, *51*, 436–441. [CrossRef] [PubMed]
22. Bateman, M.; Osborne, S.; Smith, B. Physiotherapy treatment for atraumatic recurrent shoulder instability: Updated results of the Derby Shoulder Instability Rehabilitation Programme. *J. Arthrosc. Joint Surg.* **2019**, *6*, 35–41. [CrossRef]
23. Eshoj, H.; Rasmussen, S.; Frich, L.; Hvass, I.; Christensen, R.; Boyle, E.; Lund Jensen, S.; Søndergaard, J.; Søgaard, K.; Juul-Kristensen, B. Neuromuscular Exercises Improve Shoulder Function More Than Standard Care Exercises in Patients With a Traumatic Anterior Shoulder Dislocation. *Orthop. J. Sports Med.* **2020**, *8*. [CrossRef] [PubMed]
24. Juul-Kristensen, B.; Larsen, C.M.; Eshoj, H.; Clemmensen, T.; Hansen, A.; Bo Jensen, P.; Boyle, E.; Søgaard, K. Positive effects of neuromuscular shoulder exercises with or without EMG-biofeedback, on pain and function in participants with subacromial pain syndrome—A randomised controlled trial. *J. Electromyogr. Kinesiol.* **2019**, *48*, 161–168. [CrossRef] [PubMed]
25. Baskurt, Z.; Baskurt, F.; Gelecek, N.; Özkan, M.H. The effectiveness of scapular stabilization exercise in the patients with subacromial impingement syndrome. *J. Back Musculoskelet. Rehabil.* **2011**, *24*, 173–179. [CrossRef] [PubMed]

26. DeMey, K.; Danneels, L.; Cagnie, B.; Cools, A. Scapular muscle rehabilitation exercises in overhead athletes with impingement symptoms: Effect of a 6-week training program on muscle recruitment and functional outcome. *Am. J. Sports Med.* **2012**, *40*, 1906–1915. [CrossRef] [PubMed]
27. Huang, H.Y.; Lin, J.J.; Leon Guo, Y.; Wang, W.T.; Chen, Y.-J. EMG biofeedback effectiveness to alter muscle activity pattern and scapular kinematics in subjects with and without shoulder impingement. *J. Electromyogr. Kinesiol.* **2013**, *23*, 267–274. [CrossRef] [PubMed]
28. Larsen, C.; Søgaard, K.; Chreiteh, S.; Holtermann, A.; Juul-Kristensen, B. Neuromuscular control of scapula muscles during a voluntary task in subjects with subacromial impingement syndrome. A case-control study. *J. Electromyogr. Kinesiol.* **2013**, *23*, 1158–1165. [CrossRef]
29. Worsley, P.; Warner, M.; Mottram, S.; Gadola, S.; Veeger, H.E.J.; Hermens, H.; Morrissey, D.; Little, P.; Cooper, C.; Carr, A.; et al. Motor control retraining exercises for shoulder impingement: Effects on function, muscle activation, and biomechanics in young adults. *J. Shoulder Elbow Surg.* **2013**, *22*, e11-9. [CrossRef]
30. Dorrestijn, O.; Stevens, M.; Winters, J.; van der Meer, K.; Diercks, R.L. Conservative or surgical treatment for subacromial impingement syndrome? A systematic review. *J. Shoulder Elbow Surg.* **2009**, *18*, 652–660. [CrossRef] [PubMed]
31. Struyf, F.; Cagnie, B.; Cools, A.; Baert, I.; Van Brempt, J.; Struyf, P.; Meeus, M. Scapulathoracic muscle activity and recruitment timing in patients with shoulder impingement symptoms and glenohumeral instability. *J. Electomiogr. Kinesiol.* **2014**, *24*, 277–284. [CrossRef] [PubMed]
32. Roach, K.; Budiman-Mak, E.; Songsiridej, N.; Lertratanakul, Y. Development of a Shoulder Pain and Disability Index. *Arthritis Care Res.* **1991**, *4*, 143–149. [CrossRef] [PubMed]
33. Roy, J.; MacDermid, J.; Woodhouse, L. Measuring Shoulder Function: A Systematic Review of Four Questionnaires. *Arthritid Rheum.* **2009**, *61*, 623–632. [CrossRef] [PubMed]
34. Michener, L.; Snyder, A.; Leggin, B. Responsiveness of the numeric pain rating scale in patients with shoulder pain and the effect of surgical status. *J. Sports Rehabil.* **2011**, *20*, 115–128. [CrossRef]
35. Hudak, P.; Amadio, P.; Bombardier, C. Development of an Upper Extremity Outcome Measure: The DASH (Disabilities of the Arm, Shoulder and Hand)[Corrected]. The Upper Extremity Collaborative Group (UECG). *J. Sports Rehabil.* **2011**, *20*, 115–128.
36. Ludewig, P.; Cook, T. Alterations in shoulder kinematics and associated muscle activity in people with symptoms of shoulder impingement. *Phys. Ther.* **2000**, *80*, 276–291. [CrossRef] [PubMed]
37. Lewis, L. Rotator cuff tendinopathy/subacromial impingement syndrome: Is it time for a new method of assessment? *Br. J. Sports Med.* **2009**, *43*, 259–264. [CrossRef] [PubMed]
38. Hung, C.-J.; Jan, M.-H.; Lin, Y.-F.; Wang, T.-Q.; Liu, J.-J. Scapular kinematics and impairment features for classifying patients with subacromial impingement syndrome. *Man. Ther.* **2010**, *15*, 547–551. [CrossRef]
39. Dong, W.; Goost, H.; Xang-Bo, L.; Burger, C.; Paul, C.; Wang, Z.-L.; Zhang, T.-Y.; Jiang, Z.-C.; Welle, K.; Kabir, K. Treatment for Shoulder Impingement Syndrome. A PRISMA Systematic Review and Network Meta-Analysis. *Medicine* **2015**, *94*, e510. [CrossRef]
40. Haik, M.; Alburquerque-Sendín, F.; Silva, C.; Siqueira-Junior, A.L.; Ribeiro, I.L.; Camargo, P.R. Scapular-kinematics pre-and post-thoracic thrust manipulation in individuals with and without shoulder impingement symptoms: A randomised controlled study. *J. Orthop. Sports Phys. Ther.* **2014**, *44*, 475–487. [CrossRef]
41. Tate, A.; McClure, P.; Young, I.; Salvatori, R.; Michener, L.A. Comprehensive impairment-based exercise and manual therapy intervention for patients with subacromial impingement syndrome: A case series. *J. Orthop. Sports Phys. Ther.* **2010**, *40*, 474–493. [CrossRef] [PubMed]
42. Neer, C. Anterior acromioplasty for the chronic impingement syndrome in the shoulder: A preliminary report. *J. Bone Joint Surg. Am.* **1972**, *54*, 41–50. [CrossRef] [PubMed]
43. Hawkins, R.; Kennedy, J. Impingement syndrome in athletes. *Am. J. Sports Med.* **1980**, *8*, 151–158. [CrossRef] [PubMed]
44. Jobe, F.W.; Moynes, D. Delineation of diagnostic criteria and a rehabilitation program for rotator cuff injuries. *Am. J. Sports Med.* **1982**, *10*, 336–339. [CrossRef] [PubMed]
45. Hegedus, E.J.; Goode, A.; Campbell, S.; Morin, A.; Tamaddoni, M.; Moorman, C.T., III; Cook, C. Physical examination tests of the shoulder: A systematic review with meta-analysis of individual tests. *Br. J. Sports Med.* **2008**, *42*, 80–92. [CrossRef] [PubMed]
46. Hayes, K.; Callanan, M.; Walton, J.; Paxinos, A.; Murrell, G.A.C. Shoulder instability: Management and rehabilitation. *J. Orthop. Sports Phys. Ther.* **2002**, *32*, 497–509. [CrossRef] [PubMed]
47. Hegedus, E.J.; Goode, A.; Cook, C.; Michener, L.; Myer, C.A.; Myer, D.M.; Wright, A.A. Which physical examination tests provide clinicians with the most value when examining the shoulder? Update of a systematic review with meta-analysis of individual tests. *Br. J. Sports Med.* **2012**, *46*, 964–978. [CrossRef]
48. Huskisson, E. Measurement of pain. *J. Rheumatol.* **1982**, *9*, 768–769. [CrossRef]
49. Norkin, C.; White, D. The shoulder. In *Measurement of Joint Motion: A Guide to Goniometry*, 4th ed.; F.A. Davis Company: Philadelphia, PA, USA, 2009; pp. 57–90.
50. Kolber, M.; Hanney, W. The reliability and concurrent validity of shoulder mobility measurement using a digital inclinometer and goniometer: A technical report. *Int. J. Sports Phys. Ther.* **2012**, *7*, 306–313.
51. Celik, D.; Dirican, A.; Baltaci, G.; Layman, J. Intrarater reliability of assessing strength of the shoulder and scapular muscles. *J. Sport Rehabil.* **2012**, *21*, 1–5. [CrossRef]

52. Kendall, F.; McCreary, E.; Provance, P. Upper extremity and shoulder girdle strength tests. In *Muscles: Testing and Function, with Posture and Pain*, 4th ed.; Lippincott Williams & Wilkins: Baltimore, MD, USA, 1993; pp. 235–298.
53. Shumway-Cook, A.; Woolacott, J. Motor learning and recovery of function. In *Motor Control: Theory and Practical Applications*, 2nd ed.; Lippincott Williams & Wilkins: Philadelphia, PA, USA, 2001.
54. Comerford, M.; Mottram, S. Functional stability re-training: Principles and strategies for managing mechanical dysfunction. *Man. Ther.* **2001**, *6*, 3–14. [CrossRef] [PubMed]
55. Jones, M.A.; Rivett, D.A. Clinical Reasoning: Fast and Slow Thinking in Musculoskeletal Practice. In *Clinical Reasoning in Musculoskeletal Practice*, 2nd ed.; Elsevier: Amsterdam, The Netherlands, 2019; pp. 2–31.
56. Salamh, P.; Lewis, J. It is time to put special tests for rotator cuff-related shoulder pain out to posture. *J. Orthop. Sports Phys. Ther.* **2020**, *50*, 222–225. [CrossRef]
57. Hodges, P. Pain and motor control: From the laboratory to rehabilitation. *J. Electromyogr. Kinesiol.* **2011**, *21*, 220–228. [CrossRef]
58. Glover, S. Separate visual representations in the planning and control of action. *Behav. Brain Sci.* **2004**, *27*, 3–78. [CrossRef]
59. Roy, J.; Moffet, H.; Hébert, L.; Lirette, R. Effect of motor control and strengthening exercises on shoulder function in persons with impingement syndrome: A single-subject study design. *Man. Ther.* **2009**, *14*, 180–188. [CrossRef] [PubMed]
60. Roy, J.; Moffet, H.; McFadyen, B.; Lirette, R. Impact of movement training on upper limb motor strategies in persons with shoulder impingement syndrome. *Sports Med. Arthrosc. Rehabil. Ther. Technol.* **2009**, *17*, 8. [CrossRef] [PubMed]
61. Aurin, A.; Latash, M. Directional specificity of postural muscles in feed-forward postural reactions during fast voluntary arm movements. *Exp. Brain Res.* **1995**, *103*, 323–332. [CrossRef]
62. Ludewig, P.; Reynolds, J. The association of scapular kinematics and glenohumeral joint pathologies. *J. Orthop. Sports Phys. Ther.* **2009**, *39*, 90–104. [CrossRef] [PubMed]
63. Moraes, G.; Faria, C.; Teixeira-Salmela, L. Scapular muscle recruitment patterns and isokinetic strength ratios of the shoulder rotator muscles in individuals with and without impingement syndrome. *J. Shoulder Elbow Surg.* **2008**, *17*, S48–S53. [CrossRef] [PubMed]
64. Hodges, P.W.; Tucker, K. Moving differently in pain: A new theory to explain the adaptation to pain. *Pain* **2011**, *152*, S90–S98. [CrossRef]
65. Crow, J.; Pizzari, J.; Buttifani, D. Muscle onset can be improved by therapeutic exercise: A systematic review. *Phys. Ther. Sport* **2011**, *12*, 199–209. [CrossRef] [PubMed]
66. Tsao, H.; Hodges, P. Persistence of improvements in postural strategies following motor control training in people with recurrent low back pain. *J. Electromyogr. Kinesiol.* **2008**, *18*, 559–567. [CrossRef] [PubMed]
67. Hodges, P.W.; van Dillen, L.R.; McGill, S.; Brumagne, S.; Hides, J.A.; Moseley, G.L. Integrated clinical approach to motor control interventions in low back and pelvic pain. In *Spinal Control: The Rehabilitation of Back Pain*; Churchill Livingstone: London, UK, 2013; pp. 244–306.
68. Berkovitch, Y.; Shapira, J.; Haddad, M.; Keren, Y.; Rosenberg, N. Current clinical trends in first time traumatic anterior shoulder dislocation. *Merit. Res. J. Med. Med. Sci.* **2013**, *1*, 7–13.
69. Lewis, J. Subacromial impingement syndrome: A musculoskeletal condition or a clinical illusion? *Phys. Ther. Rev.* **2011**, *16*, 388–398. [CrossRef]
70. Steuri, R.; Sattelmayer, M.; Elsig, S. Effectiveness of conservative interventions including exercise, manual therapy and medical management in adults with shoulder impingement: A systematic review and meta-analysis of RCTs. *Br. J. Sports Med.* **2017**, *18*, 1340–1347. [CrossRef]
71. Hanratty, C., McVeigh, J., Kerr, D., Basford, J.R.; Finch, M.B.; Pendleton, A.; Sim, J. The effectiveness of physiotherapy exercises in subacromial impingement syndrome: A systematic review and meta-analysis. *Semin. Arthritis. Rheum.* **2012**, *42*, 297–316. [CrossRef]
72. Haik, M.; Alburquerque-Sendín, F.; Moreira, R.; Pires, E.D.; Camargo, P.R. Effectiveness of physical therapy treatment of clearly defined subacromial pain: A systematic review of randomised controlled trials. *Br. J. Sports Med.* **2016**, *50*, 1124–1134. [CrossRef] [PubMed]
73. Bury, J.; West, M.; Chamorro-Moriana, G.; Littlewood, C. Effectiveness of scapula-focused approaches in patients with rotator cuff related shoulder pain: A systematic review and meta-analysis. *Man. Ther.* **2016**, *25*, 35–42. [CrossRef]
74. Matias, R.; Jones, M. Incorporating Biomechanical Data in the Analysis of a University Student With Shoulder Pain and Scapula Dyskinesis. In *Clinical Reasoning in Musculoskeletal Practice*; Elsevier: Amestradam, The Netherlands, 2019; pp. 483–503.
75. Ferreira, A.L.; dos Santos, C.; Matias, R. A kinematic biofeedback-assisted scapular-focused intervention reduces pain, and improves functioning and scapular dynamic control in patients with shoulder dysfunction. *Gait Posture* **2016**, *49*, 277.
76. Ekstrom, R.; Soderberg, G.; Donatelli, R. Normalization procedures using maximum voluntary isometric contractions for the serratus anterior and trapezius muscles during surface EMG analysis. *J. Electromyogr. Kinesiol.* **2005**, *15*, 418–428. [CrossRef]
77. Hermens, H.; Freriks, B.; Merletti, R.; Stegeman, D.; Blok, J.; Rau, G.; Disselhorst-Klug, C.; Hagg, G. *Eurpoean Recommendations for Surface Electromyography—Results of the Seniam Project. SENIAM 8*, 2nd ed.; Roessingh Research and Development: Enschede, The Netherlands, 1999; pp. 27–30.
78. Gamboa, H.; Matias, R.; Araújo, T.; Veloso, A. Electromyography onset detection: New methodology. *J. Biomech.* **2012**, *45*, S494. [CrossRef]

79. Bagg, S.; Forrest, W. Electromyographic study of the scapular rotators during arm abduction in the scapular plane. *Am. J. Phy. Med.* **1986**, *65*, 111–124. [PubMed]
80. Borsa, P.; Timmons, M.; Sauers, E. Scapular-positioning patterns during humeral elevation in unimpaired shoulders. *J. Athl. Trai* **2003**, *38*, 12–17. [PubMed]
81. Inman, V.; Sauders, J.; Abbott, L. Observations of the function of the shoulder joint. *Clin. Orthop. Relat. Res.* **1996**, *330*, 3–1 [CrossRef] [PubMed]
82. Pescatello, L.; Arena, R.; Riebe, D.; Thompson, P. Behavioral Theories and Strategies for promoting exercise. In *ACSM's Guideline for Exercise Testing and Prescription*, 9th ed.; Wolkers Kluwer / Lippincott Williams & Walkins: Baltimore, MD, USA, 201 pp. 355–365.
83. Baechle, T.; Earle, R. Training variation: Periodization. In *Essentials of Strength Training and Conditioning*, 2nd ed.; Human Kinetic Champaign, IL, USA, 2000; pp. 513–527.
84. Desmurget, M.; Grafton, S. Forward modeling allows feedback control for fast reaching movements. *Trends Cogn. Sci.* **2000**, 423–431. [CrossRef]

 sensors

MDPI

Article

A Wearable System with Embedded Conductive Textiles and an IMU for Unobtrusive Cardio-Respiratory Monitoring

Joshua Di Tocco [1,*,†], Luigi Raiano [2,†], Riccardo Sabbadini [1], Carlo Massaroni [1], Domenico Formica [2] and Emiliano Schena [1]

1 Unit of Measurements and Biomedical Instrumentation, Università Campus Bio-Medico di Roma, Via Alvaro del Portillo, 00128 Rome, Italy; r.sabbadini@unicampus.it (R.S.); c.massaroni@unicampus.it (C.M.); e.schena@unicampus.it (E.S.)

2 Unit of Neurophysiology and Neuroengineering of HumanTechnology Interaction (NeXT), Università Campus Bio-Medico di Roma, Via Alvaro del Portillo, 00128 Rome, Italy; l.raiano@unicampus.it (L.R.); D.Formica@unicampus.it (D.F.)

* Correspondence: j.ditocco@unicampus.it

† These authors contributed equally to this work.

Abstract: The continuous and simultaneous monitoring of physiological parameters represents a key aspect in clinical environments, remote monitoring and occupational settings. In this regard, respiratory rate (RR) and heart rate (HR) are correlated with several physiological and pathological conditions of the patients/workers, and with environmental stressors. In this work, we present and validate a wearable device for the continuous monitoring of such parameters. The proposed system embeds four conductive sensors located on the user's chest which allow retrieving the breathing activity through their deformation induced during cyclic expansion and contraction of the rib cage. For monitoring HR we used an embedded IMU located on the left side of the chest wall. We compared the proposed device in terms of estimating HR and RR against a reference system in three scenarios: sitting, standing and supine. The proposed system reliably estimated both RR and HR, showing low error averaged along subjects in all scenarios. This is the first study focused on the feasibility assessment of a wearable system based on a multi-sensor configuration (i.e., conductive sensors and IMU) for RR and HR monitoring. The promising results encourage the application of this approach in clinical and occupational settings.

Keywords: cardio-respiratory monitoring; wearable system; wearable device; smart textile; IMU; respiratory rate; heart rate

Citation: Di Tocco, J.; Raiano, L.; Sabbadini, R.; Massaroni, C.; Formica, D.; Schena, E. A Wearable System with Embedded Conductive Textiles and an IMU for Unobtrusive Cardio-Respiratory Monitoring. *Sensors* **2021**, *21*, 3018. https://doi.org/10.3390/s21093018

Communicated by: Maria de Fátima Domingues

Received: 15 March 2021
Accepted: 21 April 2021
Published: 25 April 2021

1. Introduction

Continuous, real-time and non-invasive monitoring of vital signs through wearable devices represents one of the most appealing challenges posed by the modern medicine, healthcare and occupational health [1,2]. Regarding modern medicine and healthcare, the use of unobtrusive, lightweight and comfortable wearable devices for collecting physiological signals constitutes a key aspect for improving both the monitoring in clinical settings and a remote/home monitoring of the patients [3]. In clinical settings, a continuous monitoring becomes challenging in all those wards hospitalizing patients which require particular care because they have to be connected to bulky, portable, monitoring devices and every movement around the hospital becomes thus difficult [3,4]. Outside the clinic, wearable devices have gained increased attention for the remote monitoring of the patients and healthcare, due to their intrinsic comfortably, ease of use and reduced costs [3,5–7]. Moreover, the use of wearables to monitor physiological parameters has gained attention in occupational health as well, due to the increased attention to the workers' health and safety by monitoring their condition in the era of Industry 4.0 [8]. Indeed, the monitoring of physiological parameters is beneficial to assessing physiological status, and the activities and fatigue levels of workers (e.g., muscle-skeletal and cardiovascular disorders) to improving their health, well-being and safety and thus meeting the guidelines defined by ergonomics [6,9,10].

In these scenarios, respiratory rate (RR) and heart rate (HR) have gained broad interest since they are strictly related to different physiological and pathological conditions of the patients/workers (e.g., early detection of critical events) and to different environmental stressors [6,11,12]. These vital signs can be monitored using many approaches [9,13].

In this work, we present a prototype of a novel wearable device for simultaneous monitoring of the cardio-respiratory parameters (i.e., RR and HR). The proposed system uses different sensors with respect to what has been reported in the literature and used in commercial devices, since it is based on four conductive textiles (for RR monitoring) and an IMU (for monitoring HR). These sensors were embedded within a highly integrated lightweight, comfortable and low-cost wearable device. We have tested the feasibility of the proposed device in three different scenarios to mimic conditions that can be experienced in the above-described fields. Specifically, we enrolled eight healthy volunteers and we monitored their cardio-respiratory activity, in terms of RR and HR estimation, in three different scenarios: (i) sitting (e.g., it can simulate the occupational settings of a computer worker), (ii) standing and (iii) supine position (e.g., they can simulate clinical and remote applications). This work is organized into the following sections: (i) in Section 2 we focus on the related works; (ii) in Sections 3 and 4 we describe the proposed wearable system (WS) and the experimental protocol used to assess its feasibility in monitoring RR and HR (iii) in Section 5 we describe the techniques of data analysis used to estimate RR and HR starting from the raw data recorded by the WS; (iv) Section 6 reports the results in terms of both RR and HR; (v) Section 7 deals with the discussion of the obtained results and the conclusion.

2. Related Works

The state of the art of wearable systems for RR monitoring consists of techniques based on the cyclic expansion and contraction of the rib cage during the breathing activity. Most of these systems directly measure the expansion of the rib cage by means of electrical elements that change their impedance with strain (i.e., resistive and piezoresistive sensors, capacitive sensors and inductive sensors) and fiber optic sensors [14–20]. Fiber optic sensors (e.g., fiber Bragg grating sensors) have some advantages over their electrical counterparts related to their metrological properties (high sensitivity, good accuracy and short response time), immunity from electromagnetic interference and small size, and they are most often used in this field [17,21–24]. However, the interrogation systems are usually bulky and only recently have there been commercially available portable systems, but these remain quite expensive solutions (from around 3.000 USD to 40.000 USD). When the application does not require the use of the system in a harsh environment in terms of electromagnetic field (e.g., patients monitoring during magnetic resonance scan [23,24]), the resistive, capacitive and inductive sensors may be valid alternatives due to the low prices of both the sensors and the front-end electronics, and the possibility to collect the data by wireless transmission protocol [25,26]. Among others, resistive sensors represent a convenient solution to implement reliable, accurate and low-cost assessments of breathing activity and RR [9,27]. In addition, they can be manufactured as "smart textiles"; thus, it is possible to design highly integrated solutions maximizing the comfort and minimizing the encumbrance of the system itself [28,29]. A commercially available solution for RR monitoring is the SS5LB by BIOPAC systems Inc., which transduces the chest wall deformations using a strain gauge. To allow the collection of the transduced signal, an additional component has to be purchased, increasing both the complexity and costs. Moreover, the device cannot be used in unstructured and unsupervised environments [30]. As regards HR, many techniques have been proposed to develop wearable devices. They are mainly based on electrocardiography (ECG), photoplethysmography (PPG) and the monitoring of the local mechanical vibrations provided by the heartbeat to the chest wall, in terms of accelerations (seismocardiography, SCG) [21,31] or local angular rotations (gyrocardiography, GCG) [32,33].

Specifically, monitoring the cardiac activity using chest wall induced vibrations is an appealing solution for developing highly integrated wearable systems due to the recent technological advancements that have been made in micro-electromechanical systems (MEMSs) for motion tracking that integrate tri-axial accelerometers and gyroscopes into a miniaturized inertial measurement unit (IMU) [31]. Available commercial devices for monitoring HR based on a PPG sensor proposed by Polar. Different devices have been developed to match the needs of subjects (i.e., humans and animals) when monitoring their HRs during physical activities [34,35]. One of the limitations of these devices which is crucial for the application of interest is the inability to simultaneously monitor RR and HR. There are several solutions for monitoring RR and HR by wearable systems; however, the state of the art of wearable systems for simultaneous monitoring of these two parameters consists only of a few works. In [36,37] the system was based on electrodes placed in contact with the subject's skin to monitor both ECG and breathing-induced variations of chest wall impedance during the cyclic respiration. In [22] fiber optic sensors were used for the mentioned purpose. In [38] a piezoelectric sensor was adopted to monitor SCG and breathing activity. In [26] a wearable belt embedding a capacitive sensor and two conductive textiles used as electrodes for a single lead ECG were used to monitor RR and HR simultaneously. Although this system is compliant with the scenarios presented in this study, it is characterized by a high price and having no feature to cope with sensor damage or data loss due to the sensor's failure.

3. Experimental Setup

3.1. Wearable Device

The wearable device, hereinafter referred to as *WS*, consists of two main components: the first one uses 2 elastic bands; the second one is a a custom electrical board. The elastic bands utilizes 2 sensing elements each. The sensing elements are conductive textiles laser-cut as rectangles (dimensions L $\times$ W 50 mm $\times$ 10 mm) from an A4 sheet of material (Eeontex LG-SLPA by Eeonyx Corporation). When these textiles undergo strain, their initial resistance changes according to the applied strain. In this case, the strain is provided by the expansion and contraction of the rib cage during ventilation. To retrieve the respiratory signal on the rib cage, the sensing elements are hand sewed into the elastic bands on the extremities with silver-coated yarn (mod.235/36 dtex 2-ply HC, Statex Produktions und Vertieb GmbH, Germany), whose purpose is twofold: (i) to fix the sensing element to the band and (ii) to provide the electrical contact to retrieve the sensor's output signal by connecting it to the electronic board. In addition, the elastic bands are provided with Velcro to allow the adaptability of the system to different anthropometries.

The custom electrical board has two main functions:

- To process the signal retrieved by the four conductive sensors. To accomplish this task it has 4 embedded Wheatstone bridges (1/4 bridge configuration with the sensing element connected in series with a trimmer of 50 kΩ with the other resistances of 82.5 kΩ) to transduce the conductive sensors' output (i.e., an electrical resistance) into a voltage, 2 instrumentation amplifiers (AD8426 by Analog Devices) with a set gain of 6 and a microcontroller (STM32F446RET by STMicroelectronics).
- To retrieve the cardiac activity information and the position of the subject by using a Magneto-Inertial Measuring Unit (M-IMU, LSM9DS1 by STMicroelectronics).

In addition, the board is equipped with a microSD card socket for storage the data related to respiratory activity (provided by the 4 conductive sensors) and to heart activity (provided by the IMU). All data were collected at 100 Hz. The electronics are powered by a 750 mAh Li-Po battery at 3.7 V, which guarantees autonomy of approximately 8 h. The electronic board along with the battery was placed into a custom 3D-printed TPU casing.

Figure 1 shows a schematic representation of the developed wearable system, the M-IMU axes' orientation and the reference system.

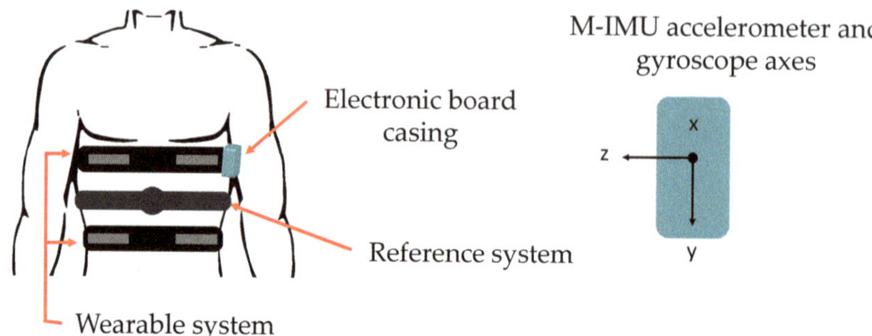

Figure 1. Schematic representations of the proposed wearable system and the reference system and their positioning on the rib cage.

The first step was to assess the response of the sensing elements by applying strain o up to 10%. We repeated 4 quasi-static trials and we calculated the calibration curve and the sensors' sensitivity. The output of the sensors (an electrical resistance) was transduced in a voltage by a voltage divider (61.9 kΩ) powered at +5 V. Therefore, the calibration curve represents the relationship between the output of the amplification stage and the applied strain. It is the well represented by a second-order polynomial ($y = 0.12 \cdot x^2 - 3.81 \cdot x$ – 59.97), as confirmed by the high value of the correlation coefficient ($R^2 > 0.99$).

3.2. *Reference System*

A reference system (Zephyr BioHarness 3.0 by Medtronic) provided the RR (collected at 25 Hz) and HR (single lead ElectroCardioGram, ECG, collected at 250 Hz).

4. Population and Experimental Protocol

To assess the performance of the proposed wearable system, we enrolled 8 healthy male volunteers (mean ± standard deviation: age—27.8 ± 2.7 years old; body mass—75.4 ± 12.2 kg; height—1.74 ± 0.08 m). Table 1 shows details regarding the subjects' ages and somatotypes.

Table 1. Age, body mass, height and body mass index (BMI) of the 8 volunteers.

Volunteer	Age [Years]	Body Mass [kg]	Height [m]	BMI [kg·m^{-2}]
V1	30	72	1.69	25.2
V2	28	76	1.80	23.5
V3	26	88	1.89	24.6
V4	27	70	1.75	22.9
V5	28	74	1.75	24.2
V6	25	98	1.77	31.9
V7	25	63	1.65	23.1
V8	33	62	1.63	23.3

Informed consent was obtained from all subjects involved in the study (protocol code 27.2(18).20 of 15/06/2020), and the principles of declaration of Helsinki and amendments were followed in all the study's steps.

Firstly, each volunteer was asked to wear the reference instrument belt on the xiphoid process line and the 2 elastic belts (one on the nipple line and one on the umbilical line). Both systems were worn in direct contact with the skin. The electronic board was positioned on the left side of the upper belt (next to the heart), in order to retrieve the cardiac activity displacements on the chest wall. Then, the volunteer was asked to perform approximately 10 s of self-paced breathing, a0 s apnea at the end of the inspiratory phase, 3 min of self-paced breathing and finally a 10 s apnea at the end of the inspiratory phase. The same protocol was applied in 3 different positions (i.e., standing, sitting and supine) for a total of 24 trials. The 9-axes M-IMU and the output of the 4 Wheatstone bridges along with the reference system parameters were collected simultaneously.

Figure 2 shows a graphical representation of the experimental setup and the protocols performed.

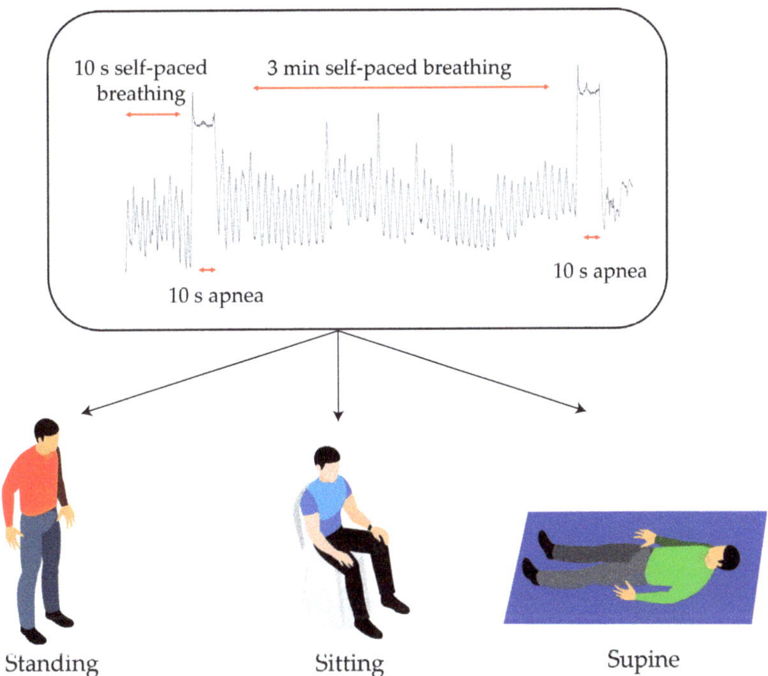

Figure 2. A schematic representation of the experimental protocol performed. The top trend represents the respiratory trial performed in the 3 tested scenarios shown in the lower part.

5. Data Analysis

The data analysis aimed at accomplishing two tasks: (i) estimating RR and HR starting from the trends of the conductive sensors' output and from the IMU; (ii) assessing the performance of the proposed wearable system by comparing the values of RR and HR estimated by the wearable system and the reference one. In this regard, we implemented both a frequency domain analysis, for estimating average RR during the trials, and a time domain analysis to estimate RR breath-by-breath) [27].

To estimate HR we considered the signals recorded by the embedded IMU, and we analyzed them using two approaches: (i) we implemented a frequency domain analysis to monitor the average HR on the whole trial (it lasted approximately 3 min); (ii) a windowed frequency domain analysis considering windows of 30 s. This solution allows investigating how HR behaves over time.

The data analysis was entirely implemented in MATLAB® for each subject and each protocol.

5.1. Respiratory Activity: Data Analysis

To assess RR we considered the signals recorded by the conductive sensors, which followed the breathing-related motions of the subjects' rib cages (see Section 3.1). Specifically, we considered the average in time of the sensors of the four recorded conductive signals hereinafter denoted as $r_{WS}(t)$. Both the conductive signal and reference signal ($r_{ref}(t)$) were filtered using a third-order Butterworth band-pass filter between 0.05 Hz and 2 Hz using zero-phase digital filtering implemented through the function *"filtfilt"* (embedded in MATLAB®). We selected 0.05 Hz as the low cut-off frequency in order to discard very slow signal variations from the recorded data; conversely, we selected 2 Hz as the high cut-off frequency, since RR is hardly above 1.5 Hz [9]. The choice of filtering the RR signals within the mentioned frequency band (i.e., from 0.05 to 2 Hz) agrees with the results reported in [27]. We aimed to filter out components not relevant for our applications while avoiding discarding any useful information recorded by the sensors [9,39].

5.1.1. Frequency Domain Analysis

For the *i*-th subject we computed the error between the RR estimated using the spectrum of $r_{ref}(t)$ ($F^{RR}_{ref,i}$) and $r_{WS}(t)$ ($F^{RR}_{WS,i}$) in each scenario as follows:

$$\tilde{F}^{RR}_{WS,i} = |F^{RR}_{ref,i} - F^{RR}_{WS,i}| \tag{1}$$

F^{RR}_{ref} and F^{RR}_{WS} correspond to the highest peak in the spectra within the range 0.1–1.5 Hz. In (1) all terms are expressed in *bpm*, denoting breaths per minute. The spectra of the signals were obtained by computing the power spectral density (PSD) considering Welch's overlapped segment averaging estimator over the duration of the trials (180 s). To that end, we used the MATLAB® function *"pwelch."* In addition, we computed the percentage version of (1) as follows:

$$\tilde{F}^{RR}_{WS\%,i} = \frac{|F^{RR}_{ref,i} - F^{RR}_{WS,i}|}{F^{RR}_{ref,i}}$$

The averages of subjects for $\tilde{F}^{RR}_{WS,i}$ and $\tilde{F}^{RR}_{WS\%,i}$ are denoted as $\tilde{F}^{RR}_{WS}$ and $\tilde{F}^{RR}_{WS\%}$, respectively.

5.1.2. Time Domain Analysis

To implement a breath-by-breath analysis we computed the breath duration ($\Delta T_{rr}[n]$) between two inspiratory peaks both considering $r_{WS}(t)$ and $r_{ref}(t)$. To that end, we implemented the following steps [28]:

- The first step was devoted to the identification of the inspiratory peaks. We used the MATLAB® function *"findpeaks"* with the inverse of the average RR (the value estimated using the frequency domain analysis) as temporal threshold; we used as amplitude threshold 50% of the *RMS* of $r_{WS}(t)$ during the entire duration of the task, and concerning $r_{ref}(t)$ we used as the amplitude threshold 40% of its RMS.
  We used two different amplitude thresholds to optimize the detection of the peaks. After this step, we visually inspected the detected peaks and eventually removed those not related to the end of inspiratory phase. This correction was performed on the data collected by the reference system and by the wearable system, mainly in the supine position, and it was needed due to the different morphologies of the signals which are affected by the position assumed by the subject.
- The second step was devoted to computing the period of each breathing act, $\Delta T_{rr}[n]$. This parameter was considered as the time elapsed between two consecutive peaks. This analysis was performed for both the wearable system ($\Delta T_{rr,WS}$) and the reference one ($\Delta T_{rr,ref}$) (see Figure 3);
- The third step was devoted to computing the RR for the *n*-th breath as $\frac{60}{\Delta T_{rr}[n]}$, for $r_{WS}(t)$ ($f^{RR}_{WS}[n]$) and $r_{ref}(t)$ ($f^{RR}_{ref}[n]$).

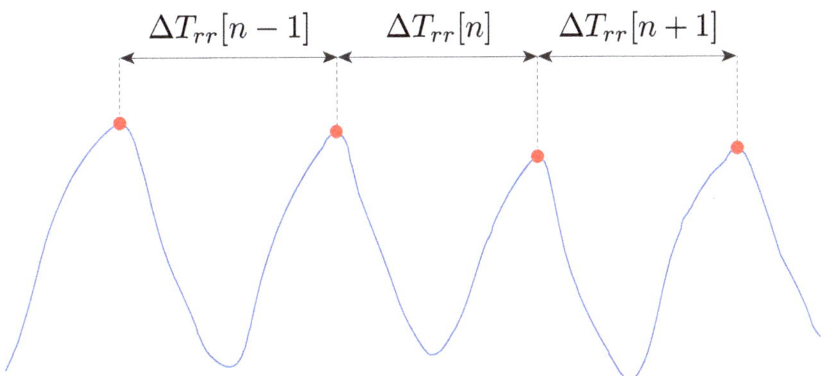

Figure 3. A schematic representation of the breathing act period $\Delta T_{rr}[n]$. The blue line represents $r_{WS}(t)$, and the red circles represent the identified respiratory peaks.

To compare the *WS* with the reference system in the time domain for the *i*-th subject and each protocol, we computed the mean absolute error ($MAE_{WS,i}$) as follows:

$$MAE_{WS,i} = \frac{1}{N_{breaths}} \sum_{n=1}^{N_{breaths}} |f_{ref}^{RR}[n] - f_{WS}^{RR}[n]| \quad (2)$$

In (2), $N_{breaths}$ denotes the number of breaths identified in the *i*-th subject and the specific scenario. In addition, we computed the percentage version of (2) as follows:

$$MAE_{WS\%,i} = \frac{1}{N_{breaths}} \sum_{n=1}^{N_{breaths}} \frac{|f_{ref}^{RR}[n] - f_{WS}^{RR}[n]|}{f_{ref}^{RR}[n]}$$

The averages of $MAE_{WS,i}$ and $MAE_{WS\%,i}$ over all subjects are denoted as MAE_{WS} and $MAE_{WS\%}$, respectively.

A method specifically proposed to test the feasibility of a new measuring system for monitoring physiological parameters has been proposed in this study. Indeed, we performed Bland–Altman analysis [10] considering all the RR values collected by the proposed system (i.e., $f_{WS}^{RR}[n]$) and by the reference one (i.e., $f_{ref}^{RR}[n]$). This analysis was performed considering all the 8 volunteers in all the three scenarios. As recommended in [40], we computed the following parameters:

- $f_{RR}mean[n]$, calculated as the average value between $f_{ref}^{RR}[n]$ and $f_{WS}^{RR}[n]$;
- Δf_{RR}, calculated as the difference between $f_{ref}^{RR}[n]$ and $f_{WS}^{RR}[n]$;
- Mean of the differences (*MOD*), calculated as the mean of the difference between $f_{ref}^{RR}[n]$ and $f_{WS}^{RR}[n]$;
- Limits Of agreement (*LOAs*), calculated as $MOD \pm (1.96 \cdot STD(\Delta f_{RR}))$.

5.2. Cardiac Activity: Data Analysis

According to Figure 1, to monitor the cardiac activity we considered the following signals:

- $s_{a_x}(t)$, which denotes the acceleration along x-axis of the M-IMU;
- $s_{g_x}(t)$ denoting the angular rotation around x-axis of the M-IMU;

Firstly, we band-pass filtered the two signals from 0.7 Hz to 20 Hz in order to remove or minimize bias, breathing activity-related signal and high frequency noise [41]. The choice of this filtering frequency band allowed preserving the informative content related to SCG [33,42].

Subsequently, in order to enhance the effect of the heart beat on recorded signals, we computed the Hilbert transform of $s_{a_x}(t)$ and $s_{g_x}(t)$. It is typically used in SCG and GCG data analysis [43], and given a generic signal $s(t)$, its Hilbert transform is defined as follows:

$$\hat{s}(t) = \frac{1}{\pi} \int_{-\infty}^{+\infty} \frac{s(\xi)}{t - \xi} d\xi, \tag{3}$$

The outcome of the Hilbert transform, i.e., $\hat{s}(t)$, is a complex signal containing in its real part the copy of $s(t)$ and in its imaginary part a 90 deg phase shift of $s(t)$ itself. Assuming that the heart-beat activity ($h(t)$) is hidden and only its modulation can be measured, it is possible to model the recorded signal ($s(t)$) as follows [44]:

$$s(t) = h(t)cos(2\pi f_0 t) + \epsilon(t) \tag{4}$$

In (4), $cos(2\pi f_0 t)$ denotes the modulating term [44], while $\epsilon(t)$ denotes additive noise. Therefore, according to the effect of (3) on the input signal, it is possible to extract $h(t)$ as follows:

$$h(t) = \sqrt{(\Re(\hat{s}(t)))^2 + (\Im(\hat{s}(t)))^2}, \tag{5}$$

denoting $\Re(\hat{s}(t))$ and $\Im(\hat{s}(t))$ the real part and the imaginary part of $\hat{s}(t)$, respectively.

To estimate HR we considered the following signals related to the *WS*:

- $h_{a_x}(t)$, denoting the heart-beat activity estimated considering $s_{a_x}(t)$;
- $h_{g_x}(t)$, denoting the heart-beat activity estimated considering $s_{g_x}(t)$;

All the above-mentioned signals were further filtered using a zero-phase shift band pass filter from 0.7 to 5 Hz, in order to remove bias and obtain the heart-beat envelope (<5 Hz) [43].

The ECG signal recorded by the reference system and band-pass filtered from 0.7 to 20 Hz is denoted as $h_{ref}(t)$.

5.2.1. Frequency Domain Analysis

For the i-th subject we computed the error between the HR estimated using the spectrum of $h_{ref}(t)$ ($F^{HR}_{ref,i}$) and $h_{WS}(t)$ ($F^{HR}_{WS,i}$) in each scenario as follows:

$$\tilde{F}^{HR}_{WS,i} = |F^{HR}_{ref,i} - F^{HR}_{WS,i}| \tag{6}$$

F^{HR}_{ref} and F^{HR}_{WS} correspond to the highest peaks in the spectra within the range 0.7–4 Hz of the signals collected by the reference system and the wearable device, respectively. Thus, F^{HR}_{WS} was calculated by considering either h_{a_x} or h_{g_x}. As for the RR analysis (described in Section 5.1.1), the spectra of the signals were computed by considering the power spectral density (PSD) using a Welch's overlapped segment averaging estimator over the entire duration of the trials (180 s). In addition, we computed the percentage version of (6) as follows:

$$\tilde{F}^{HR}_{WS\%,i} = \frac{|F^{HR}_{ref,i} - F^{HR}_{WS,i}|}{F^{HR}_{ref,i}}.$$

The averages of subjects of $\tilde{F}^{HR}_{WS,i}$ and $\tilde{F}^{HR}_{WS\%,i}$ are denoted as $\tilde{F}^{HR}_{WS}$ and $\tilde{F}^{HR}_{WS,i}$, respectively.

5.2.2. Windowed Frequency Domain Analysis

To further investigate the HR estimation capabilities of the proposed *WS*, we implemented a new frequency domain analysis, considering 30 s lasting windows to compute the PSD.

To that end, we considered only $h_{g_x}(t)$, being the most reliable according to Section 6.2.2, and we computed its spectrum by using the the MATLAB® *"pwelch"* function with a Hamming window of 30 s with an overlap between segments of 50%.

Similarly to (6), considering the i-th subject and the k-th window, we computed the error between the HR estimated using the spectrum of $h_{ref}(t)$ ($F^{HR}_{ref,ik}$) and $h_{g_x}(t)$ ($F^{HR}_{g_x,ik}$) in each scenario as follows:

$$\tilde{f}^{HR}_{g_x,ik} = |F^{HR}_{ref,ik} - F^{HR}_{g_x,ik}| \tag{7}$$

In addition, we computed the percentage version of (7) as follows:

$$\tilde{f}^{HR}_{g_x\%,ik} = \frac{|F^{HR}_{ref,ik} - F^{HR}_{g_x,ik}|}{F^{HR}_{ref,ik}}$$

The averages of $\tilde{f}^{HR}_{g_x,ik}$ and $\tilde{f}^{HR}_{g_x\%,ik}$ for windows are denoted as $\tilde{f}^{HR}_{g_x,i}$ and $\tilde{f}^{HR}_{g_x\%,i}$, respectively; their averages for subjects are denoted as $\tilde{f}^{HR}_{g_x}$ and $\tilde{f}^{HR}_{g_x\%}$.

6. Results

6.1. Respiratory Activity

6.1.1. Frequency Domain Analysis

The frequency domain analysis allowed us to estimate the average RR during the entire duration for each volunteer in each scenario (i.e., standing, sitting and supine). An example of a signal spectrum for a representative subject is presented in Figure 4 which shows the normalized PSD (nPSD), computed by dividing the amplitude of the spectrum by its maximum peak value, of both the reference system and the *WS* in all scenarios.

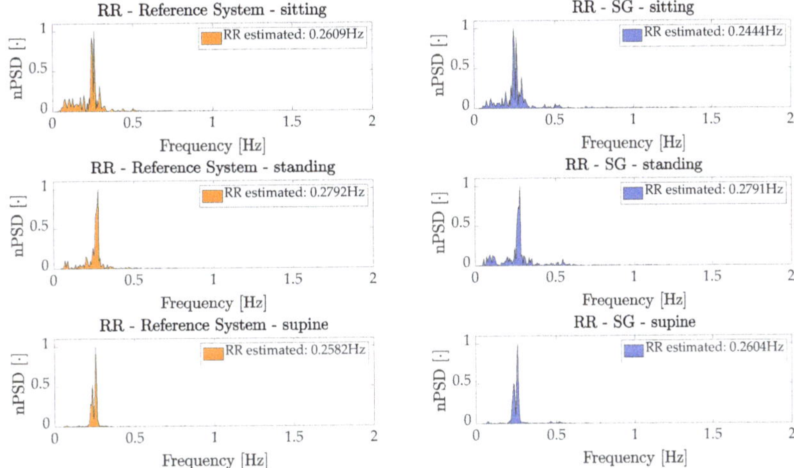

Figure 4. nPSD of r_{ref} (left) and r_{WS} (right) a representative subjects in each scenario.

Tables 2 and 3 report the values of $\tilde{F}^{RR}_{WS,i}$ and $\tilde{F}^{RR}_{WS\%,i}$ in the upper part and their average along subjects, i.e., $\tilde{F}^{RR}_{WS}$ and $\tilde{F}^{RR}_{WS\%}$, respectively, in the lower part in all three scenarios. The worst case is related to subject 8 during the scenario "supine," in which the system apparently failed in estimating the average RR. This might have been caused by a too low or absent pre-strain on the sensing elements due to the supine position. However, if such a value is discarded the average error in the supine scenario is equal to 0.14 bpm.

Table 2. The absolute error of RR estimated for each volunteer, considering the three scenarios. The average value of the mentioned error for every subject is also shown.

Volunteer	$\tilde{F}^{RR}_{WS,i}$ [bpm]-Sitting	$\tilde{F}^{RR}_{WS,i}$ [bpm]-Standing	$\tilde{F}^{RR}_{WS,i}$ [bpm]-Supine
1	0.99	0.01	0.01
2	0.06	2.39	0.36
3	0.07	0.06	0.01
4	0.02	0.01	0.03
5	0.04	0.00	0.13
6	0.01	0.01	0.01
7	0.11	0.27	0.34
8	0.04	0.01	22.57
Average	0.17	0.35	2.95

Table 3. The percentage of absolute error of RR estimated for each volunteer, considering the three scenarios. The average value of the mentioned error for each of the subjects is also shown.

Volunteer	$\tilde{F}^{RR}_{WS\%,i}$ [%]-Sitting	$\tilde{F}^{RR}_{WS\%,i}$ [%]-Standing	$\tilde{F}^{RR}_{WS\%,i}$ [%]-Supine
1	6.78	0.05	0.87
2	0.40	25.43	2.16
3	0.95	0.72	0.08
4	0.09	0.10	0.13
5	0.33	0.01	1.19
6	0.06	0.08	0.04
7	0.72	1.47	2.58
8	0.32	0.07	66.94
Average	1.21	3.49	9.25

6.1.2. Time Domain Analysis

The behaviours of $r_{ref}(t)$ and $r_{WS}(t)$ over time are reported in Figure 5 in all scenarios for a representative subject.

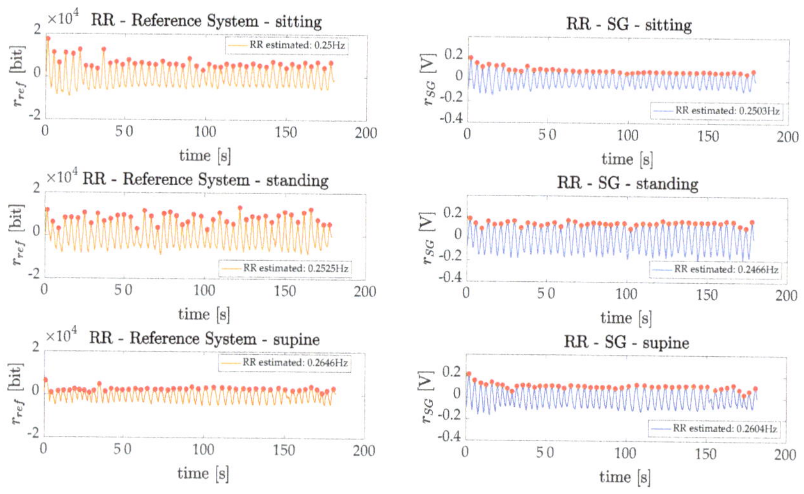

Figure 5. r_{ref} (left) and r_{WS} (right) plotted over time for all scenarios using a representative subject. Peaks selected using the method presented in Section 5 were superimposed on the signals (red circles).

The values of $MAE_{WS,i}$ and $MAE_{WS\%,i}$ in all scenarios are reported in the upper part of Tables 4 and 5, and their averages for subjects (MAE_{WS} and $MAE_{WS\%}$) in the lower part.

Table 4. MAE_{WS} of RR estimated for each volunteer, considering the three scenarios. The average value of the mentioned error for each subject is also shown.

Volunteer	$MAE_{WS,i}$ [bpm]-Sitting	$MAE_{WS,i}$ [bpm]-Standing	$MAE_{WS,i}$ [bpm]-Supine
1	0.24	0.61	1.16
2	0.30	0.28	1.08
3	0.22	0.39	1.11
4	0.24	0.33	3.14
5	0.11	0.11	1.92
6	0.18	0.25	0.87
7	0.36	0.23	1.78
8	0.05	0.23	0.07
Average	0.21	0.30	1.39

Table 5. $MAE_{WS\%}$ of RR estimated for each volunteer, considering the three scenarios. The average value of the mentioned error for each subject is also shown.

Volunteer	$MAE_{WS\%,i}$ [%]-Sitting	$MAE_{WS\%,i}$ [%]-Standing	$MAE_{WS\%,i}$ [%]-Supine
1	1.52	3.17	7.72
2	1.44	2.25	7.41
3	2.56	3.59	9.08
4	1.36	1.89	15.00
5	0.91	0.81	18.09
6	1.15	1.64	5.65
7	2.25	1.29	12.57
8	0.37	1.93	0.57
Average	1.45	2.07	9.51

Concerning the Bland–Altman analysis, the values of MODs and LOAs estimated for each scenario are reported in Table 6 and depicted in Figure 6.

Table 6. Results of the Bland–Altman analysis of speed for the three tested scenarios.

Scenario	MOD [bpm]	LOA-Upper [bpm]	LOA-Lower [bpm]
Sitting	−0.0039	0.9391	−0.9470
Standing	0.0186	1.6753	−1.6392
Supine	−0.2948	4.9264	−5.5160

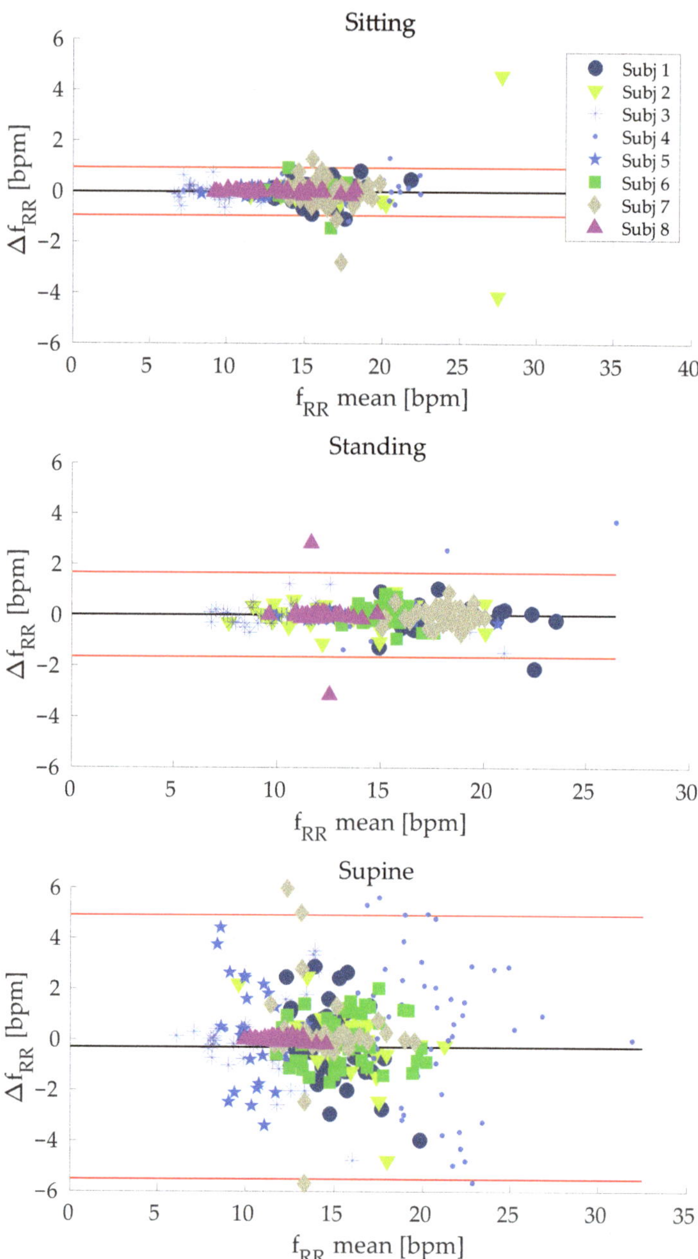

Figure 6. Plot related to Bland-Altman analysis for each scenario. Each plot contains all breath-by-breath RR values estimated for each subject. $\Delta f_{RR} = f_{ref}^{RR}[n] - f_{WS}^{RR}[n]$, and $f_{mean}^{RR} = \frac{f_{ref}^{RR}[n]+f_{WS}^{RR}[n]}{2}$, for n-th breath estimated.

6.2. Cardiac Activity

6.2.1. Frequency Domain Analysis

According to Section 5, we have estimated HR considering $h_{a_x}(t)$ and $h_{g_x}(t)$. The best results were obtained considering $h_{g_x}(t)$, and the standing scenario was the worst case. The results related to $\tilde{F}^{HR}_{WS,i}$ and $\tilde{F}^{HR\%}_{WS,i}$ are reported in Tables 7 and 8, respectively.

Table 7. Error between average HR estimated using r_{ref} and r_{WS} from $h_{a_x}(t)$ and $h_{g_x}(t)$ signals over subjects ($\tilde{F}^{HR}_{WS,i}$) for each scenario.

Volunteer	$\tilde{F}^{HR}_{WS,i}$ [bpm]-Sitting	$\tilde{F}^{HR}_{WS,i}$ [bpm]-Standing	$\tilde{F}^{HR}_{WS,i}$ [bpm]-Supine
$h_{g_x}(t)$			
1	0.56	2.92	0.89
2	3.83	0.05	0.18
3	5.19	5.97	0.03
4	0.28	13.07	0.09
5	0.13	1.98	0.06
6	0.30	0.23	0.28
7	0.18	6.20	0.02
8	0.25	0.08	0.80
Average	1.34	3.81	0.29
$h_{a_x}(t)$			
1	0.41	32.80	0.06
2	3.83	0.05	0.18
3	8.45	24.55	0.03
4	6.29	0.73	2.23
5	0.13	0.02	0.06
6	3.04	9.93	0.28
7	0.16	36.21	0.02
8	0.25	6.24	0.12
Average	2.82	13.82	0.37

Table 8. Percentage of error between average HR estimated using r_{ref} and r_{WS} from $h_{a_x}(t)$ and $h_{g_x}(t)$ signals over subjects ($\tilde{F}^{HR}_{WS\%,i}$) for each scenario.

Volunteer	$\tilde{F}^{HR}_{WS\%,i}$ [%]-Sitting	$\tilde{F}^{HR}_{WS\%,i}$ [%]-Standing	$\tilde{F}^{HR}_{WS\%,i}$ [%]-Supine
$h_{g_x}(t)$			
1	0.83	6.44	1.53
2	5.54	0.06	0.28
3	7.96	7.37	0.06
4	0.41	17.98	0.14
5	0.21	2.69	0.10
6	0.35	0.23	0.37
7	0.31	11.21	0.04
8	0.30	0.09	1.14
Average	1.99	5.75	0.46

Table 8. *Cont.*

Volunteer	$\tilde{F}^{HR}_{WS\%,i}$ [%]-Sitting	$\tilde{F}^{HR}_{WS\%,i}$ [%]-Standing	$\tilde{F}^{HR}_{WS\%,i}$ [%]-Supine
$h_{a_x}(t)$			
1	0.61	72.34	0.11
2	5.54	0.06	0.28
3	12.96	30.31	0.06
4	9.16	1.00	3.59
5	0.21	0.03	0.10
6	3.53	9.85	0.37
7	0.28	65.49	0.04
8	0.30	7.01	0.17
Average	4.07	23.26	0.59

An example of the spectra obtained considering $h_{g_x}(t)$ is presented in Figure 7, which refers to a representative subject.

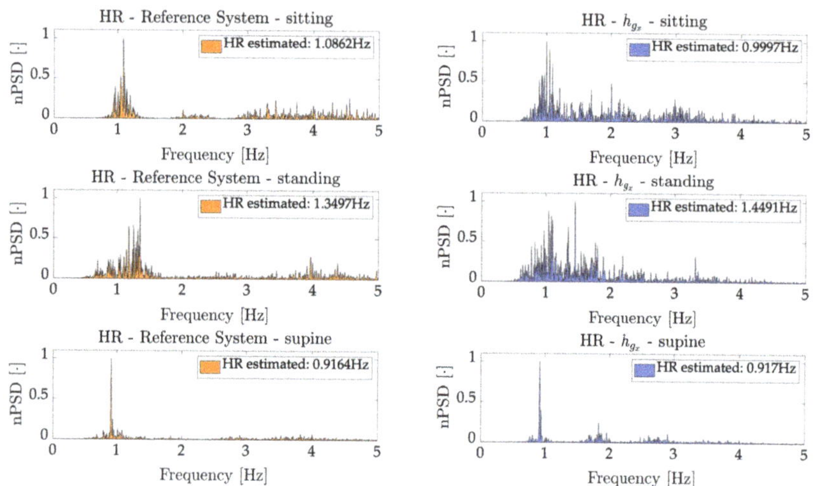

Figure 7. PSD of h_{ref} (left) and h_{g_x} (right) for a representative subject and all scenarios.

6.2.2. Windowed Frequency Domain Analysis

The results obtained for $\tilde{f}^{HR}_{g_x,i}$ (i.e., the error between HR estimated using h_{ref} and h_{g_x} for *i*-th subject averaged along the 30 s time windows considered to compute the spectra) and $\tilde{f}^{HR}_{g_x}$ (i.e., $\tilde{f}^{HR}_{g_x,i}$ averaged along subject) are reported for each scenario in Table 9. Conversely, results related to $\tilde{f}^{HR}_{g_x\%,i}$ are reported in Table 10.

Table 9. Values of $\tilde{f}^{HR}_{g_x,i}$ obtained for all subjects in the three experimental scenarios.

Volunteer	$\tilde{f}^{HR}_{g_x,i}$ [bpm]-Sitting	$\tilde{f}^{HR}_{g_x,i}$ [bpm]-Standing	$\tilde{f}^{HR}_{g_x,i}$ [bpm]-Supine
1	5.45	27.27	0.18
2	0.00	27.45	0.67
3	13.45	16.18	1.00
4	8.60	4.60	5.20
5	0.20	0.00	0.00
6	0.20	35.60	28.73
7	0.80	15.20	0.00
8	0.20	0.20	0.40
Average	3.61	15.81	4.52

Table 10. Values of $\tilde{f}^{HR}_{g_x\%,i}$ obtained for all subjects in the three experimental scenarios.

Volunteer	$\tilde{f}^{HR}_{g_x\%,i}$ [%]-Sitting	$\tilde{f}^{HR}_{g_x\%,i}$ [%]-Standing	$\tilde{f}^{HR}_{g_x\%,i}$ [%]-Supine
1	8.02	53.99	0.29
2	0.00	16.99	1.08
3	20.96	20.91	1.98
4	13.03	6.33	8.01
5	0.31	0.00	0.00
6	0.23	30.01	13.42
7	1.30	22.56	0.00
8	0.29	0.24	0.58
Average	5.52	18.88	3.17

7. Discussion and Conclusions

In this study, we presented a prototype of an unobtrusive and multiparametric wearable system for continuous monitoring of RR and HR. The feasibility of the system has been assessed in different static positions (i.e., sitting, standing and supine), simulating clinical and remote/home monitoring scenarios, and an occupational setting—specifically, a computer worker sitting at a desk. Continuously monitoring those parameters can provide useful information on the health status of an individual, including insights on upcoming potentially critical conditions, and can improve workers' conditions in terms of health, well-being and safety [1–3]. Indeed, although HR is a well established parameter for evaluating an individual's critical critical state, RR is mostly neglected. Indeed, RR is directly affected by the effort made (e.g., physical activity, load handling), the surrounding environment and the psycho-physical state. Thus, a system capable of jointly monitoring breathing activity and cardiac activity may be beneficial to providing comprehensive assessments of the mentioned conditions [3,12,45].

As shown in Figure 1, the proposed wearable system embeds four conductive textiles sewed into two elastic bands located on the chest wall of the user (pulmonary rib cage and abdomen) for RR monitoring and an inertial measurement unit (IMU) integrated within a custom and compact PCB (located on the left side of pulmonary rib cage) for retrieving HR in terms of SCG and GCG.

Concerning the breathing activity, we monitored both average RR, by means of a frequency domain analysis (Section 5.1.1), and RR breath-by-breath, through a time domain analysis (Section 5.1.2). In both cases we considered $r_{sg}(t)$ band-pass filtered using a zero phase shift filter and we compared the estimated RR with the one estimated by the reference system ($r_{ref}(t)$). According to Section 6.1.1, the average RR estimated by the proposed *WS* can be considered as reliable, since the errors obtained were, on average, fractions of the breath-per-minute in sitting and standing tasks. The average error was ~3 bpm when considering the supine task, which corresponds to an average percentage error of ~9%. Such results were confirmed in the time domain analysis (Section 6.1.2). Indeed, the MAE_{WS} obtained for sitting and standing are fractions of the breaths per minute.

Additionally, in the worst case scenario (represented by the supine task) acceptable errors were obtained (~1.5 bpm, corresponding to a $MAE_{WS\%}$ of ~9.5%). The same behavior was obtained considering Bland–Altman analysis (see Table 6 and Figure 6). Although the MOD value was acceptable in all the cases, the LOAs were high especially for the supine position. A possible explanation for the worse results obtained during supine task may lie in an undesired interaction between the elastic bands with the support used let subjects lay down. Probably, in this configuration the elastic band stretched, thereby worsening the signals recorded by conductive sensors. The use of bands with the rear part being stiff instead of being elastic might solve this issue. However, this is just a speculation and future investigations will be devoted to study such aspects.

As regards HR, we considered the signals recorded by IMU. Firstly, we employed the Hilbert Transform to enhance the contribute of the heartbeat on the recorded signals as already used in similar applications [43,44]. This technique allowed us to obtain the heartbeat envelope relative to the x-axis of the accelerometer (h_{a_x}) and the x-axis of the gyroscope (h_{g_x}). Considering such signals, we implemented a frequency domain analysis to estimate an average HR during the entire duration of the trials. Afterwards, we estimated the average HR on 30 s time windows to better assess the capabilities of the proposed device. In both cases, the results were compared with respect the ECG recorded by the reference system. Regarding the average HR, estimated considering the entire duration of the trials, h_{g_x} and h_{a_x} returned similar results in sitting and supine tasks, while h_{g_x} (average error of 3.81 bpm, corresponding to a percentage error of 5.75%) prominently outperformed h_{a_x} (average error of 13.82 bpm, corresponding to a percentage error of 23.26%) considering the standing task. This is likely due to the higher sensitivity of the accelerometer to the body motions, which are higher in standing being the subjects less constrained than in sitting and supine. As expected, the best results were obtained in the supine scenario (average error of 0.29 bpm and percentage error of 0.46% considering h_{g_x}, while 0.37 bpm and 0.59% considering h_{a_x}), where most of the movements detected by the IMU are due to heartbeat, once the respiration has been filtered out. Considering the average HR on a 30 s time window, we considered only h_{g_x}, on the basis the better results obtained in the above-mentioned frequency domain analysis, which allowed obtaining error (averaged along subjects) of ~3.5 bpm (~5.5%), ~4.5 bpm (~3.2%) and ~16 bpm (~18.9%) in sitting supine and standing tasks respectively.

A few studies have investigated to simultaneously monitor breathing and cardiac activities, and the proposed system show error in line with the systems reported in literature [22,36–38]. Results presented in [22] show errors smaller than ~2% and ~6%; however fiber Bragg grating sensors were used, which require more expensive and bulky systems to retrieve the signals, and above all, the HR were estimated during apnea. In [36], where the authors used a belt embedding textile electrodes for recording ECG and breathing activity through impedance variation of the chest wall. They showed errors of ~2% concerning RR estimation and better results in terms of HR estimation. Despite the very good results obtained, the main drawback of this solution lays in the contact required between the electrodes and the skin of the subjects and the need to continue guarantee a low impedance at the contact points. A similar approach was proposed by [37]; however, no performance comparisons with a reference system were presented. In [38] the system proposed is based on a single piezoelectric sensors which allowed the authors to obtain errors of (in average) ~10% and ~6% for RR and HR respectively. To conclude the comparison, the main advantage of our solution lays in the ease of use, simple and low cost electronics required and high wearability and comfort, which does not require direct contact with the skin or further adjustments of the sensors after they are worn by the user. Moreover, because of the presence of IMU within the device, it is possible to exploit their sensitivity on body motion artifacts to further improve the estimation of RR, similarly to [20,46].

The present work is mainly focused on the design of the device and the technology, rather than implementing or assessing robust and efficient algorithms to remove motion artifact from recorded signals during the everyday life. However, we reckon that the problem of motion artifact removal should be taken into account since in a real-life scenario, movements of the subjects can occur. To overcome this concern, different solutions have been proposed for both RR [20,28,47,48] and regarding HR [33,41,49–51]. Future works will be devoted to defining a tailored approach on our system, by combining the two different sensor technologies embedded (i.e., textile strain sensors and an M-IMU) to develop a sensor-fusion algorithm to remove motion artifacts occurring during real life.

Future works will be devoted to further improve the proposed system to enhance its capabilities of HR and RR estimation. Indeed, in its present version, the sensors are sewed on two elastic bands which, despite being comfortable for an ease removal, introduce an additional compliance thus reducing the sensitivity of the conductive sensors. This contingency does not allow them to reliably catch the SCG activity. Therefore, as a future work we are planning on sewing the sensors directly onto an elastic t-shirt (i.e., sportswear) to reduce as much as possible additional compliant elements between the sensors and the user, aiming at investigating whether conductive sensors allow also reliably and robustly monitoring HR, as much as they do with respect RR. Moreover, this may lead to a lower system complexity, to slightly improve its cost and to provide a more comfortable system. In addition, we will test the improved system on a larger population, including females and pathological subjects, to evaluate its potential use also in clinical settings. Since we tested the device on only male subjects, we can just speculate that the use of the proposed device might not be of any discomfort on female subjects. Taking into account what most women wear during sport activities (i.e., sport bras), the use of the upper band of the proposed device should not be of relevant discomfort, since they also allow being regulated in length thanks to the provided Velcro. We are convinced that with the improved device this potential discomfort will be avoided. In addition, the respiratory movements will be hardly detected by the upper band due to the presence of the breast. However, this hurdle will be overcome thanks to the presence of the second band. Moreover, we will evaluate the system performances during different daily living activities which result to be more challenging but represent a typical use of the proposed system. Finally, it is worth noting that we used basic data analysis techniques; therefore, more sophisticated analyses (e.g., based on machine learning methods) may allow improving the estimation of respiratory and cardiac parameters.

Author Contributions: Conceptualization, J.D.T., L.R., R.S., C.M., D.F. and E.S.; manufacturing of the wearable device, J.D.T., L.R. and R.S.; data processing, J.D.T. and L.R.; funding acquisition, C.M., D.F. and E.S.; investigation, J.D.T., L.R. and R.S.; methodology, J.D.T., L.R., R.S., C.M., D.F. and E.S.; supervision, C.M., D.F., and E.S.; validation, J.D.T. and L.R.; writing—original draft, J.D.T. and L.R.; writing—review and editing, J.D.T., L.R., R.S., C.M., D.F. and E.S. All authors have read and agreed to the published version of the manuscript.

Funding: This work has been funded by the Italian National Institute for Insurance against Accidents at Work (INAIL) in the framework of BRIC 2018 (SENSE-RISC project, number ID10/2018), H2020/ICT European project CONBOTS ("CONnected through roBOTS: physically coupling humans to boost handwriting and music learning," grant agreement number 871803, call topic ICT-09-2019-2020) and POR FESR LAZIO 2014-2020 (SMiLe, "Sistema indossabile di Monitoraggio di parametri fisiologici per il benessere della persona e la prevenzione di malattie Lavoro-correlate").

Institutional Review Board Statement: The study was conducted according to the guidelines of the Declaration of Helsinki, and approved by the Ethics Committee of Università Campus Bio-Medico di Roma (protocol code 27.2(18).20 of 15/06/2020).

Informed Consent Statement: Informed consent was obtained from all subjects involved in the study.

Data Availability Statement: The data presented in this study are available on request from the corresponding author. The data are not publicly available due to privacy reason.

Conflicts of Interest: The authors declare no conflict of interest.

Abbreviations

The following abbreviations are used in this manuscript:

WS	Wearable System
RR	Respiratory Rate
HR	Heart Rate
PSD	Power Spectral Density
nPSD	Normalized Power Spectral Density
PCB	Printed Circuit Board
ADC	Analog Digital Converter
M-IMU	Magneto-Inertial Measurement Unit
IMU	Inertial Measurement Unit
BMI	Body Mass Index
RMS	Root Mean Square
MOD	Mean of Difference
LOA	Limit of Agreement
ECG	Electrocardiography
SCG	Seismocardiography
GCG	Gyrocardiography

References

1. Sparks, K.; Faragher, B.; Cooper, C.L. Well-being and occupational health in the 21st century workplace. *J. Occup. Organ. Psychol* **2001**, *74*, 489–509. [CrossRef]
2. Romero, D.; Mattsson, S.; Fast-Berglund, Å.; Wuest, T.; Gorecky, D.; Stahre, J. Digitalizing occupational health, safety and productivity for the operator 4.0. In Proceedings of the IFIP International Conference on Advances in Production Management Systems, Seoul, Korea, 26–30 August 2018; pp. 473–481.
3. Dunn, J.; Runge, R.; Snyder, M. Wearables and the medical revolution. *Personal. Med.* **2018**, *15*, 429–448. [CrossRef] [PubMed]
4. Kroll, R.R.; McKenzie, E.D.; Boyd, J.G.; Sheth, P.; Howes, D.; Wood, M.; Maslove, D.M. Use of wearable devices for post-discharge monitoring of ICU patients: A feasibility study. *J. Intensive Care* **2017**, *5*, 1–8. [CrossRef]
5. Yetisen, A.K.; Martinez-Hurtado, J.L.; Ünal, B.; Khademhosseini, A.; Butt, H. Wearables in medicine. *Adv. Mater.* **2018**, *30*, 1706910 [CrossRef]
6. Dias, D.; Paulo Silva Cunha, J. Wearable health devices—vital sign monitoring, systems and technologies. *Sensors* **2018**, *18*, 2414 [CrossRef]
7. Patel, S.; Park, H.; Bonato, P.; Chan, L.; Rodgers, M. A review of wearable sensors and systems with application in rehabilitation. *J. Neuroeng. Rehabil.* **2012**, *9*, 1–17. [CrossRef]
8. Rüßmann, M.; Lorenz, M.; Gerbert, P.; Waldner, M.; Justus, J.; Engel, P.; Harnisch, M. Industry 4.0: The future of productivity and growth in manufacturing industries. *Boston Consult. Group* **2015**, *9*, 54–89.
9. Massaroni, C.; Nicolò, A.; Lo Presti, D.; Sacchetti, M.; Silvestri, S.; Schena, E. Contact-based methods for measuring respiratory rate. *Sensors* **2019**, *19*, 908. [CrossRef] [PubMed]
10. Wood, W. Toward developing new occupational science measures: An example from dementia care research. *J. Occup. Sci.* **2005**, *12*, 121–129. [CrossRef]
11. van Loon, K.; van Zaane, B.; Bosch, E.J.; Kalkman, C.J.; Peelen, L.M. Non-invasive continuous respiratory monitoring on general hospital wards: A systematic review. *PLoS ONE* **2015**, *10*, e0144626. [CrossRef]
12. Nicolò, A.; Massaroni, C.; Schena, E.; Sacchetti, M. The Importance of Respiratory Rate Monitoring: From Healthcare to Sport and Exercise. *Sensors* **2020**, *20*, 6396. [CrossRef]
13. Chan, M.; Estève, D.; Fourniols, J.Y.; Escriba, C.; Campo, E. Smart wearable systems: Current status and future challenges. *Artif. Intell. Med.* **2012**, *56*, 137–156. [CrossRef] [PubMed]
14. Huang, C.T.; Shen, C.L.; Tang, C.F.; Chang, S.H. A wearable yarn-based piezo-resistive sensor. *Sens. Actuators A Phys.* **2008**, *141*, 396–403. [CrossRef]
15. Kundu, S.K.; Kumagai, S.; Sasaki, M. A wearable capacitive sensor for monitoring human respiratory rate. *Jpn. J. Appl. Phys.* **2013**, *52*, 04CL05. [CrossRef]
16. Wu, D.; Wang, L.; Zhang, Y.T.; Huang, B.Y.; Wang, B.; Lin, S.J.; Xu, X.W. A wearable respiration monitoring system based on digital respiratory inductive plethysmography. In Proceedings of the 2009 Annual International Conference of the IEEE Engineering in Medicine and Biology Society, St. Paul, MN, USA, 2–6 September 2009; pp. 4844–4847.
17. Presti, D.L.; Massaroni, C.; Leitão, C.S.J.; Domingues, M.D.F.; Sypabekova, M.; Barrera, D.; Floris, I.; Massari, L.; Oddo, C.M.; Sales, S.; et al. Fiber Bragg Gratings for medical applications and future challenges: A review. *IEEE Access* **2020**, *8*, 156863–156888. [CrossRef]
18. Massaroni, C.; Di Tocco, J.; Presti, D.L.; Longo, U.G.; Miccinilli, S.; Sterzi, S.; Formica, D.; Saccomandi, P.; Schena, E. Smart textile based on piezoresistive sensing elements for respiratory monitoring. *IEEE Sens. J.* **2019**, *19*, 7718–7725. [CrossRef]

19. Di Tocco, J.; Massaroni, C.; Raiano, L.; Formica, D.; Schena, E. A wearable system for respiratory and pace monitoring in running activities: A feasibility study. In Proceedings of the 2020 IEEE International Workshop on Metrology for Industry 4.0 & IoT, Rome, Italy, 3–5 June 2020; pp. 44–48.
20. Raiano, L.; Di Tocco, J.; Massaroni, C.; Di Pino, G.; Schena, E.; Formica, D. Clean-Breathing: A Novel Sensor Fusion Algorithm Based on ICA to Remove Motion Artifacts from Breathing Signal. In Proceedings of the 2020 IEEE International Workshop on Metrology for Industry 4.0 & IoT, Rome, Italy, 3–5 June 2020; pp. 734–739.
21. Presti, D.L.; Massaroni, C.; Di Tocco, J.; Schena, E.; Formica, D.; Caponero, M.A.; Longo, U.G.; Carnevale, A.; D'Abbraccio, J.; Massari, L.; et al. Cardiac monitoring with a smart textile based on polymer-encapsulated FBG: Influence of sensor positioning. In Proceedings of the 2019 IEEE International Symposium on Medical Measurements and Applications (Memea), Istanbul, Turkey, 23–28 June 2019; pp. 1–6.
22. Lo Presti, D.; Romano, C.; Massaroni, C.; D'Abbraccio, J.; Massari, L.; Caponero, M.A.; Oddo, C.M.; Formica, D.; Schena, E. Cardio-respiratory monitoring in archery using a smart textile based on flexible fiber Bragg grating sensors. *Sensors* **2019**, *19*, 3581. [CrossRef]
23. Dziuda, L.; Skibniewski, F.W.; Krej, M.; Baran, P.M. Fiber Bragg grating-based sensor for monitoring respiration and heart activity during magnetic resonance imaging examinations. *J. Biomed. Opt.* **2013**, *18*, 057006. [CrossRef]
24. Dziuda, Ł.; Krej, M.; Skibniewski, F.W. Fiber Bragg grating strain sensor incorporated to monitor patient vital signs during MRI. *IEEE Sens. J.* **2013**, *13*, 4986–4991. [CrossRef]
25. Merritt, C.R. Electronic Textile-Based Sensors and Systems for Long-Term Health Monitoring. Ph.D. Thesis, NC State University, Raleigh, NC, USA, 2008.
26. BioHarness. Available online: https://www.zephyranywhere.com/ (accessed on 19 April 2021).
27. Massaroni, C.; Di Tocco, J.; Bravi, M.; Carnevale, A.; Presti, D.L.; Sabbadini, R.; Miccinilli, S.; Sterzi, S.; Formica, D.; Schena, E. Respiratory monitoring during physical activities with a multi-sensor smart garment and related algorithms. *IEEE Sens. J.* **2019**, *20*, 2173–2180. [CrossRef]
28. Raiano, L.; Di Tocco, J.; Massaroni, C.; Di Pino, G.; Schena, E.; Formica, D. A PCA-based method to select the number and the body location of piezoresistive sensors in a wearable system for respiratory monitoring. *IEEE Sens. J.* **2020**, *21*, 6847–6855. [CrossRef]
29. Dinh, T.; Nguyen, T.; Phan, H.P.; Nguyen, N.T.; Dao, D.V.; Bell, J. Stretchable respiration sensors: Advanced designs and multifunctional platforms for wearable physiological monitoring. *Biosens. Bioelectron.* **2020**, *166*, 112460. [CrossRef]
30. SS5LB respiratory effort transducer - Available online: https://www.biopac.com/wp-content/uploads/SS5LB.pdf (accessed on 19 April 2021).
31. Inan, O.T.; Migeotte, P.F.; Park, K.S.; Etemadi, M.; Tavakolian, K.; Casanella, R.; Zanetti, J.; Tank, J.; Funtova, I.; Prisk, G.K.; et al. Ballistocardiography and seismocardiography: A review of recent advances. *IEEE J. Biomed. Health Inform.* **2014**, *19*, 1414–1427. [CrossRef] [PubMed]
32. D'Mello, Y.; Skoric, J.; Xu, S.; Roche, P.J.; Lortie, M.; Gagnon, S.; Plant, D.V. Real time cardiac beat detection and heart rate monitoring from combined seismocardiography and gyrocardiography. *Sensors* **2019**, *19*, 3472. [CrossRef]
33. Yang, C.; Tavassolian, N. Combined seismo-and gyro-cardiography: A more comprehensive evaluation of heart-induced chest vibrations. *IEEE J. Biomed. Health Inform.* **2017**, *22*, 1466–1475. [CrossRef]
34. Optical Heart Rate Sensors. Available online: https://www.polar.com/en/products/heart-rate-sensors(accessed on 19 April 2021).
35. Horse Heart Rate Sensor. Available online: https://www.polar.com/en/products/heart-rate-sensors (accessed on 19 April 2021).
36. Piuzzi, E.; Pisa, S.; Pittella, E.; Podestà, L.; Sangiovanni, S. Wearable Belt With Built-In Textile Electrodes for Cardio—Respiratory Monitoring. *Sensors* **2020**, *20*, 4500. [CrossRef] [PubMed]
37. Trindade, I.G.; Machado da Silva, J.; Miguel, R.; Pereira, M.; Lucas, J.; Oliveira, L.; Valentim, B.; Barreto, J.; Santos Silva, M. Design and evaluation of novel textile wearable systems for the surveillance of vital signals. *Sensors* **2016**, *16*, 1573. [CrossRef] [PubMed]
38. Allataifeh, A.; Al Ahmad, M. Simultaneous piezoelectric noninvasive detection of multiple vital signs. *Sci. Rep.* **2020**, *10*, 1–13. [CrossRef] [PubMed]
39. da Costa, T.D.; Vara, M.D.F.F.; Cristino, C.S.; Zanella, T.Z.; Neto, G.N.N.; Nohama, P. Breathing monitoring and pattern recognition with wearable sensors. In *Wearable Devices—the Big Wave of Innovation*; Books on Demand: Norderstedt, Germany, 2019.
40. Bland, J.M.; Altman, D. Statistical methods for assessing agreement between two methods of clinical measurement. *Lancet* **1986**, *327*, 307–310. [CrossRef]
41. Kaisti, M.; Tadi, M.J.; Lahdenoja, O.; Hurnanen, T.; Saraste, A.; Pänkäälä, M.; Koivisto, T. Stand-alone heartbeat detection in multidimensional mechanocardiograms. *IEEE Sens. J.* **2018**, *19*, 234–242. [CrossRef]
42. Taebi, A.; Mansy, H.A. Time-frequency distribution of seismocardiographic signals: A comparative study. *Bioengineering* **2017**, *4*, 32. [CrossRef]
43. Tadi, M.J.; Lehtonen, E.; Hurnanen, T.; Koskinen, J.; Eriksson, J.; Pänkäälä, M.; Teräs, M.; Koivisto, T. A real-time approach for heart rate monitoring using a Hilbert transform in seismocardiograms. *Physiol. Meas.* **2016**, *37*, 1885. [CrossRef] [PubMed]
44. Xie, Q.; Li, Y.; Wang, G.; Lian, Y. An unobtrusive system for heart rate monitoring based on ballistocardiogram using Hilbert transform and viterbi decoding. *IEEE J. Emerg. Sel. Top. Circuits Syst.* **2019**, *9*, 635–644. [CrossRef]

45. Nicolò, A.; Girardi, M.; Sacchetti, M. Control of the depth and rate of breathing: Metabolic vs. non-metabolic inputs. *J. Physio* **2017**, *595*, 6363. [CrossRef] [PubMed]
46. Massaroni, C.; Di Tocco, J.; Presti, D.L.; Schena, E.; Bressi, F.; Bravi, M.; Miccinilli, S.; Sterzi, S.; Longo, U.G.; Berton, A.; et a Influence of motion artifacts on a smart garment for monitoring respiratory rate. In Proceedings of the 2019 IEEE Internationa Symposium on Medical Measurements and Applications (MeMeA), Istanbul, Turkey, 26–28 June 2019; pp. 1–6.
47. Siqueira, A.; Spirandeli, A.F.; Moraes, R.; Zarzoso, V. Respiratory Waveform Estimation From Multiple Accelerometers: A Optimal Sensor Number and Placement Analysis. *IEEE J. Biomed. Health Inform.* **2018**, *23*, 1507–1515. [CrossRef] [PubMed]
48. Loo, N.; Chiew, Y.; Tan, C.; Arunachalam, G.; Ralib, A.; Mat-Nor, M.B. A machine learning model for real-time asynchronou breathing monitoring. *IFAC-PapersOnLine* **2018**, *51*, 378–383. [CrossRef]
49. Shafiq, G.; Veluvolu, K.C. Surface chest motion decomposition for cardiovascular monitoring. *Sci. Rep.* **2014**, *4*, 1–9. [CrossRef
50. Faust, O.; Hagiwara, Y.; Hong, T.J.; Lih, O.S.; Acharya, U.R. Deep learning for healthcare applications based on physiologica signals: A review. *Comput. Methods Programs Biomed.* **2018**, *161*, 1–13. [CrossRef]
51. Rim, B.; Sung, N.J.; Min, S.; Hong, M. Deep learning in physiological signal data: A survey. *Sensors* **2020**, *20*, 969. [CrossRef]

 *sensors* 

Communication

Validity of Hip and Ankle Worn Actigraph Accelerometers for Measuring Steps as a Function of Gait Speed during Steady State Walking and Continuous Turning

Lucian Bezuidenhout [1,*], Charlotte Thurston [1], Maria Hagströmer [1,2] and David Moulaee Conradsson [1,3]

1 Department of Neurobiology, Care Sciences and Society, Division of Physiotherapy, Karolinska Institutet, 141 83 Stockholm, Sweden; charlotte.thurston@ki.se (C.T.); maria.hagstromer@ki.se (M.H.); david.m.conradsson@ki.se (D.M.C.)
2 Academic Primary Health Care Centre, Region Stockholm, 104 31 Stockholm, Sweden
3 Women's Health and Allied Health Professionals Theme, Medical Unit Occupational Therapy and Physiotherapy, Karolinska University Hospital, 171 64 Stockholm, Sweden
* Correspondence: lucian.bezuidenhout@ki.se; Tel.: +46-76-094-5156

Citation: Bezuidenhout, L.; Thurston, C.; Hagströmer, M.; Moulaee Conradsson, D. Validity of Hip and Ankle Worn Actigraph Accelerometers for Measuring Steps as a Function of Gait Speed during Steady State Walking and Continuous Turning. *Sensors* **2021**, *21*, 3154. https://doi.org/10.3390/s21093154

Academic Editor: Maria de Fátima Domingues

Received: 19 March 2021
Accepted: 29 April 2021
Published: 1 May 2021

Abstract: This study aimed to investigate the accuracy and reliability of hip and ankle worn Actigraph GT3X+ (AG) accelerometers to measure steps as a function of gait speed. Additionally, the effect of the low frequency extension filter (LFEF) on the step accuracy was determined. Thirty healthy individuals walked straight and walked with continuous turns in different gait speeds. Number of steps were recorded with a hip and ankle worn AG, and with a Stepwatch (SW) activity monitor positioned around the right ankle, which was used as a reference for step count. The percentage agreement, interclass correlation coefficients and Bland–Altmann plots were determined between the AG and the reference SW across gait speeds for the two walking conditions. The ankle worn AG with the default filter was the most sensitive for step detection at >0.6 m/s, whilst accurate step detection for gait speeds < 0.6 m/s were only observed when applying the LFEF. The hip worn AG with the default filter showed poor accuracy (12–78%) at gait speeds < 1.0 m/s whereas the accuracy increased to >87% for gait speeds < 1.0 m/s when applying the LFEF. Ankle worn AG was the most sensitive to measure steps at a vast range of gait speeds. Our results suggest that sensor placement and filter settings need to be taken into account to provide accurate estimates of step counts.

Keywords: accelerometers; Bland–Altman plots; gait speed; interclass correlation coefficient; low frequency extension filter; Stepwatch

1. Introduction

Walking is the most common form and used marker of physical activity, where the number of steps per day is associated with health, e.g., cardiovascular health, dementia and future mortality risk [1–3]. Accelerometry is an established method for measuring steps [4], with the Actigraph GT3X+ (AG; ActiGraph Corp.) being one of the most commonly used accelerometers. AG is a small triaxial accelerometer (dimensions: 4.6 cm × 3.3 cm × 1.5 cm; weight: 19 g) that can be worn on different body positions (e.g., wrist, ankle and hip) and has a dynamic range of ±6G (1G = 9.81 m/s^2). Additionally, AG has a long battery life and can continuously measure physical activity for up to six weeks at 30 Hz [5,6]. For the AG software to detect a step, the bandpass filtered vertical component (y-axis) accelerometer signal must exceed a proprietary amplitude threshold and cross the zero axis (i.e., positive and negative values) of the proprietary amplitude threshold [7].

Accelerometer device placement and gait speed are known to influence the step detection accuracy of accelerometers [8,9]. The hip is the most commonly used placement for measuring walking since the hip accurately reflects the center of mass of the body. However, since slow walking generally has low acceleration amplitudes and the amplitude decreases from the ground upwards, the signal measured at the hip might not be sufficient to cross the proprietary threshold for detecting a step. Previous studies have shown AG to accurately detect steps at a gait speed of >1.0 m/s [5,10], whereas few studies have investigated the sensitivity of AG to detects steps at different ranges of gait speeds < 1.0 m/s. Slow walking speeds have been linked to various movement disorders and high risk of morbidity and mortality [11]. Therefore, it is important to be able to accurately detect steps at low gait speeds (<1.0 m/s) to be able to identify individuals who have an increased risk of deteriorating health. This is especially important for people with disability (i.e., stroke or Parkinson's) or the elderly, who often have compromised gait speeds. Hergenroeder et al. [1] showed the hip worn AG percentage agreement with observed step count to be <52% at gait speed $\leq$ 1.0 m/s and >85% at >1.0 m/s in older adults [1]. While it has been suggested that the accuracy of accelerometers to detect a step is inversely proportional to gait speed [5,8,12], it is unclear at which gait speed the sensitivity of hip and ankle worn AG starts decreasing. Furthermore, previous studies exploring the step detection accuracy of accelerometers have assessed straight walking, i.e., steady speed without changing the walking direction [1,13–15]. However, walking in everyday life is rarely performed during steady state; in fact, most walking bouts in daily life include four steps or less [16] and 50% of the steps executed each day incorporate turning steps [17]. Therefore, it is important that validation of accelerometer step detection accuracy also incorporates non-steady state walking (e.g., turning).

To take into consideration low frequency/amplitude movements, AG developed a low frequency extension filter (LFEF), which increases the sensitivity of the accelerometer signal at low intensity movements by decreasing the proprietary amplitude threshold. This allows for more accurate step detection during slow walking (<1.0 m/s) [6,14,18]. Although the LFEF has shown to improve step count accuracy during slow walking, studies by Wallen et al. [18], Toth et al. [19] and Feito et al. [13,14] showed a significant overestimation in daily steps taken in the free-living environment. Little is known at which range of gait speeds the LFEF should be applied to optimize the sensitivity of step detection. While previous studies [8,9,20] have shown the accuracy of AG to determine steps to be poor during slow walking, no previous studies have explored at what level the accuracy drops when walking straight and performing continuous turns.

This study aspired to determine the validity of the ankle and hip worn AG in healthy adults in a controlled environment before testing the validity of the accelerometers in people with a disability (e.g., those with neurological diseases) or elderly people. The aim of this study was to determine the validity of hip and ankle worn AG accelerometer for measuring steps compared to step counts measured with a reference ankle sensor in healthy adults. We especially investigated the accuracy of the AG accelerometer to detect steps across a large span of gait speeds during straight walking and walking with continuous turns. Additionally, we observed the step detection accuracy of the AG using the default filter (i.e., AG-DF) compared to the LFEF (i.e., AG-LFEF).

2. Materials and Methods

2.1. Study Participants

Thirty healthy participants (14 males, mean age $\pm$ standard deviation: 42 $\pm$ 13 years) with no ongoing or recent medical conditions affecting their gait participated in this cross-sectional study. The study was approved by the Regional Board of Ethics in Stockholm (2017/1626-31 and 2018/2524-32) and all participants gave written informed consent prior to participation.

2.2. Data Collection

Participants attended one gait assessment including two different walking tasks; walking straight and walking with continuous turns, since these conditions reflect different walking patterns occurring in everyday life. Prior to assessment, participants were fitted with two AG's and one Stepwatch (SW) activity monitor. SW has been shown to be sensitive to detect steps over a range of gait speeds, especially during slow walking [21]. The position of the accelerometers was guided by Webber and St. John [8] with the AG accelerometers attached around the right hip (above the iliac crest) and left ankle (proximal to the lateral malleolus), and the SW reference sensor positioned on the right ankle (proximal to the lateral malleolus) [8]. The AG devices recorded time series acceleration data at a sampling rate of 30 Hz whereas the SW sensor records the number of steps per one second epochs. For the SW sensor, the participants height were entered into the Modus Health software and the following configuration was used: "no quick stepping", "slow walking speed", "rarely varying pace" and "gentle leg motion" [8]. The AG and SW devices utilize the local computer time to initialize the device timestamp. In order to synchronize the AG and SW devices, the local time on the local computer were reset to coordinated universal time approximately 15 min before the start of data collection for each participant. We are aware of the potential problems that might occur during data synchronization (i.e., clock drifting), therefore we validated the start and stop times for each trail by; (1) asking the participants to stand still between 10 and 20 s before and after each trail to delineate the devices step onset and offset; (2) manually recording the start and stop times from the local computer and (3) comparing the devices step onset and offset times to the manually recorded start and stop time for each trail.

For straight walking, participants were asked to walk straight for a distance of 40 m with a 180 degree turn around a cone after 20 m (Figure 1A). For the walking with continuous turning trial, participants were instructed to walk through a maze (34 m distance) consisting of an equal distribution of 45 and 90 degree turns to the right and left (Figure 1B). Participants were instructed to start in their self-selective comfortable gait speed and gradually decrease their speed after each trail in order to achieve a good distribution of speeds for both walking conditions. Each participant performed between 10 and 15 trials per walking condition and we aimed to measure gait speeds between 0.2 and 1.6 m/s for straight walking and between 0.2 and 1.0 m/s for continuous turning. The time taken to complete each trail was measured with a stopwatch and immediately entered into a prearranged Microsoft Excel spreadsheet to calculate the average gait speed for each trail. The narrow gait speed range for turning reflects the nature of walking and continuous changing the direction, which often occurs at lower speeds. The start and stop time, manually counted steps and gait speeds were recorded for each trail by a trained investigator.

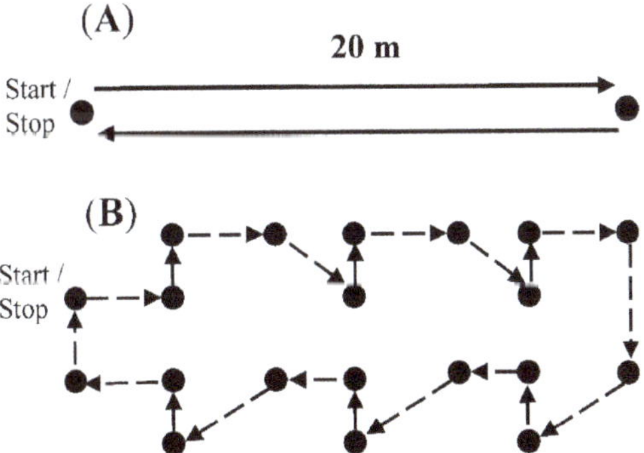

Figure 1. (**A**) Straight ahead walking. (**B**) Continuous turning. Each circle represents a cone and the direction of walking is shown by the arrows.

2.3. Data Analysis

The raw acceleration signal was bandpass filtered between 0.25 and 2.5 Hz to attenuate noise and artifacts and to extract physical movements [22]. Subsequently, the raw signal was digitized by a 12-bit analog to digital (A/D converter) at 30 Hz, which allows for 4096 levels of both positive and negative accelerations measurements (i.e., $2^{12} = 4096$). Since the acceleration is both in the positive (acceleration towards the earth surface) and negative (acceleration away from the earth surface) direction, no movement (zero acceleration) is associated with the center of the A/D scale (i.e., A/D = 4096/2 = 2048). The positive and negative acceleration, which is proportional to the vector component of acceleration deviates the signal from the center value (zero acceleration). Subsequently, the AG software calculated the difference between the measured A/D reading and the center value and thereby only retaining the magnitude of the acceleration and removing the sign [23]. The magnitude of the AG signal is then converted to counts, where a count is defined as the acceleration amplitude crossing some AG proprietary amplitude threshold. The AG count data were then summed to one second epochs, i.e., sum of 30 counts per second, and exported to excel using the ActiLife 6 software (version 6.13.4). The number of steps was obtained using both the DF (0.25–2.5 Hz) [14] and the LFEF [14,24]. Subsequently, the AG step data were imported to Matlab where the start and stop times were used to delineate the total number of steps for each walking and turning trial. The SW data were exported to excel using the Modus Health (StepWatch4 RE 1.1.6) software. The SW output yields the timestamp at which each step was taken. The start and stop time were used to sum the number of steps for each trail of the two walking conditions. The total number of steps for each trail was grouped in four gait speed groups of 0.4 m/s for straight walking (i.e., 0.2–0.6 m/s, >0.6–1.0 m/s, >1.0–1.4 m/s and >1.4 m/s) and two gait speed groups for walking with continuous turning (0.2–0.6 m/s and >0.6–1.0 m/s). The gait speeds were stratified into 0.4 m/s in order to obtain a relatively good distribution of gait speeds samples within each group and to resemble gait speed categories of individuals who are dependent on others in activities in daily living and also have rehabilitation needs [25].

Statistical analyses were carried out using IBM SPSS v.27 software and the level of significance was set to 5%. In line with previous findings [8,9], the mean percentage agreement and two-way random inter class correlation coefficient ($ICC_{2,1}$) between the manually counted steps and the SW activity monitor across all gait speeds was >99% and 0.99, respectively, for both straight walking and continuous turning. We therefore used SW as a reference for measuring steps in our study. The percentage agreement between the SW step count and the ankle and hip worn AG step count using the DF and the LFEF were calculated as: (AG step count/SW step count) × 100) for the gait speed groups [1,9]. The interrater reliability between the SW monitor and the AG devices was determined using the two-way random $ICC_{2,1}$ for both filter settings. The strength of the ICC was classified as follows: <0.50 = poor; 0.50–0.75 = moderate; 0.75–0.9 = good and >0.90 = excellent [26,27]. Additionally, Bland–Altmann plots were used to describe the mean percentage bias between the SW and the AG devices [28,29]. We defined the mean percentage bias as: (the difference between the number of steps between AG and SW/mean number of steps between AG and SW) × 100 [29]. We also defined a mean percentage bias of <10% to be an acceptable agreement between the SW and AG devices.

3. Results

3.1. Number of Steps and Time Spent in Each Gait Speed Group for Straight Walking and Walking with Continuous Turning

For straight walking there was a lower mean number of steps and time spent in the >1.0 m/s gait speed groups (>1.0–1.4 m/s: 30 steps and 34 s and >1.4 m/s: 26 steps and 25 s) compared to <1.0 m/s gait speed groups (0.2–0.6 m/s: 54 steps and 103 s and >0.6–1.0 m/s: 36 steps and 51 s). For continuous turning the mean number of steps and time spent in the 0.2–0.6 m/s gait speed group was also higher (55 steps and 91 s) compared to gait speeds between >0.6 and 1.0 m/s (39 steps and 45 s). The lower time spent and the higher number of steps for walking with continuous turns compared to straight walking for the same gait speed ranges are indicative of the shorter walking distance (i.e., 34 m) and the nature of walking and turning, which often requires shorter steps.

3.2. Straight Walking

The mean percentage agreement for the ankle worn AG-DF was high (≥92%) at gait speeds > 0.6 m/s (Table 1) but dropped to 71% for gait speeds < 0.6 m/s. The ankle worn AG-DF showed moderate to good reliability ($ICC_{2,1}$: 0.70–0.85) with the SW activity monitor at gait speeds > 0.6 m/s (Table 2) whereas the reliability was poor ($ICC_{2,1}$ < 0.29) at gait speeds < 0.6 m/s. The ankle worn AG-LFEF showed a percentage agreement of ≥94% and moderate to excellent reliability ($ICC_{2,1}$: 0.70–0.97) across all represented gait speeds. For the hip worn AG-DF, the percentage agreement was ≥92% at gait speeds > 1.0 m/s, with the percentage agreement decreasing rapidly at gait speeds < 1.0 m/s (Table 1). When the AG-LFEF was applied to the hip worn sensor, the percentage agreement was ≥96% for gait speeds > 0.6–1.4 m/s. While the hip worn AG-DF showed poor to moderate reliability ($ICC_{2,1}$: 0.00–0.58) with the SW across all gait speeds, the agreement was moderate to good at all represented gait speeds when the LFEF was applied (Table 2).

Table 1. Mean percentage agreement (standard deviation) between SW activity monitor and ankle and hip worn AG using the DF and LFEF for different gait speeds during straight walking and walking with continuous turns.

Gait Speed (m/s)	Straight Walking				Continuous Turning			
	Ankle		Hip		Ankle		Hip	
	DF	LFEF	DF	LFEF	DF	LFEF	DF	LFEF
0.2–0.6	71 (19)	96 (7)	12 (17)	87 (22)	82 (17)	96 (5)	25 (23)	88 (15)
>0.6–1.0	96 (5)	97 (5)	69 (25)	97 (5)	95 (4)	97 (4)	78 (13)	96 (5)
>1.0–1.4	96 (4)	97 (4)	92 (7)	97 (4)	-	-	-	-
>1.4	92 (8)	94 (8)	94 (12)	96 (11)	-	-	-	-

Table 2. $ICC_{2,1}$ between SW activity monitor and the ankle and hip worn AG using the DF and the LFEF for different gait speeds during straight walking and walking with continuous turning.

Gait Speed (m/s)	Straight Ahead Walking				Continuous Turning			
	Ankle		Hip		Ankle		Hip	
	DF	LFEF	DF	LFEF	DF	LFEF	DF	LFEF
0.2–0.6	0.29	0.97	0.00	0.50	0.28	0.97	0.00	0.58
>0.6–1.0	0.79	0.83	0.00	0.86	0.89	0.93	0.12	0.88
>1.0–1.4	0.85	0.86	0.58	0.87	-	-	-	-
>1.4	0.70	0.70	0.42	0.57	-	-	-	-

The Bland–Altman plots (Figure 2) showed a low mean percentage bias (−3.0–−8.0%) for both the AG-DF and AG-LFEF at gait speeds between >0.6 and >1.4 m/s (Figure 2F–H) and between 1.0 and >1.4 m/s (Figure 2C,D) for the ankle and hip worn AG, respectively. The mean percentage bias was high for the hip worn AG-DF for gait speeds ranging between 0.2 and 1.0 m/s (−40.0–−160.0%; Figure 2A,B) compared to the AG-LFEF. The high negative mean percentage bias is indicative of the number of steps being significantly underestimated by the AG sensors.

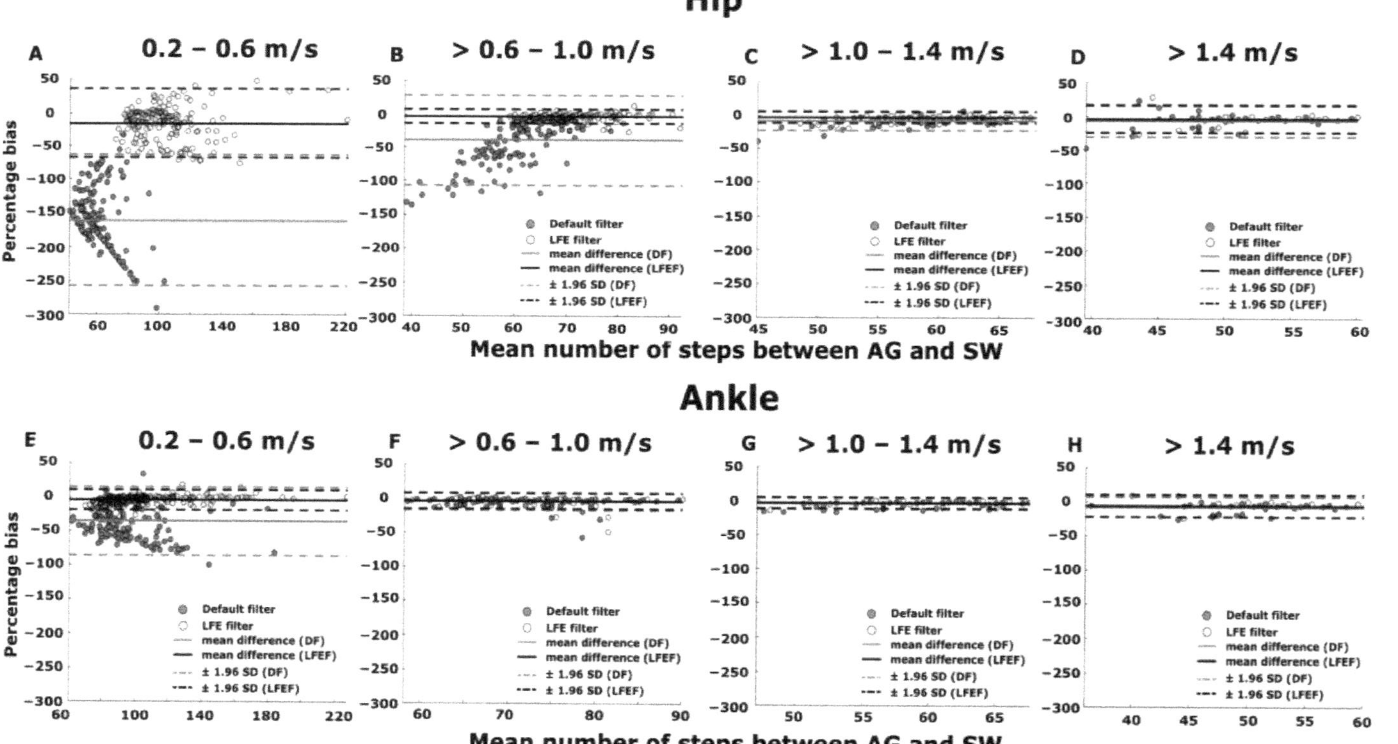

Figure 2. Bland–Altman plots for percentage bias between the hip worn AG and SW vs. the mean number of steps between the hip worn AG and SW using the DF and LFEF for gait speeds between (**A**) 0.2–0.6 m/s. (**B**) >0.6–1.0 m/s. (**C**) >1.0–1.4 m/s. (**D**) > 1.4 m/s during straight walking. Bland–Altman plots for percentage bias between the ankle worn AG and SW vs. the mean number of steps between the ankle worn AG and SW using the DF and LFEF for gait speeds between (**E**) 0.2–0.6 m/s. (**F**) >0.6–1.0 m/s. (**G**) >1.0–1.4 m/s. (**H**) > 1.4 m/s during straight walking. The grey and white filled circles represent the DF and LFEF, respectively. The grey and black solid line represents the mean percentage bias for the DF and the LFEF, respectively. The grey and black dashed line represents the ± 95% limits of agreement.

3.3. Walking with Continuous Turning

The percentage agreement for the ankle worn AG-DF was 95% for gait speeds between >0.6 and 1.0 m/s and 82% for gait speeds < 0.6 m/s (Table 1). The ankle worn AG-DF also showed good reliability ($ICC_{2,1}$ = 0.89) at gait speeds between >0.6 and 1.0 m/s whereas the reliability was poor ($ICC_{2,1}$ = 0.28) at gait speeds < 0.6 m/s (Table 2). When the AG-LFEF was applied to the ankle, the percentage agreement was >96% and the level of reliability was excellent ($ICC_{2,1}$: 0.93–0.97) across all represented gait speeds. The hip worn AG-DF showed a percentage agreement of 78% at gait speeds between >0.6 and 1.0 m/s but dropped drastically for gait speeds < 0.6 m/s. The reliability was poor ($ICC_{2,1}$: 0.00–0.12) for all gait speeds using the hip worn AG-DF. Conversely, when the hip worn AG-LFEF was applied, the percentage agreement was >88% and the reliability was moderate to good ($ICC_{2,1}$: 0.58–0.88) for gait speeds ranging 0.2–1.0 m/s.

The Bland–Altman plots for walking with continuous turns showed a low mean percentage bias (−3.0––5.0%; Figure 3C,D) for both ankle worn AG-DF and AG-LFEF for all represented gait speeds (0.2–1.0 m/s) except when using the ankle worn AG-DF at gait speeds between 0.2 and 0.6 m/s (−21.0%; Figure 3C). The mean percentage bias was low for the hip worn AG-LFEF (−4.0%) for gait speeds between >0.6 and 1.0 m/s and otherwise high at all represented gait speeds (−13––127%; Figure 3A,B).

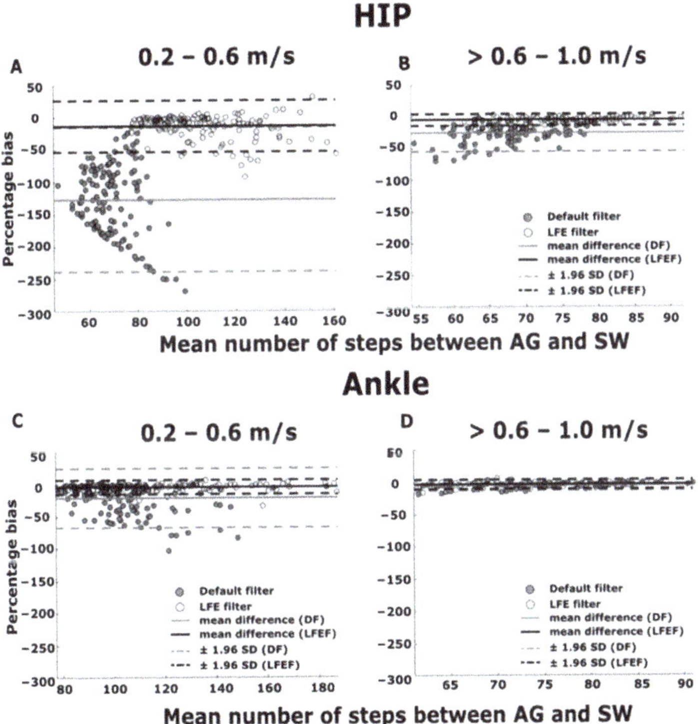

Figure 3. Bland–Altman plots for percentage bias between the hip worn AG and SW vs. the mean number of steps between the hip worn AG and SW using the DF and LFEF for gait speeds between (**A**) 0.2–0.6 m/s. (**B**) >0.6–1.0 m/s during continuous turning. Bland–Altman plots for percentage bias between the ankle worn AG and SW vs. the mean number of steps between the ankle worn AG and SW using the DF and LFEF for gait speeds between (**C**) 0.2–0.6 m/s. (**D**) >0.6–1.0 m/s during continuous turning. The grey and white filled circles represent the DF and LFEF, respectively. The grey and black solid line represents the mean percentage bias for the DF and the LFEF, respectively. The grey and black dashed line represents the ± 95% limits of agreement.

4. Discussion

The purpose of this study was to determine the accuracy of ankle and hip worn AG accelerometers to detect steps using the DF and LFEF as a function of gait speed during steady state walking and continuous turning. The results showed the ankle worn AG-DF to be the most sensitive for step detection at gait speeds > 0.6 m/s, whilst accurate step detection for gait speeds < 0.6 m/s were only observed when applying the LFEF. The hip worn AG-DF showed poor accuracy (12–78%) at gait speeds < 1.0 m/s whereas the accuracy increased to >87% for gait speeds < 1.0 m/s when applying the LFEF. Walking straight in a steady state or while walking with continuous turns did not impact the sensitivity of AG to detect step counts.

While previous studies [8,9,20] have shown the accuracy of AG to determine steps to be poor during slow walking, no previous study have explored at what level the accuracy drops when walking straight and performing continuous turns. In line with previous findings by Treacy et al. (2017), our results showed the hip worn AG-DF to have acceptable percentage agreement with SW for gait speeds > 1.0 m/s but significantly under counts the number of steps at gait speeds between 0.2 and 1.0 m/s for both straight walking and continuous turning. Irrespective of walking condition, the ankle worn AG-DF showed poor accuracy for step detection at gait speeds < 0.6 m/s and an increased accuracy and level of reliability for gait speeds between >0.6 and 1.4 m/s. These findings are in agreement with Treacy et al. (2017), Klaasen et al. (2016), Weber and St. John (2016) and Hergenroeder et al. (2018) who found that ankle worn accelerometers (i.e., SW, Fitbit and AG ankle) are generally more accurate when compared to sensors placed at the hip. This is likely due to the criteria of the AG step detection algorithm, which depends on a signal amplitude threshold of the vertical acceleration to determine if a step is taken. The ground impact and the signal amplitude picked up by the AG generally decreases from the distal to proximal placement (i.e., higher at ankle compared to the hip) [20], which is a plausible explanation for why ankle worn accelerometers overall show greater accuracy for step detection than hip worn sensors. In contrast, the acceleration signal at the hip during slow speeds is most likely not sufficient to register a step using the existing algorithms developed for AG [9].

Walking is an important marker for health where the number of steps per day is often associated with cardiovascular health [1]. Moreover, slow walking speeds has been linked with various movement disorders and high risk of morbidity and mortality [11]. Therefore, it is important to be able to accurately detect steps at low gait speeds to be able to identify health risk. This is especially important for people with disability, who often have compromised gait speeds. Previous studies [8,30] have shown improved accuracy for detecting steps using AG when applying the LFEF. For example, Weber and St. John [8] compared the accuracy of hip and ankle worn AG to an ankle reference sensor in older adults during straight walking [8]. Their results showed the absolute percentage error decreased from 47% to <3% for the ankle worn AG and from 96% to 19% for the hip worn AG when applying the LFEF. In our study, the overall step detection sensitivity for hip and ankle worn AG improved while applying the LFEF, especially at gait speeds < 0.6 m/s. Currently, there is no consensus regarding which gait speeds and (or) frequency movement the LFEF should be applied at. Our results suggest the LFEF should be used at gait speeds < 0.60 m/s for the ankle worn AG and <1.0 m/s for the hip worn AG. On the other hand, walking in daily life often occur in different speed ranges with most people varying their gait speed depending on the purpose of taking steps. Therefore, using the DF and LFEF is a bit more challenging for populations that walk both slow and at normal speeds. In line with this, it is worth noting, that previous studies [18,19,24,31], which recorded accelerometer data over a few days in daily living, have shown the LFEF to overestimate the number of steps taken over a day. Since the LFEF increases the sensitivity at low frequency movements, the AG might be prone to falsely detect steps during stationary movements not related to walking and especially the hip worn AG. Therefore, exploring and redefining the accelerometer amplitude cut points at low frequency movements to negate potential overestimation in daily living is warranted. We suggest that future work should also entail validating the present findings in daily living and to determine a cut-off gait velocity at which the LFEF should be used.

Study limitations include the relatively small sample size including healthy adults that walked slower than their self-selective gait speeds. It is unclear whether healthy controls walking at slower gait speeds reflect the walking pattern of individuals who walk slowly due to old age or disability (e.g., stroke or Parkinson's). Therefore, future work entails validating the findings in these populations. The study measured steps over a relative short distance and measuring a longer walking distance could result in an increased reliability. On the other hand, the variability of the gait pattern among healthy adults is low and approximately 30 steps has shown to be sufficient for reliable measures of spatial and temporal gait parameters [32]. Therefore, we do not believe our result would have been different if we had assessed a longer duration of walking. Finally, our study included relatively small number of data points at gait speeds > 1.0 m/s. This could result in misrepresentation of the percentage agreement and interclass correlation coefficient at gait speeds > 1.0 m/s. Still, previous studies has shown good reliability of accelerometers at gait speeds > 1.0 m/s [5].

5. Conclusions

Our results showed the hip worn AG device to have poor agreement and reliability at gait speeds of <1.0 m/s, with the LFEF drastically increasing the accuracy of the step count at gait speeds between 0.2 and 1.0 m/s. The ankle worn AG showed the highest accuracy at gait speeds > 0.60 m/s, however at gait speeds < 0.60 m/s the LFEF needs to be applied to heavily negate underestimating. Walking straight in a steady state or while walking with continuous turns did not impact the sensitivity of AG to detect step counts.

Author Contributions: Conceptualization, L.B. and D.M.C.; Methodology, L.B. and D.M.C.; Validation, L.B. and D.M.C.; Formal analysis, L.B., D.M.C. and M.H.; Investigation, L.B. and C.T.; Writing—Original Draft Preparation, L.B. and D.M.C.; Writing—Review and Editing, L.B., C.T., M.H. and D.M.C.; Visualization, L.B.; Supervision, D.M.C.; Project Administration, L.B.; Funding acquisition, D.M.C. All authors have read and agreed to the published version of the manuscript.

Funding: The research was funded by the Norrbacka-Eugenia foundation, Promobilia foundation, NEURO Sweden and the Swedish Stroke Association.

Institutional Review Board Statement: The study was conducted according to the guidelines of the Declaration of Helsinki and approved by the Regional Board of Ethics in Stockholm (2017/1626-31 and 2018/2524-32).

Informed Consent Statement: Informed consent was obtained from all subjects involved in the study

Data Availability Statement: The data used in this study are available from the authors upon request.

Acknowledgments: The authors would like to thank all the participants for their valuable time and efforts.

Conflicts of Interest: The authors report no conflict of interest.

References

1. Hergenroeder, A.L.; Gibbs, B.B.; Kotlarczyk, M.P.; Kowalsky, R.J.; Perera, S.; Brach, J.S. Accuracy of Objective Physical Activity Monitors in Measuring Steps in Older Adults. *Gerontol. Geriatr. Med.* **2018**, *4*, 2333721418781126. [CrossRef]
2. Tudor-Locke, C.; Craig, C.L.; Aoyagi, Y.; Bell, R.C.; A Croteau, K.; De Bourdeaudhuij, I.; Ewald, B.; Gardner, A.W.; Hatano, Y.; Lutes, L.D.; et al. How many steps/day are enough? For older adults and special populations. *Int. J. Behav. Nutr. Phys. Act.* **2011**, *8*, 80. [CrossRef] [PubMed]
3. Middleton, A.; Fritz, S.L.; Lusardi, M. Walking speed: The functional vital sign. *J. Aging Phys. Act.* **2015**, *23*, 314–322. [CrossRef] [PubMed]
4. Karas, M.; Bai, J.; Strączkiewicz, M.; Harezlak, J.; Glynn, N.W.; Harris, T.; Zipunnikov, V.; Crainiceanu, C.; Urbanek, J.K. Accelerometry Data in Health Research: Challenges and Opportunities. *Stat. Biosci.* **2019**, *11*, 210–237. [CrossRef]
5. Storti, K.L.; Pettee, K.K.; Brach, J.S.; Talkowski, J.B.; Richardson, C.R.; Kriska, A.M. Gait Speed and Step-Count Monitor Accuracy in Community-Dwelling Older Adults. *Med. Sci. Sports Exerc.* **2008**, *40*, 59–64. [CrossRef]

6. Campos, C.; DePaul, V.G.; Knorr, S.; Wong, J.S.; Mansfield, A.; Patterson, K.K. Validity of the ActiGraph activity monitor for individuals who walk slowly post-stroke. *Top. Stroke Rehabil.* **2018**, *25*, 295–304. [CrossRef]
7. John, D.; Morton, A.; Arguello, D.; Lyden, K.; Bassett, D. "What Is a Step?" Differences in How a Step Is Detected among Three Popular Activity Monitors That Have Impacted Physical Activity Research. *Sensors* **2018**, *18*, 1206. [CrossRef]
8. Webber, S.C.; John, P.D.S. Comparison of ActiGraph GT3X+ and StepWatch Step Count Accuracy in Geriatric Rehabilitation Patients. *J. Aging Phys. Act.* **2016**, *24*, 451–458. [CrossRef]
9. Treacy, D.; Hassett, L.; Schurr, K.; Chagpar, S.; Paul, S.S.; Sherrington, C. Validity of Different Activity Monitors to Count Steps in an Inpatient Rehabilitation Setting. *Phys. Ther.* **2017**, *97*, 581–588. [CrossRef]
10. Ingebrigtsen, J.; Stemland, I.; Christiansen, C.; Skotte, J.; Hanisch, C.; Krustrup, P.; Holtermann, A. Validation of a Commercial and Custom Made Accelerometer-Based Software for Step Count and Frequency during Walking and Running. *J. Ergon.* **2013**, *3*, 1000119. [CrossRef]
11. Lugade, V.; Fortune, E.; Morrow, M.; Kaufman, K. Validity of using tri-axial accelerometers to measure human movement—Part I: Posture and movement detection. *Med. Eng. Phys.* **2014**, *36*, 169–176. [CrossRef] [PubMed]
12. Grant, P.M.; Dall, P.M.; Mitchell, S.L.; Granat, M.H. Activity-Monitor Accuracy in Measuring Step Number and Cadence in Community-Dwelling Older Adults. *J. Aging Phys. Act.* **2008**, *16*, 201–214. [CrossRef]
13. Feito, Y.; Bassett, D.R.; Thompson, D.L. Evaluation of Activity Monitors in Controlled and Free-Living Environments. *Med. Sci. Sports Exerc.* **2012**, *44*, 733–741. [CrossRef] [PubMed]
14. Feito, Y.; Garner, H.R.; Bassett, D.R. Evaluation of ActiGraph's Low-Frequency Filter in Laboratory and Free-Living Environments. *Med. Sci. Sports Exerc.* **2015**, *47*, 211–217. [CrossRef]
15. Sandroff, B.M.; Motl, R.W.; Pilutti, L.A.; Learmonth, Y.C.; Ensari, I.; Dlugonski, D.; Klaren, R.E.; Balantrapu, S.; Riskin, B.J. Accuracy of StepWatch™ and ActiGraph Accelerometers for Measuring Steps Taken among Persons with Multiple Sclerosis. *PLoS ONE* **2014**, *9*, e93511. [CrossRef] [PubMed]
16. Orendurff, M.S. How humans walk: Bout duration, steps per bout, and rest duration. *J. Rehabil. Res. Dev.* **2008**, *45*, 1077–1090. [CrossRef] [PubMed]
17. Glaister, B.C.; Bernatz, G.C.; Klute, G.K.; Orendurff, M.S. Video task analysis of turning during activities of daily living. *Gait Posture* **2007**, *25*, 289–294. [CrossRef] [PubMed]
18. Wallén, M.B.; Nero, H.; Franzén, E.; Hagströmer, M. Comparison of two accelerometer filter settings in individuals with Parkinson's disease. *Physiol. Meas.* **2014**, *35*, 2287–2296. [CrossRef] [PubMed]
19. Toth, L.P.; Park, S.; Springer, C.M.; Feyerabend, M.D.; Steeves, J.A.; Bassett, D.R. Video-Recorded Validation of Wearable Step Counters under Free-living Conditions. *Med. Sci. Sports Exerc.* **2018**, *50*, 1315–1322. [CrossRef] [PubMed]
20. Klassen, T.D.; Simpson, L.A.; Lim, S.B.; Louie, D.R.; Parappilly, B.; Sakakibara, B.M.; Zbogar, D.; Eng, J.J. "Stepping Up" Activity Poststroke: Ankle-Positioned Accelerometer Can Accurately Record Steps During Slow Walking. *Phys. Ther.* **2016**, *96*, 355–360. [CrossRef]
21. Svarre, F.R.; Jensen, M.M.; Nielsen, J.; Villumsen, M. The validity of activity trackers is affected by walking speed: The criterion validity of Garmin Vivosmart((R)) HR and StepWatch() 3 for measuring steps at various walking speeds under controlled conditions. *PeerJ* **2020**, *8*, e9381. [CrossRef] [PubMed]
22. Brønd, J.C.; Andersen, L.B.; Arvidsson, D. Generating ActiGraph Counts from Raw Acceleration Recorded by an Alternative Monitor. *Med. Sci. Sports Exerc.* **2017**, *49*, 2351–2360. [CrossRef]
23. Tryon, W.W.; Williams, R. Fully proportional actigraphy: A new instrument. *Behav. Res. Methods Instrum. Comput.* **1996**, *28*, 392–403. [CrossRef]
24. Feito, Y.; Hornbuckle, L.M.; Reid, L.A.; Crouter, S.E. Effect of ActiGraph's low frequency extension for estimating steps and physical activity intensity. *PLoS ONE* **2017**, *12*, e0188242. [CrossRef] [PubMed]
25. Fritz, S.; Lusardi, M. White paper: "walking speed: The sixth vital sign". *J. Geriatr. Phys. Ther.* **2009**, *32*, 46–49. [CrossRef] [PubMed]
26. Bobak, C.A.; Barr, P.J.; O'Malley, A.J. Estimation of an inter-rater intra-class correlation coefficient that overcomes common assumption violations in the assessment of health measurement scales. *BMC Med. Res. Methodol.* **2018**, *18*, 93. [CrossRef]
27. Koo, T.K.; Li, M.Y. A Guideline of Selecting and Reporting Intraclass Correlation Coefficients for Reliability Research. *J. Chiropr. Med.* **2016**, *15*, 155–163. [CrossRef] [PubMed]
28. Giavarina, D. Understanding Bland Altman analysis. *Biochem. Med.* **2015**, *25*, 141–151. [CrossRef] [PubMed]
29. Negrini, F.; Gasperini, G.; Guanziroli, E.; Vitale, J.A.; Banfi, G.; Molteni, F. Using an Accelerometer-Based Step Counter in Post-Stroke Patients: Validation of a Low-Cost Tool. *Int. J. Environ. Res. Public Health* **2020**, *17*, 3177. [CrossRef]
30. Korpan, S.M.; Schafer, J.L.; Wilson, K.C.; Webber, S.C. Effect of ActiGraph GT3X+ Position and Algorithm Choice on Step Count Accuracy in Older Adults. *J. Aging Phys. Act.* **2015**, *23*, 377–382. [CrossRef] [PubMed]
31. Barreira, T.V.; Brouillette, R.M.; Foil, H.C.; Keller, J.N.; Tudor-Locke, C. Comparison of Older Adults' Steps per Day Using an NL-1000 Pedometer and Two GT3X+ Accelerometer Filters. *J. Aging Phys. Act.* **2013**, *21*, 402–416. [CrossRef] [PubMed]
32. Galna, B.; Lord, S.; Rochester, L. Is gait variability reliable in older adults and Parkinson's disease? Towards an optimal testing protocol. *Gait Postur.* **2013**, *37*, 580–585. [CrossRef] [PubMed]

MDPI

Article

A Smart Walker for People with Both Visual and Mobility Impairment

Nafisa Mostofa [1], Christopher Feltner [1], Kelly Fullin [2], Jonathan Guilbe [1], Sharare Zehtabian [1], Salih Safa Bacanlı [1], Ladislau Bölöni [1] and Damla Turgut [1,*]

[1] Department of Computer Science, University of Central Florida, Orlando, FL 32816, USA; nafisa.mostofa@knights.ucf.edu (N.M.); chris.feltner@knights.ucf.edu (C.F.); jguilbe@knights.ucf.edu (J.G.); sharare.zehtabian@knights.ucf.edu (S.Z.);bacanli@knights.ucf.edu (S.S.B.); Ladislau.Boloni@ucf.edu (L.B.)

[2] Department of Computer Science, Ashland University, Ashland, OH 44805, USA; kellyfullin@gmail.com

* Correspondence: damla.turgut@ucf.edu

Citation: Mostofa, N.; Feltner, C.; Fullin, K.; Guilbe, J.; Zehtabian, S.; Bacanlı, S.S.; Bölöni, L.; Turgut, D. A Smart Walker for People with Both Visual and Mobility Impairment. *Sensors* **2021**, *21*, 3488. https://doi.org/10.3390/s21103488

Academic Editors: Ayman Radwan, Andrea Sciarrone and Maria de Fátima Domingues

Received: 4 April 2021
Accepted: 10 May 2021
Published: 17 May 2021

Abstract: In recent years, significant work has been done in technological enhancements for mobility aids (smart walkers). However, most of this work does not cover the millions of people who have *both* mobility and visual impairments. In this paper, we design and study four different configurations of smart walkers that are specifically targeted to the needs of this population. We investigated different sensing technologies (ultrasound-based, infrared depth cameras and RGB cameras with advanced computer vision processing), software configurations, and user interface modalities (haptic and audio signal based). Our experiments show that there are several engineering choices that can be used in the design of such assistive devices. Furthermore, we found that a holistic evaluation of the end-to-end performance of the systems is necessary, as the quality of the user interface often has a larger impact on the overall performance than increases in the sensing accuracy beyond a certain point.

Keywords: smart walker; obstacle detection; aging; rehabilitation

1. Introduction

While a significant amount of research and industry interest targets mobility aids at the elderly and disabled, these efforts are often not applicable to people who have specific comorbidities. A particularly widespread group of such patients have both visual and mobility impairments. For instance, according to the World Health Organization, there are an estimated 1.3 billion people globally living with some form of visual impairment [1]. As this population ages, they will also require mobility assistance at least at the rate of the people with a healthy vision, estimated to be about 16% for individuals 65 years of age or older [2]. A significant challenge, however, is that the assistive technologies developed for people with normal eyesight often cannot be used by people with visual impairments. This is partially due to the fact that the user interfaces often rely on visual feedback. Furthermore, people with visual impairment already need some type of assistive technology to navigate their environment—the use of two distinct devices would lead to an unacceptable cognitive overload.

The research described in this paper focuses on devices that simultaneously help people with their visual and mobility impairments, and at the same time also use user interfaces which are appropriate to the capabilities of the user and do not lead to cognitive overload. As there has been very little work done on devices with this particular combination of capabilities (some examples include [3,4]), it is not clear what type of path planning and obstacle detection technologies are appropriate in these settings (ultrasound, vision, infrared and or structured/light technologies). It is also uncertain what type of user interaction is the least distracting (sound, haptic or high-contrast visual) and what the content and frequency of the communication with the user should be. In conclusion, one of our objectives was the exploration of the design space: we investigated a variety of obstacle detection and user interaction technologies, and evaluated them under various scenarios. The four obstacle detection techniques we used are (a) ultrasonic distance sensors; (b) 2D observable light camera input processed with state-of-the-art deep learning-based computer vision algorithms; (c) depth images from an RGB-D camera with minimal post-processing; and (d) the processing of a 3D point cloud acquired from an RGB-D camera. For the user interface, we investigated the use of audible and haptic signals.

Beyond the investigation of the technologies involved, we need to keep in mind that the output of our work needs to be an assistive device that can be deployed and used by people who have mobility and visual impairments. We implemented the technological solutions in the form of a smart walker, and we also took into consideration several practical design requirements. The walkers need to be affordable, they should preferably be autonomous, and do not require a network connection or cloud computation. Furthermore, the user interface is very important, as it needs to convey information about obstacles and proposed avoidance strategies without distracting the user.

The contributions of this paper are as follows:

- We described the designs of several smart walker configurations that can use various technologies for obstacle detection and user interaction.
- We described four different approaches for obstacle detection, based on different sensing techniques (ultrasound, RGB cameras, and RGB-D depth cameras). Correspondingly, we described the appropriate processing techniques that can transform the sensor output to a signal identifying the detected obstacles.
- We validated the proposed object detection techniques in a series of real-world experiments.
- We studied the user interaction techniques based on audio versus haptic notifications.

The remainder of the paper is organized as follows. In Section 2, we review the related work. In Section 3, we described the proposed approaches in detail. The evaluations of the proposed approaches through various configurations are given in Section 4 and we conclude in Section 5.

2. Related Work

One of the earliest implementations for a smart walker and smart cane for elderly users with impaired mobility were the PAMM designs implemented at MIT in the late 1990s (Dubowsky et al. [5]). These devices were essentially small mobile robots augmented with handles. They used a sonar array for obstacle detection, an upward pointing camera for localization using ceiling mounted markers and a force/torque sensor mounted on the handle for user control. While limited by the technology available at the time, these designs remained influential, and outlined the research directions which are now being pursued by many researchers.

MacNamara and Lacey [3] proposed a wheeled walker (rollator) targeted towards aged people who also have visual impairments. Similarly to our work, this system was designed to detect obstacles in the path of the user and communicate their presence using two techniques: audio feedback and a feedback that uses motors to align the wheels in an obstacle-free direction.

This work led to an early commercial implementation in the form of the motorized rollator Guido by Haptica [6]. In addition to a more polished commercial design, this system specifically focused on older blind people. It used sonar and laser ranging devices to avoid obstacles and a SLAM algorithm to build a map of its environment, allowing it to guide the user to a predetermined destination. One of the challenges faced by Guido as a commercial product launched in the early 2000s was the high cost of devices such as LIDARs.

Zehtabian et al. [7] described an IoT-augmented four-legged walker, which uses sensors to track and visualize the weight distribution over its legs during use. This facilitates proper walker usage for rehabilitation and assists physicians in checking their patients' rehabilitation progress.

In a follow-up paper, Khodadadeh et al. [8] processes the walker's data stream using a deep neural network-based classifier. This allows the detection of unsafe usage that could hinder a patient's rehabilitation.

Paulo et al. [9] implemented ISR-AIWALKER, a robotic walker using computer vision as the primary human–machine interface modality. This contrasts with previous approaches that primarily relied on force sensing. The walker was also augmented with multimodal sensing capabilities that allow it to analyze and classify the gait of the user. In follow-up work, Garrote et al. [4] augmented the ISR-AIWALKER with robot-assisted navigation targeted towards users who lack a dexterous upper limb or have visual impairments. The walker uses reinforcement learning algorithms to learn a behavior that fuses user intent and the environmental sensing of the obstacles. Whenever obstacles are detected, the system adds corrections to the movement in order to avoid collisions.

Kashyap et al. [10] developed a self-driven smart walker by augmenting the rear wheels of a rollator with motors. The user interacts with the rollator using voice commands, such as "go to (room name)". The authors evaluated several off-the-shelf, LIDAR-based simultaneous localization and mapping (SLAM) implementations for mapping and navigation. The system also has a fall detection algorithm that prevents the rollator from moving away if the user falls.

Bhatlawande et al. [11] proposed a system where ultrasonic sensors are attached to a belt and a pair of glasses worn by the user. The system detects and identifies obstacles and indicates an obstacle-free path using audio feedback.

Orita et al. [12] implemented a device that augments the white cane used by visually impaired people with a Microsoft Kinect camera connected to a laptop and uninterruptible power supply in a backpack. Information about the lack of obstacles was communicated to the user through vibration feedback. A small experimental study using blindfolded subjects had shown that the device helped users navigate an indoor environment.

Pham et al. [13] devised a system to provide a blind user feedback about drops, objects, walls, and other potential obstacles in the environment. The system relies on a Kinect sensor worn by the user, with the output processed by a laptop computer in a backpack. Feedback to the user is provided using a Tongue Display Unit, a sensory substitution device.

Panteleris and Argyros [14] investigated the challenges of vision-based SLAM arising in the use of the c-Walker [15], a smart rollator with a Kinect sensor as an RGB-D camera. Rollator users normally move in environments with many other people around, thus the SLAM algorithm must consider a large number of independently moving objects.

Viegas et al. [16] described a system which allows a four-legged smart walker to collect data about the load the users put on each walker leg using load cells and the relative position of the user to the walker using a LIDAR. This information is transmitted using Bluetooth to an external device and can be used to guide the users in the correct use of the device and prevent dangerous situations.

Ramadhan [17] described a wrist-wearable system that allows visually impaired persons to navigate public places and seek assistance if needed. The system contains a suite of sensors, haptic interaction modules as well as a GPS module and has the ability to communicate over cellular networks.

Kim and Cho [18] performed a user study about the challenges encountered by the users of several types of commercial smart canes with obstacle detection capabilities, and compared them with the traditional white canes. The output of the customer interviews was used to advance design guidelines for improved smart canes.

Feltner et al. [19] and Mostofa et al. [20] designed walkers targeted at visually impaired people. These works describe the early versions of the walker configurations described in the next section.

3. Proposed Approaches

The most widely used assistive devices for mobility are the cane, four wheeled rollators, and four legged walkers. These devices are simple and intuitive for most people and, by physical construction, they support and stabilize the users. In order to extend the benefits of these devices to people who are both mobility *and* visually impaired, they need to be augmented with additional capabilities. A relatively wide number of choices exist with regards to the type of sensing and processing capabilities that can be deployed. To explore this design space, we started from a standard four-wheeled rollator with a basket and seat. We implemented four different augmentation configurations across a range of sensor types, processing hardware and software, and user interface techniques.

3.1. User Interaction

Assistive devices in general, and devices for visually impaired users in particular have special user interface (UI) requirements. Graphical user interfaces, the most widely used techniques to convey information to the user, are not applicable. UIs for assistive devices must be robust to environmental noise, not require significant cognitive effort from the user and reduce the chance of misunderstood signals. Given the capabilities of the augmented rollator, there are two distinct messages that the UI must convey. The *obstacle detected* message warns the users that they would hit an obstacle if they continue on the current trajectory. A signal of increased urgency can be used to convey the proximity of the obstacle. The *navigational guidance* message provides a recommendation to the user to turn towards the left or right in order to avoid the obstacle.

For our rollator, we chose to implement and compare two UI methods: a voice-based user interface that conveys the information through spoken messages, and a haptic feedback that is enacted through coin vibration motors attached to the handles of the walker. Both modalities can convey both the obstacles detected and the navigational guidance messages, as well as their various gradations. In the case of the voice interface, this is conveyed through the content and tone of the voice. For the haptic interface, the intensity of the vibration corresponds to the proximity of the obstacle, while the vibration in the left or right handle conveys the direction of the recommended turn.

3.2. (A) Ultrasonic Sensors

In this configuration, we removed the basket of the walker and attached nine HC-SR04 [21] ultrasonic sonar distance sensors to the lower front crossbar of the walker. We used a Raspberry Pi 3b+ device to drive the sensors, as well as collect and interpret the results.

The HC-SR04 sensor operates as follows. A ultrasonic sound wave, above the frequency of human hearing, is generated as a trigger. If there is an object in the sensor's path, the sound wave bounces back to the sensor as an echo and is captured by the sensor. By measuring the time taken by sound to travel to the obstacle and back, we can calculate the distance to the obstacle. In practice, this sensor is limited to a viewing angle of 14 degrees and can operate to a distance of up to 400 cm–s with an accuracy of 3 mm. In order to cover the range of obstacles of interest to the user of the walker, we attached seven of these sensors across the width of the walker. They were also angled slightly towards the floor in order to detect obstacles immediately in the front of the walker. In addition, to detect obstacles to the left and right of the walker, we attached one outward angled sensor to each side of the walker (see Figure 1). One technical challenge we encountered was that the sound waves from the multiple sensors interfered with each other, which is possible for one sensor to detect a delayed echo from a wave started by a different sensor. To avoid this, in our implementation the sensors perform their detection sequentially.

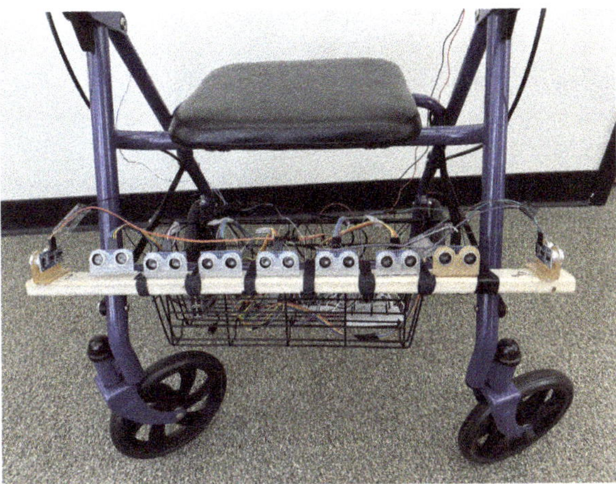

Figure 1. The rollator configured with nine HC-SR04 ultrasonic sensors. Seven sensors are facing forward to capture obstacles in a wide area in front of the walker. To facilitate navigational guidance, two sensors (one on the left and the other on the right) are capturing obstacles.

The outcome of the detection step is an array of distance values corresponding to the directions covered by the individual sensors. The first use of these data is the obstacle detection message: the user will be warned if there is an obstacle closer than 200 cm–s in the direction of movement. This message is triggered if any of the central sensors detect an obstacle in this range. In addition to this, the arrangement of the sensors also allows us to recommend the user a navigational guidance for the avoidance of the obstacle. For instance, if there are obstacles detected in the front and to the right of the walker, the system will recommend the user to move towards the left.

3.3. (B) RGB Camera with Deep Learning-Based Computer Vision Algorithms

In recent years, the significant reduction in the cost of video cameras, together with the advances of deep learning-based computer vision algorithms made it possible to detect obstacles based solely on video information, without using a dedicated distance or depth sensors. To investigate the feasibility of such an approach, in this configuration, we mounted a forward-facing Logitech C270HD webcam on the top crossbar of the walker. We used deep learning-based computer vision algorithms to process the video stream. The algorithms required the full-featured version of the Tensorflow library, exceeding the capabilities of a Raspberry Pi device. In our experiments, we used a laptop computer placed on the seating area of the walker.

Recent research on deep learning-based object detection systems created a number of approaches that can detect objects of specific types in images in real time. Some of these approaches include R-CNN, Fast R-CNN, Faster R-CNN, YOLO and others. While the training of such systems requires non-trivial computational capabilities, many pre-trained neural networks are available in the public domain. For instance, the Object Detection API of the TensorFlow library provides a simple programmatic interface to a choice of several different pre-trained networks. In our experiments, we used a pre-trained network with the Faster R-CNN region proposal network [22] with the Inception Resnet V4 [23] model trained on the Open Images data set [24]. By applying this object detector to the video frames captured through the camera driver of OpenCV, we obtained a collection of bounding boxes on the image, together with the tentative label and a confidence value. While the pre-trained network handles and detects a wider variety of objects, we only retain objects detected with the labels relevant to our application, such as doors, cars and chairs (see Figure 2).

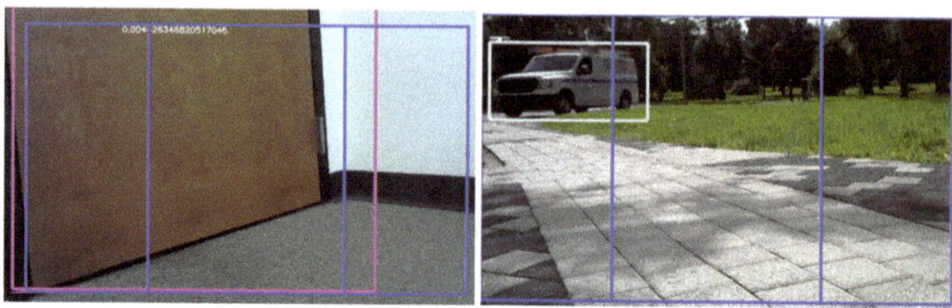

Figure 2. Object detection of door and car with TensorFlow.

Many objects of interest to the user have relatively uniform sizes. Thus, for a given camera with a fixed focal length, the size of a bounding box in the image can allow us to approximate the distance to the object. For instance, if a chair occupies approximately half of the field of view of a camera with the 45° view angle, the distance to the chair will be approximately 2–3 ft. Note that this approximation critically depends on the correct identification of the object—a car with a similar size bounding box would be much farther away. This approach, while limited in absolute precision, in practice allows us to develop a software implementation that provides sufficient accuracy for the purposes of obstacle detection and navigational messages.

We identify the central region of the visual area, which is where the walker is currently heading. However, comparing the bounding box of the obstacle to this region, we can identify whether the obstacle is in this region, and also if it is on the left or right side of the region. This allows us to generate the appropriate navigational messages.

3.4. (C) Depth Camera with Direct Processing of the Depth Image

In this configuration, we mounted a Microsoft Kinect RGB-D depth camera to the lower front crossbar of the walker. As with the ultrasonic sensors in configuration (A), the Kinect was angled towards the floor. The device was connected to a Raspberry Pi and powered through a dedicated 12 V DC rechargeable battery. Figure 3—top shows the depth component of a captured image, with darker colors representing closer distances. The areas in black are locations where the camera was unable to determine the location of the point.

In order to implement the functionality required by the walker while relying only on the limited computational capabilities of a Raspberry Pi, we chose to implement an algorithm based on an idea from Ortigosa et al. [25]. We started from the central vertical stripe of the depth image. To reduce the noise and limit the data to be processed, we performed a pre-processing step by calculating the average of the values on the center stripe, skipping the black pixels for which no value was available. This created a one-dimensional array of the size of the height of the image (see Figure 3—bottom). If there is no obstacle in front of the walker, the pixels at the bottom will have the smallest distance values, gradually and smoothly increasing to the top of the image. Obstacles will be indicated by a sudden increase in the slope of this array. On the other hand, a negative slope indicates a drop in the elevation such as the beginning of a staircase or a street curb.

The advantage of this approach is that it only requires an averaging of a central area, followed by an iteration over a one-dimensional array while tracking the slope. This computational load is well within the reach of most IoT devices. A disadvantage of the approach is that by focusing only on the central stripe, it cannot take into consideration the available space to the left and right. Thus, a system using this algorithm can only perform obstacle detection, not navigational guidance.

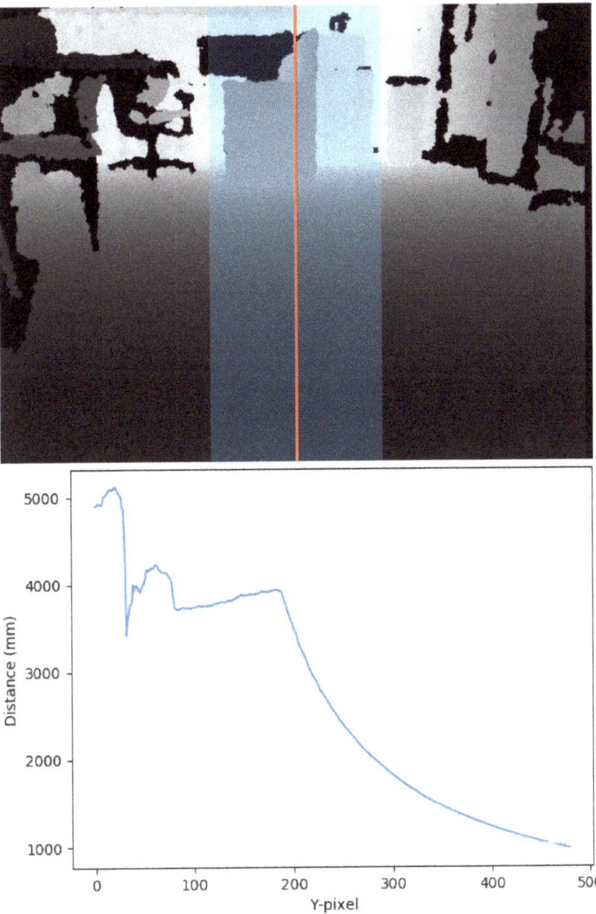

Figure 3. Top: The depth component of an image recorded by a Kinect camera in configuration (C). Darker colors show closer points, with black pixels representing points for which depth information is not available. The blue central column is the area processed for obstacle detection. Bottom: The one-dimensional array extracted from the depth map. Pixel 500 refers to the bottom of the image. The smooth increase in distance from 500 to 200 shows an approximately 4 m free area in front of the walker, with a drop starting after that.

3.5. (D) Depth Camera with Point Cloud Model and Processing

With the Kinect mounted as in case (C), for this configuration we used more complex algorithms that build a point cloud of the scene as an intermediate step. These algorithms required more computational power than available in the Raspberry Pi 3, thus, as in case (B), we used a laptop computer to process the input and to power the Kinect device. Figure 4 shows the rollator in this configuration.

We are extracting a point cloud from the Kinect device, and processing it through an approach similar to that of Pham et al. [13]. In order to focus on the processing of the data relevant to the user of the walker, we performed a series of pre-processing steps using the Point Cloud Library [26]. We removed the points from the point cloud that were not immediately relevant to the user of the walker, as well as points that lie outside the Kinect's range of accuracy. Second, we noticed that the initially captured point cloud contained hundreds of thousands of points. The user of the walker does not need such a detailed *spatial* resolution for obstacle avoidance. However, the time needed to process this information would reduce the *temporal* resolution of the obstacle detector, introducing unacceptable delays into the obstacle notification and navigation guidance. To solve this problem, we down-sampled the point cloud using a voxel grid filter to a spatial resolution of 1 cm.

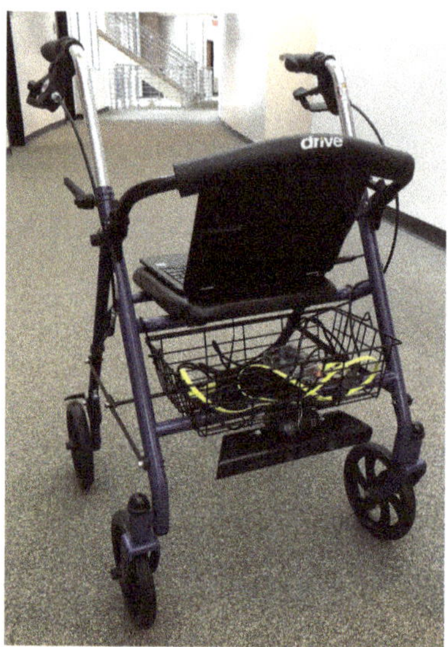

Figure 4. The smart walker in configuration D.

The next step is to extract a model of the floor in the point cloud which allows us to interpret other points as belonging to obstacles. We use the random sampling and consensus (RANSAC) algorithm to determine the coefficients of a large horizontal plane within the point cloud, with the inlier points being considered as part of the floor. If no such plane could be found, this means that either the walker is at the edge of a sudden drop (such as a staircase) or that a large, close obstacle obstructed the majority of the camera view. In both cases, the user is warned about an immediate obstacle at close range.

After the plane of the floor was extracted, the outlier points were considered to be part of the obstacles. We used the techniques described by Li et al. [27] to iterate through all outlier points and determine the distance from the obstacles. The stages of processing the point cloud are shown in Figure 5.

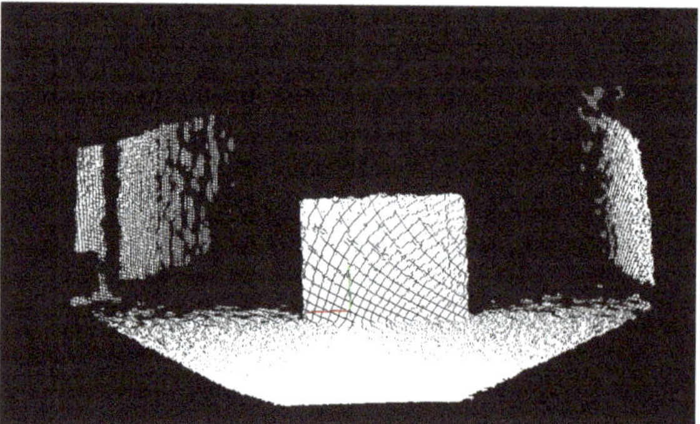

Figure 5. *Cont.*

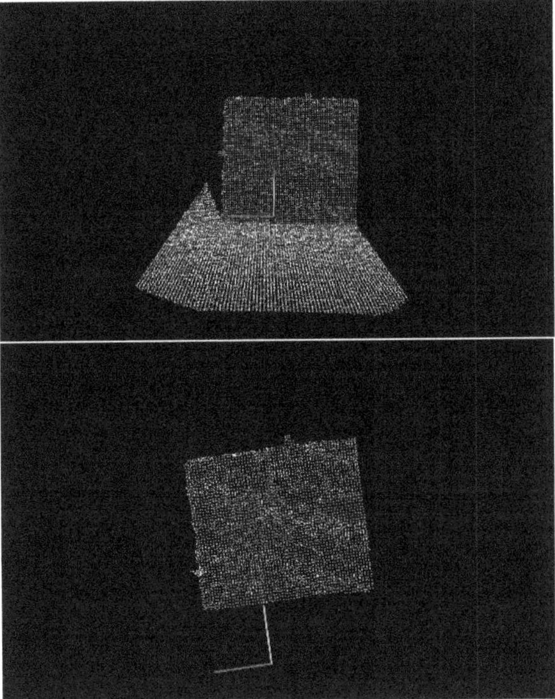

Figure 5. The stages of processing the point cloud: (top) the high-resolution point cloud extracted from the Kinect sensor; (middle) the point cloud after the elimination of non-relevant points and downsampling; and (bottom) the points of an obstacle, after the floor plane was identified and removed from the image using RANSAC.

4. Results

There is no single criteria that can be used to compare the various configurations of the smart walker. Obviously, devices with higher computational power and more capable sensors should yield better performance when measured in localization accuracy. For instance, an RGB-D camera that returns both an image and a depth value for every pixel will inevitably yield a high-quality model of the environment on which complex path planning algorithms can be applied. However, a sensor of this complexity has both a higher cost and more costly processing requirements.

Sensors with significantly less capabilities, such as the ultrasonic sensors that return only a single numerical value for a depth (configuration (A)), or sensors that collect only an image (configuration (B)), will yield a walker with less capabilities in an absolute sense. However, such a walker can still provide useful services to the user, and be better in terms of its dimensions of cost, robustness and reliability compared to walkers with more complex sensors (such as configurations (C) and (D)).

Thus, we are going to evaluate our designs in two different ways. Objective tests for obstacle detection measure the performance of the sensor paired with a specific processing algorithm. These tests do not involve the participation of a human user.

End-to-end usability tests measure the utility provided by a walker to the user for obstacle avoidance and navigation. Such tests holistically consider the entire system, including sensor performance, processing latency, the quality, and types of feedback given to the user.

The objective of these experiments was to investigate the capabilities of the augmented walker. These tests consider a human user performing specific navigation tasks with the configured walker. As these experimental systems do not yet meet the criteria of safety for human subjects research with elderly and disabled people, in these experiments we used healthy volunteers, from 20 to 30 years old, from the research group. The subjects were instructed to put some of their weight on the walker to model mobility impairment and were blindfolded to model visual impairment.

4.1. Obstacle Detection: Configuration (A) versus (B)

In this set of experiments, we compared the two configurations that do not have an RGB-D sensor ((A) and (B)) in their ability to detect and estimate the distance to an obstacle placed in front of the walker.

For the experiments with the (A) and (B) configurations, we chose a set of obstacles to experiment with based on the following considerations. We included both indoor and outdoor obstacles, as these sensor types can function in both environments (this is not possible for (C) and (D)). We also included both obstacles that raise above the level of the floor, and obstacles that represent a drop (such as the stairs, curb and swimming pool). Finally, we added some obstacles that test the ability of the computer vision system in configuration (B) to identify obstacles that cannot be distinguished by a simple distance sensor like in configuration (A). For instance, the computer vision system can distinguish between a door (which can be opened, thus treated differently by the user) from a wall.

What we are interested in here is whether these more limited sensors are able to capture a variety of obstacles that a typical user might encounter. Table 1 shows the results of a series of experiments we performed with a variety of obstacles in a household environment. The ground truth have been obtained through direct measurement from the sensor to the closest point of the obstacle. Some of the conclusions we can draw from these experiments are as follows. Both configurations (A) (ultrasonic sensor) and (B) (camera processed through computer vision) were able to detect all the obstacles in these experiments, and returned the correct "no obstacle" answer in the empty hallway. The ultrasonic sensor, which is an active sensor specialized on the distance measurement, obtained the sub-centimeter accuracy, an operational parameter of this sensor type. As expected, the values obtained from the processing of the camera image were less accurate. The camera, as a passive image sensor, does not directly capture any depth information, as all values are inferred only from relative image sizes. Nevertheless, we conclude that the accuracy of both sensors was sufficient for making a decision about whether the user should be notified of the obstacle or not. As a note, with regards to the quality of this notification, the computer vision system was able to identify the type of obstacle (e.g., person, wall or car), while the depth sensor can only detect the distance, albeit at a higher accuracy.

Table 1. Comparing the accuracy of configurations A and B for measuring the distance (cm) to several indoors (upper part) and outdoors (lower part) obstacle types.

Obstacle	**Ground Truth**	**Config (A)**	**Config (B)**
Empty hallway	N/A	No obstacle	No obstacle
Wall	91.9	92.3	82.9
Door	53.7	54.6	40.3
Person	110.3	110.2	120.9
Stairs	51.8	52.3	54.9
Backpack	42.5	42.6	39.5
Curb	87.3	86.9	100.3
Car	132.6	131.3	111.2
Swimming pool	49.3	50.1	68.3

4.2. Obstacle Detection: Configuration (C) versus (D)

In this series of experiments, we compared the two configurations of the walker that use the same Kinect RGB-D camera as the sensor. As discussed in the previous section, the difference between these configurations is the processing algorithms: (C) uses a simple central stripe averaging technique that can be implemented on a Raspberry Pi 3 device, while (D) uses a more complex approach based on creating and processing a point cloud.

A general observation with the use of the Kinect sensor is that due to the fact that it uses an infrared emitter and sensor, it does not work in bright sunshine, when the infrared rays from the sun confuse the measurements. Configurations (A) and (B) do not suffer from this problem. For the experiments with (C) and (D) configurations, the nature of the sensor did not allow for outdoor experiments, which restricted the use of obstacles in the indoor setups. At the same time, the low-resolution point cloud representations are not suitable for certain types of obstacle classifications such as between a wall and a door. On the other hand, the 3D representations allow us to distinguish between obstacles that represent a drop or a raise in the floor. To verify that our algorithms can identify these situations, we added experiments to test for this, looking at the bottom of a stairwell and a drop in the floor level.

The measurements for these obstacle types are shown in Table 2. We found that for these measurements, the approach (D) slightly under-estimated the distance to the obstacles, while (C) slightly overestimated them. However, for all cases, the obstacles and drops were detected correctly, and the errors were small enough not to affect in any measurable way the experience of the user.

We concluded from this experiment that if the only goal was the detection of the obstacles in front of the walker, the much simpler algorithm using configuration (C) is sufficient. However, we note that (C) ignores the obstacles outside of the center stripe, and thus it cannot provide navigational guidance.

Table 2. Comparing the accuracy of configurations C and D for measuring the distance (cm) to several likely obstacle types.

Obstacle	Ground Truth	Config (C)	Config (D)
Empty hallway	N/A	No obstacle	No obstacle
Wall	128	130	120
Drop	N/A	Detected	Detected
Bottom of stairwell	125	129	119
Item in path	182	187	172

4.3. Blindfolded Navigation (Configuration A versus B)

As devices need to be deployed to the user, the most useful evaluation is that of measuring the way in which they impact the user's daily routine. We are less interested, for instance, in the high precision measurement of the distance to an obstacle, than the fact that the obstacle has been detected, and the user had successfully avoided the collision and navigated to their destination with the help of the walker. To evaluate the performance of the augmented walkers along this dimension, we performed a series of experiments that tested the navigational guidance of the walkers for users with severe visual impairment.

In these environments, we considered three setups which involved the same walker frame, but with different configurations and ways of interaction with the user. BASELINE involved the walker without any of the augmentations, serving only as a mobility aid. A+H was the walker in Configuration (A) with a haptic user interface for signaling the navigational guidance. B+A was the walker in configuration (B) with an audio feedback. To model severe visual impairment, the users (healthy volunteers) were blindfolded.

We chose these setups to investigate some representative choices at various technology levels. Thus, BASELINE is a "no-tech", traditional assistive device. A+H is the "mid-tech" choice, it uses a relatively simple sensor technology, with a low dimensional output (seven dimensions), simple user interface based on a binary haptic actuator. As a perceived complexity, the code driving this system can be measured in about one hundred lines of Python code, without relying on external libraries. The B+A setup is the "high-tech" choice: the input technology is the video input at a resolution of 1024 $\times$ 1024 corresponding to a dimensionality of one million. It is processed through a deep learning system combining several technologies (ResNet, Inception, FasterRCNN) with the number of parameters exceeding 10 million. It also uses a high-level, voice-based output. Naturally, many other combinations could be investigated. However, a full exploration of the possible pairings are beyond the scope of this paper.

We performed two types of experiments: simple navigation from source to destination point and more complex indoor navigation experiments. In both types of experiments we measured the number of time the users collided with the obstacles. There were two major situations that led to hit obstacles. In the first type, the user was already moving when the notification was issued, and the momentum of the movement led to hitting the obstacle. In the second type, a notification about an obstacle led the user to change their trajectory, and this led to immediately hitting another obstacle, different from the one about which it had been notified. We conjecture that the first type of collision could be mitigated with a faster overall process of the detection–decision–notification cycle. The second type would require the system to have a more sophisticated navigation and user prediction model, which would take into account more obstacles in the scene and model the user's likely reaction to the navigation command. We have not encountered a situation where the obstacle detector would have missed one of the obstacles in the scene.

The detection range of the HC-SR04 ultrasonic sensor is 4 m, the one of the Kinect sensor is about 3.5 m, and the one based on a camera is basically unlimited, in practice extending to the nearest occlusion. We judge that these ranges are sufficient for obstacle detection in a rollator scenario. The practical problem, however, is that it is impractical for a system to make a notification when it sees an obstacle four meters away—in any scenario there is always going to be some kind of obstacle that is far out. The main challenge, as we noted above, is not the detection of the obstacle, but making the decision to notify the user in the right way and at the right time. If we notify the user too early, we risk spamming them about obstacles they will not hit anyhow, while if we notify them too late, the user is already committed to a move and will collide with the obstacle despite the notification.

4.3.1. Simple Navigation

The simple navigation task involved the user moving from a source to a destination location, on a trajectory that would be a straight line in the absence of obstacles. The experiments involved navigating an environment with various obstacle densities. We used both an indoor and an outdoor environment, with the type of obstacles suitable to the setting. The experiments considered four levels of obstacle densities: empty, low, medium and high. We used the same ten obstacles, but the higher the obstacle density, the closer the obstacles were to each other leading to a more complex navigation task. Note that even for the empty setting, the user had to avoid collision with walls for the indoor environment and curbs and cars for the outdoor environment. For each experiment, we counted the number of obstacles the user collided with during the movement, as well as the time it took to navigate from the source to the destination.

Figure 6—top shows the percentage of the obstacles hit under various configurations, as an average of two trials. The results show that while there is considerable individual variation, overall, the more densely packed the obstacles were, the lower the number of the collisions and the faster the traversal. We conjecture that the main reason for this is that when the obstacles were closely clustered, the user could traverse the area with the obstacles very carefully and slowly to avoid them, but it could speed up on the rest of the trajectory. However, when the same number of obstacles was distributed in the whole area (in this case, with lower density) the user was more likely to be taken by surprise by an obstacle.

Comparing the three setups, the B+A approach consistently showed the lowest number of collisions. The A+H approach was in general better than the baseline, with one outlier for the medium-density outdoor setting. Empirical observation led us to conjecture that this was due to a "snowball" effect: if a user hits several obstacles in close succession, it is likely that they will hit other ones as well, perhaps as a result of disorientation.

These results validate the fact that the walkers need to be evaluated as a holistic system. While both configurations correctly detected the obstacles, with in fact the ultrasonic sensor in A+H being the more accurate, the overall results in setup B+A were better. We conjecture that the reason for this is that the users were not accustomed to navigate based on the haptic feedback, with the audio signals offering a clearer guidance. In addition, the vision-based sensing combined with audio output allowed the walker to identify the obstacle. This information was not available in setup A+H. This information could be used by the user for a more successful navigation.

This finding matches other studies that investigate the cognitive load of audio and haptic feedback in assistive systems. For instance, Martinez et al. [28] found that blind people prefer haptic feedback over audio feedback for short range navigation tasks, but prefer audio feedback for other tasks such as orientation, communication and alerts.

Figure 6—bottom shows the measured values for the time to reach the destination.

For the indoor environment, we found that the higher the density of the obstacles, that is, the closer the obstacles were clustered together, the faster the traversal time, as the trajectory contains large stretches where no obstacle was present, allowing the user to speed up. However, we found that, in general, there is little difference between the different walker configurations in the time taken to navigate the trajectory.

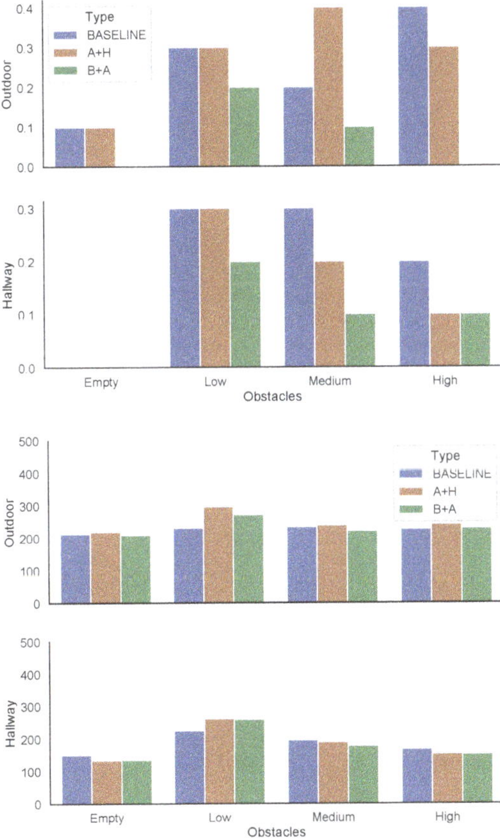

Figure 6. (top) Efficacy of the navigational guidance system for configurations A and B, measured as the percentage of the obstacles in the environment that were hit during the navigation (lower is better); and (bottom) time needed to perform a navigation task (lower is better).

4.3.2. Complex Indoor Navigation Task

The most frequently encountered navigation task by a mobility and visually challenged person is navigating their personal environment: this involves moving from the bedroom to bathroom or from the front room to the kitchen. In contrast to the fixed source–destination pairs we considered in the previous section, these navigation tasks are more complex: they involve finding paths, going through doors and maneuvering around furniture (see Figure 7).

To investigate the impact of the different walker setups for this task, we measured the user's navigation of four different paths in a house. The experiments were repeated with the BASELINE, A+H and B+A configurations. The time it took to navigate these paths and the number of obstacles with which they collided is shown in Table 3. In this environment we found that both augmented configurations A+H and B+A allowed the user to complete the navigation tasks, both faster and with a lower number of collisions compared to the BASELINE. There was no significant difference between the two augmented configurations.

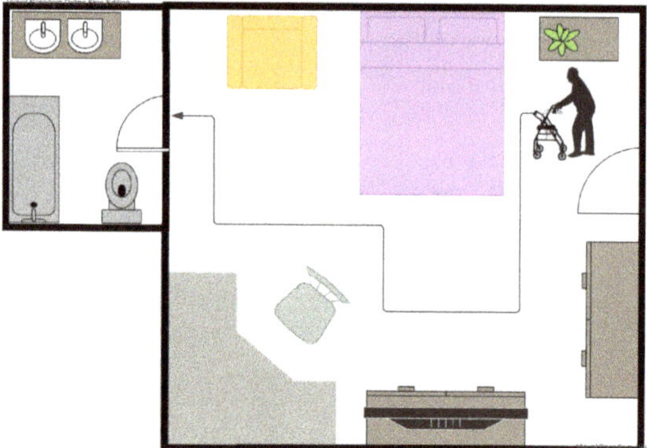

Figure 7. A complex navigation task in an indoor environment. The disabled user needs to navigate from the bedroom to the bathroom, avoid obstacles such as the bed and the chair, and must find and open the appropriate doors.

Table 3. Experimental results for the complex navigation tasks.

Type of Aid	Scenario	Time	Obstacles Hit
BASELINE	front door to kitchen	2:39	3
	kitchen to bedroom	1:31	4
	bedroom to bathroom	1:28	3
	patio	3:32	3
A+H	front door to kitchen	1:52	1
	kitchen to bedroom	1:12	2
	bedroom to bathroom	1:04	1
	patio	2:31	2
B+A	front door to kitchen	1:50	1
	kitchen to bedroom	1:10	1
	bedroom to bathroom	1:09	2
	patio	2:23	0

5. Conclusions

In this paper, we described and studied several prototypes for a smart walker specialized for people with both visual and mobility impairments. As a first conclusion, we found that there are multiple, very different choices of sensors that can ultimately ensure a similar user experience. Active sensors such as ultrasonic distance sensors or infrared depth cameras achieve the best accuracy in localizing obstacles. However, recent advances in computer vision, in particular object detection and recognition, allow passive, inexpensive cameras to achieve accuracy that is sufficient for the purposes of such a walker. In addition, computer vision systems can provide additional functionality such as identifying and naming the type of obstacle encountered by the user. Another conclusion of our experiments is that the performance of such a walker needs to be evaluated in a holistic way—the accuracy and reliability of the sensor, the type of user interaction used (such as haptic or audio), the friendliness and clarity of the user interaction and the low latency all contribute to the overall performance of the walker. Not every configuration justifies the additional cost of the technology. In particular, it is not enough that there is a sensor that detects the obstacle—we also need to find a way to convey it to the user in a way that triggers the right real-time obstacle avoidance behavior.

Author Contributions: Conceptualization, L.B. and D.T.; methodology, N.M., K.F., C.F., S.Z., S.S.B. and J.G.; software, N.M., C.F., S.Z., S.S.B. and J.G.; validation, N.M., C.F., S.Z. and J.G.; formal analysis, N.M., C.F., S.Z., J.G., L.B. and D.T.; investigation, N.M., C.F., S.S.B. and J.G.; resources, D.T., S.Z. and S.S.B.; data curation, N.M., C.F., K.F., S.Z., J.G. and S.S.B.; writing—original draft preparation, N.M., S.Z., D.T. and L.B.; writing—review and editing, L.B., D.T. and N.M.; visualization, N.M., C.F., J.G., L.B. and D.T.; supervision, D.T., L.B., S.Z. and S.S.B.; project administration, D.T. and L.B.; funding acquisition, D.T. and L.B. All authors have read and agreed to the published version of the manuscript.

Funding: This research was funded by National Science Foundation REU program award numbers 1560302 and 1852002.

Institutional Review Board Statement: Not applicable.

Informed Consent Statement: Not applicable.

Data Availability Statement: Not applicable.

Acknowledgments: Any opinions, findings, and conclusions and recommendations expressed in this material are those of the author(s) and do not necessarily reflect the views of the National Science Foundation.

Conflicts of Interest: The authors declare no conflict of interest.

References

1. Blindness and Vision Impairment. Available online: https://www.who.int/news-room/fact-sheets/detail/blindness-and-visual-impairment (accessed on 12 October 2019).
2. Gell, N.M.; Wallace, R.B.; Lacroix, A.Z.; Mroz, T.M.; Patel, K.V. Mobility device use in older adults and incidence of falls and worry about falling: Findings from the 2011–2012 National Health and Aging Trends Study. *J. Am. Geriatr. Soc.* **2015**, *63*, 853–859. [CrossRef]
3. MacNamara, S.; Lacey, G. A smart walker for the frail visually impaired. In Proceedings of the IEEE International Conference on Robotics and Automation (ICRA-2000), San Francisco, CA, USA, 24–28 April 2000; Volume 2, pp. 1354–1359.
4. Garrote, L.; Paulo, J.; Nunes, U.J. Reinforcement Learning Aided Robot-Assisted Navigation: A Utility and RRT Two-Stage Approach. *Int. J. Soc. Robot.* **2020**, *12*, 689–707. [CrossRef]
5. Dubowsky, S.; Genot, F.; Godding, S.; Kozono, H.; Skwersky, A.; Yu, H.; Yu, L.S. PAMM—A robotic aid to the elderly for mobility assistance and monitoring: A "helping-hand" for the elderly. In Proceedings of the IEEE International Conference on Robotics and Automation (ICRA-2000), San Francisco, CA, USA, 24–28 April 2000; Volume 1, pp. 570–576.
6. Lacey, G.J.; Rodriguez-Losada, D. The evolution of Guido. *IEEE Robot. Autom. Mag.* **2008**, *15*, 75–83. [CrossRef]
7. Zehtabian, S.; Khodadadeh, S.; Pearlman, R.; Willenberg, B.; Kim, B.; Turgut, D.; Bölöni, L.; Ross, E.A. Supporting rehabilitation prescription compliance with an IoT-augmented four-legged walker. In Proceedings of the 2nd Workshop on AI for Aging, Rehabilitation and Independent Assisted Living (ARIAL-18), Stockholm, Sweden, 13–15 July 2018.
8. Khodadadeh, S.; Zehtabian, S.; Guilbe, J.; Pearlman, R.; Willenberg, B.; Kim, B.; Ross, E.A.; Bölöni, L.; Turgut, D. Detecting unsafe use of a four-legged walker using IoT and deep learning. In Proceedings of the IEEE International Conference on Communications (ICC-2019), Shanghai, China, 20–24 May 2019.
9. Paulo, J.; Peixoto, P.; Nunes, U.J. ISR-AIWALKER: Robotic Walker for Intuitive and Safe Mobility Assistance and Gait Analysis. *IEEE Trans. Hum. Mach. Syst.* **2017**, *47*, 1110–1122. [CrossRef]
10. Kashyap, P.; Saleh, M.; Shakhbulatov, D.; Dong, Z. An Autonomous Simultaneous Localization and Mapping Walker for Indoor Navigation. In Proceedings of the IEEE 39th Sarnoff Symposium, Newark, NJ, USA, 24–25 September 2018; pp. 1–6.

11. Bhatlawande, S.S.; Mukhopadhyay, J.; Mahadevappa, M. Ultrasonic spectacles and waist-belt for visually impaired and blind person. In Proceedings of the IEEE National Conference on Communications (NCC-2012), Kharagpur, India, 3–5 February 2012; pp. 1–4.
12. Orita, K.; Takizawa, H.; Aoyagi, M.; Ezaki, N.; Shinji, M. Obstacle detection by the Kinect cane system for the visually impaired. In Proceedings of the IEEE/SICE International Symposium on System Integration, Kobe, Japan, 15–17 December 2013; pp. 115–118.
13. Pham, H.H.; Le, T.L.; Vuillerme, N. Real-time obstacle detection system in indoor environment for the visually impaired using Microsoft Kinect sensor. *J. Sens.* **2016**, *2016*, 3754918. [CrossRef]
14. Panteleris, P.; Argyros, A.A. Vision-based SLAM and moving objects tracking for the perceptual support of a smart walker platform. In Proceedings of the Workshop on Assistive Computer Vision and Robotics (ACVR 2014) in Conjunction with ECCV 2014, Zurich, Switzerland, 6–7 September 2014; pp. 407–423.
15. Rizzon, L.; Moro, F.; Passerone, R.; Macii, D.; Fontanelli, D.; Nazemzadeh, P.; Corrà, M.; Palopoli, L.; Prattichizzo, D. c-Walker: A cyber-physical system for ambient assisted living. In *Applications in Electronics Pervading Industry, Environment and Society*; Springer: Berlin/Heidelberg, Germany, 2016; pp. 75–82.
16. Viegas, V.; Dias Pereira, J.; Postolache, O.; Girão, P.S. Monitoring walker assistive devices: A novel approach based on load cells and optical distance measurements. *Sensors* **2018**, *18*, 540. [CrossRef] [PubMed]
17. Ramadhan, A.J. Wearable Smart System for Visually Impaired People. *Sensors* **2018**, *18*, 843. [CrossRef] [PubMed]
18. Kim, S.Y.; Cho, K. Usability and design guidelines of smart canes for users with visual impairments. *Int. J. Des.* **2013**, *7*, 99–110.
19. Feltner, C.; Guilbe, J.; Zehtabian, S.; Khodadadeh, S.; Bölöni, L.; Turgut, D. Smart Walker for the Visually Impaired. In Proceedings of the IEEE ICC 2019, Shanghai, China, 20–24 May 2019; pp. 1–6.
20. Mostofa, N.; Fullin, K.; Zehtabian, S.; Bacanli, S.S.; Bölöni, L.; Turgut, D. IoT-Enabled Smart Mobility Devices for Aging and Rehabilitation. In Proceedings of the IEEE ICC 2020, Dublin, Ireland, 7–11 June 2020; pp. 1–6.
21. Ultrasonic Ranging Module HC—SR04. Available online: https://www.mouser.com/datasheet/2/813/HCSR04-1022824.pdf (accessed on 12 October 2019).
22. Ren, S.; He, K.; Girshick, R.B.; Sun, J. Faster R-CNN: Towards Real-Time Object Detection with Region Proposal Networks. *arXiv* **2015**, arXiv:1506.01497.
23. Szegedy, C.; Ioffe, S.; Vanhoucke, V. Inception-v4, Inception-ResNet and the Impact of Residual Connections on Learning. *arXiv* **2016**, arXiv:1602.07261.
24. Kuznetsova, A.; Rom, H.; Alldrin, N.; Uijlings, J.; Krasin, I.; Pont-Tuset, J.; Kamali, S.; Popov, S.; Malloci, M.; Kolesnikov, A.; et al. The Open Images Dataset V4: Unified image classification, object detection, and visual relationship detection at scale. *IJCV* **2020**, *128*, 1956–1981. [CrossRef]
25. Ortigosa, N.; Morillas, S.; Peris-Fajarnés, G. Obstacle-free pathway detection by means of depth maps. *J. Intell. Robot. Syst.* **2011**, *63*, 115–129. [CrossRef]
26. Rusu, R.B.; Cousins, S. 3D is here: Point Cloud Library (PCL). In Proceedings of the IEEE International Conference on Robotics and automation (ICRA-2011), Shanghai, China, 9–13 May 2011; pp. 1–4.
27. Li, B.; Zhang, X.; Muñoz, J.P.; Xiao, J.; Rong, X.; Tian, Y. Assisting blind people to avoid obstacles: An wearable obstacle stereo feedback system based on 3D detection. In Proceedings of the IEEE International Conference on Robotics and Biomimetics (ROBIO-2015), Zhuhai, China, 6–9 December 2015; pp. 2307–2311.
28. Martinez, M.; Constantinescu, A.; Schauerte, B.; Koester, D.; Stiefelhagen, R. Cognitive evaluation of haptic and audio feedback in short range navigation tasks. In Proceedings of the International Conference on Computers for Handicapped Persons, Paris, France, 9–11 July 2014; pp. 128–135.

MDPI
St. Alban-Anlage 66
4052 Basel
Switzerland
Tel. +41 61 683 77 34
Fax +41 61 302 89 18
www.mdpi.com

Sensors Editorial Office
E-mail: sensors@mdpi.com
www.mdpi.com/journal/sensors

www.ingramcontent.com/pod-product-compliance
Lightning Source LLC
LaVergne TN
LVHW071007170726
843515LV00006B/1098

* 9 7 8 3 0 3 6 5 2 8 1 2 0 *